Biochemistry and Physiology of Anaerobic Bacteria

Springer
New York
Berlin
Heidelberg
Hong Kong
London
Milan
Paris
Tokyo

Lars G. Ljungdahl
Michael W. Adams   Larry L. Barton
James G. Ferry   Michael K. Johnson

Editors

Biochemistry and Physiology of Anaerobic Bacteria

With 71 Illustrations

 Springer

Lars G. Ljungdahl
Department of Biochemistry and
   Molecular Biology
University of Georgia
Athens, GA 30602
USA
larsljd@bmb.uga.edu

Michael W. Adams
Department of Biochemistry and
   Molecular Biology
University of Georgia
Athens, GA 30602
USA
adams@bmb.uga.edu

Larry L. Barton
Department of Biology
University of New Mexico
Albuquerque, NM 87131
USA
barton@unm.edu

James G. Ferry
Department of Biochemistry and
   Molecular Biology
Pennsylvania State University
University Park, PA 16801
USA
jpf3@psu.edu

Michael K. Johnson
Department of Chemistry
Center for Metalloenzyme Studies
University of Georgia
Athens, GA 30602
USA
johnson@chem.uga.edu

QR
89
. 5
.B55
2003

Library of Congress Cataloging-in-Publication Data
Biochemistry and physiology of anaerobic bacteria / editors, Lars G. Ljungdahl . . . [et al.].
      p. cm.
   Includes bibliographical references and index.
   ISBN 0-387-95592-5 (alk. paper)
      1. Anaerobic bacteria. I. Ljungdahl, Lars G.
   QR89.5 .B55 2003
   579.3′149—dc21   2002036546

ISBN 0-387-95592-5   Printed on acid-free paper.

© 2003 Springer-Verlag New York, Inc.

Printed in the United States of America.

9 8 7 6 5 4 3 2 1   SPIN 10893900

www.springer-ny.com

Springer-Verlag   New York Berlin Heidelberg
A member of BertelsmannSpringer Science+Business Media GmbH

*To the memory of Harry D. Peck, Jr. (1927–1998)
professor, founder, and chairman of the
Department of Biochemistry at the University of
Georgia and pioneer in studies of sulfate-reducing
bacteria and hydrogenases.*

Preface

During the last thirty years, there have been tremendous advances within all realms of microbiology. The most obvious are those resulting from studies using genetic and molecular biological methods. The sequencing of whole genomes of a number of microorganisms having different physiologic properties has demonstrated their enormous diversity and the fact that many species have metabolic abilities previously not recognized. Sequences have also confirmed the division of prokaryotes into the domains of Archaea and bacteria. Terms such as hyper- or extreme thermopiles, thermophilic alkaliphiles, acidophiles, and anaerobic fungi are now used throughout the microbial community. With these discoveries has come a new realization about the physiological and metabolic properties of microoganisms. This, in turn, has demonstrated their importance for the development, maintenance, and sustenance of all life on Earth. Recent estimates indicate that the amount of prokaryotic biomass on Earth equals—and perhaps exceeds—that of plant biomass. The rate of uptake of carbon by prokaryotic microorganisms has also been calculated to be similar to that of uptake of carbon by plants. It is clear that microorganisms play extremely important and typically dominant roles in recycling and sequestering of carbon and many other elements, including metals.

Many of the advances within microbiology involve anaerobes. They have metabolic pathways only recently elucidated that enable them to use carbon dioxide or carbon monoxide as the sole carbon source. Thus they are able to grow autotrophically. These pathways differ from that of the classical Calvin Cycle discovered in plants in the mid-1900s in that they lead to the formation of acetyl-CoA, rather than phosphoglycerate. The new pathways are prominent in several types of anaerobes, including methanogens, acetogens, and sulfur reducers. It has been postulated that approximately twenty percent of the annual circulation of carbon on the Earth is by anaerobic processes. That anaerobes carry out autotrophic type carbon dioxide fixation prompted studies of the mechanisms by which they conserve energy and generate ATP. It is now clear that the pathways of autotrophic carbon dioxide fixation involve hydrogen metabolism and that they are coupled to electron transport and generation of ATP by chemiosmosis. Enzymes catalyzing the metabolism of carbon dioxide, hydrogen, and other materials for building cell material and for electron transport are now intensely studied in anaerobes. Almost without exception, these enzymes depend on metals such as iron, nickel, cobalt, molybdenum, tungsten, and selenium. This pertains also to electron carrying proteins like cytochromes, several types of iron-sulfur and flavoproteins. Much present knowledge of electron transport and phosphorylation in anaerobic microorganisms has been obtained from studies of sulfate reducers. More recent investigations with methanogens and acetogens corroborate the findings obtained with the sulfate reducers, but they also demonstrate the diversity of mechanisms and pathways involved.

This book stresses the importance of anaerobic microorganisms in nature and relates their wonderful and interesting metabolic properties to the fascinating enzymes that are involved. The first two chapters by H. Gest and H.G. Schlegel, respectively, review the recycling of elements and the diversity of energy resources by anaerobes. As mentioned above, hydrogen metabolism plays essential roles in many anaerobes, and there are several types of hydrogenase, the enzyme responsible for catalyzing the oxidation and production of this gas. Some contain nickel at their catalytic sites, in addition to iron-sulfur clusters, while others contain only iron-sulfur clusters. They also vary in the types of compounds that they use as electron carriers. The mechanism of activation of hydrogen by enzymes is discussed by Simon P.J. Albracht, and the activation of a purified hydrogenase from *Desulfovibrio vulgaris* and its catalytic center by B. Hanh Huynh, P. Tavares, A.S. Pereira, I. Moura, and J.G. Moura. The biosynthesis of iron-sulfur clusters, which are so prominent in most hydrogenases, formate and carbon monoxide dehydrogenases, nitrogenases, many other reductases, and several types of electron carrying proteins, is explored by J.N. Agar, D.R. Dean, and M.K. Johnson. R.J. Maier, J. Olson, and N. Mehta write about genes and proteins involved in the expression of nickel dependent hydrogenases. Genes and the genetic manipulations of *Desulfovibrio* are examined by J.D. Wall and her research associates. In Chapter 8, G. Voordouw discusses the function and assembly of electron transport complexes in *Desulfovibrio vulgaris*. In the next chapter Richard Cammack and his colleagues introduce eukaryotic anaerobes, including anaerobic fungi and their energy metabolism. They explore the role of the hydrogenosome, which in the eukaryotic anaerobes replaces the mitochondrion. A rather new aspect related to anerobic microorganisms is the observation that they exhibit some degree of tolerance toward oxygen. They typically lack the known oxygen stress enzymes superoxide dismutase and catalase, but they contain novel iron-containing protein including hemerythrin-like proteins, desulfoferrodoxin, rubrerythrin, new types of rubredoxins, and a new enzyme termed superoxide reductase. D.M. Kurtz, Jr., discusses in Chapter 10 these proteins and proposes that they function in the defense toward oxygen stress in anaerobes and microaerophiles. Over six million tons of methane is produced biologically each year, most of it from acetate, by methanogenic anaerobes. J.G. Ferry describes in Chapter 11 that reactions include the activation of acetate to acetyl-CoA, which is cleaved by acetyl-CoA synthase. The methyl group is subsequently reduced to methane, and the carbonyl group is oxidized to carbon dioxide. The pathway is similar but reverse of that of acetyl-CoA synthesis by acetogens, but it involves cofactors unique to the methane-producing Archaea. Selenium has been found in several enzymes from anaerobes including species of clostridia, acetogens, and methanogens. In Chapter 12, W.T. Self has summarized properties of selenoenzymes, that are divided into three groups. The first constitutes amino acid reductases that utilize glycine, sarcosine, betaine, and proline. In these and also in the second group, which includes formate dehydrogenases, selenium is present as selenocysteine. Selenocysteine is incorporated into the polypeptide chain via a special seryl-tRNA and selenophosphate. The third group of selenoenzymes is selenium-molybdenum hydroxylases found in purinolytic clostridia. The nature of the selenium in this group has yet to be determined. Chapters 13 and 14 deal with acetogens, which produce anaerobically a trillion kilograms of acetate each year by carbon dioxide fixation via the acetyl-CoA pathway. H.L. Drake and K. Küsel highlight the diversity of acetogens and their ecological roles. A. Das and L.G. Ljungdahl discuss evidence that the acetyl-CoA pathway of carbon dioxide fixation is coupled with electron transport and ATP generation. In addition, they present some data showing how acetogens can deal with oxydative stress. In Chapter 15, D.P. Kelly discusses the biochemical features common to both anaerobic sulfate reducing bacteria and aerobic thiosulfate oxidizing thiobacilli. His chapter is also a tribute to Harry Peck. The last three chapters are devoted to the reduction by anaerobic bacteria of metals, metalloids and nonessential elements. L.L. Barton, R.M. Plunkett, and B.M. Thomson in their review point out the geochemical importance these reductions, which involve both metal cations and metal anions. J. Wiegel, J. Hanel, and K. Aygen describe the isolation of recently discovered chemolithoautotrophic thermophilic iron(III)-reducers from geothermally heated sediments and water samples of hot springs. They propose that these bacteria are ancient and were involved in formation of iron deposits during the Precambrian era. The last chapter is a discussion of electron flow in ferrous bioconversion by E.J. Laishley and R.D. Bryant. They visualize a model for biocorrosion by sulfate-reducing bacteria that involves both iron and nickel-iron hydrogenases, high molecular cytochrome, and electron transport using sulfate as an acceptor.

Lars G. Ljungdahl
Michael W. Adams
Larry L. Barton
James G. Ferry
Michael K. Johnson

Contents

Contributors

JEFFREY N. AGAR
Department of Chemistry, Center for Metalloenzyme Studies, University of Georgia, Athens, GA 30602, USA

SIMON P.J. ALBRACHT
Department of Biochemistry, E.C. Slater Institute, University of Amsterdam, NL-1018 TV Amsterdam, The Netherlands

KAYA AYGEN
Department of Microbiology, Center for Biological Resource Recovery, University of Georgia, Athens, GA 30602, USA

LARRY L. BARTON
Department of Biology, University of New Mexico, Albuquerque, NM 87131, USA

R.D. BRYANT
Department of Biological Sciences, University of Calgary, Calgary, Alberta T2N lN4, Canada

RICHARD CAMMACK
Division of Life Sciences, King's College, London SE1 9NN, UK

LAURENCE CASALOT
Department of Biochemistry, University of Missouri-Columbia, Columbia, MO 65211, USA

AMARESH DAS
Department of Biochemistry and Molecular Biology, Center for Biological Resource Recovery, University of Georgia, Athens, GA 30602, USA

DENNIS R. DEAN
Department of Biochemistry, Virginia Institute of Technology, Blacksburg, VA 24061, USA

HAROLD L. DRAKE
Department of Ecological Microbiology, BITOEK, University of Bayreuth, 95440 Bayreuth, Germany

JAMES G. FERRY
Department of Biochemistry and Molecular Biology, Pennsylvania State University, University Park, PA 16801, USA

HOWARD GEST
Department of History and Philosophy of Science, Department of Biology, Photosynthetic Bacteria Group, Indiana University, Bloomington, IN 47405, USA

TARA GIBLIN
28024 Marguerite Parkway, Mission Viejo, CA 92692, USA

JUSTIN HANEL
Department of Microbiology, Center for Biological Resource Recovery, University of Georgia, Athens, GA 30602, USA

BOI HANH HUYNH
Department of Physics, Emory University, Atlanta, GA 20322, USA

CHRISTOPHER L. HEMME
Department of Biochemistry, University of Missouri-Columbia, Columbia, MO 65211, USA

DAVID S. HORNER
Department of Zoology, Molecular Biology Unit, Natural History Museum, London SW7 5BD, UK. *Current address:* Department of Physiology and General Biochemistry, University of Milan, 20133 Milan, Italy

MICHAEL K. JOHNSON
Department of Chemistry, Center for Metalloenzyme Studies, University of Georgia, Athens, GA 30602, USA

DONOVAN P. KELLY
Department of Biological Sciences, University of Warwick, Coventry CV4 7AL, UK

JAROSLAV KULDA
Department of Parasitology, Charles University, 128 44 Prague 2, Czech Republic

DONALD M. KURTZ, JR.
Department of Chemistry, Center for Metalloenzyme Studies, University of Georgia, Athens, GA 30602, USA

KIRSTEN KÜSEL
Department of Ecological Microbiology, BITOEK, University of Bayreuth, 95440 Bayreuth, Germany

E.J. LAISHLEY
Department of Biological Sciences, University of Calgary, Calgary, Alberta T2N 1N4, Canada

LARS G. LJUNGDAHL
Department of Biochemistry and Molecular Biology, University of Georgia, Athens, GA 30602, USA

DAVID LLOYD
School of Pure and Applied Biology, University of Wales, Cardiff CF1 3TL, UK

R.J. MAIER
Department of Microbiology, Center for Biological Resource Recovery, University of Georgia, Athens, GA 30602, USA

N. MEHTA
Department of Microbiology, Center for Biological Resource Recovery, University of Georgia, Athens, GA 30602, USA

ISABEL MOURA
Departamento de Química e Centro de Química Fina e Biotecnologia, Faculdade de Ciências e Tecnologia, Universidade Nova de Lisboa, 2825-114 Caparica, Portugal

JOSÉ J.G. MOURA
Departamento de Química e Centro de Química Fina e Biotecnologia, Faculdade de Ciências e Tecnologia, Universidade Nova de Lisboa, 2825-114 Caparica, Portugal

J. OLSON
Department of Microbiology, Center for Biological Resource Recovery, University of Georgia, Athens, GA 30602, USA

ALICE S. PEREIRA
Departamento de Química e Centro de Química Fina e Biotecnologia, Faculdade de Ciências e Tecnologia, Universidade Nova de Lisboa, 2825-114 Caparica, Portugal

RICHARD M. PLUNKETT
Department of Biology, University of New Mexico, Albuquerque, NM 87131, USA

BARBARA RAPP-GILES
Department of Biochemistry, University of Missouri-Columbia, Columbia, MO 65211, USA

JOSEPH A. RINGBAUER, JR.
Department of Biochemistry, University of Missouri-Columbia, Columbia, MO 65211, USA

HANS GÜNTER SCHLEGEL
Institut für Mikrobiologie der Georg-August-Universität, 37077 Göttingen, Germany

WILLIAM T. SELF
Laboratory of Biochemistry, National Heart, Lung and Blood Institute, National Institutes of Health, Bethesda, MD 20892, USA.

PEDRO TAVARES
Departamento de Química e Centro de Química Fina e Biotecnologia, Faculdade de Ciências e Tecnologia, Universidade Nova de Lisboa, 2825-114 Caparica, Portugal

BRUCE M. THOMSON
Department of Civil Engineering, University of New Mexico, Albuquerque, NM 87131, USA

MARK VAN DER GIEZEN
Department of Zoology, Molecular Biology Unit, Natural History Museum, London SW7 5BD, UK. *Current address:* School of Biological Sciences, Royal Holloway, University of London, Egham, Surrey TW2O OEX, UK

GERRIT VOORDOUW
Department of Biological Sciences, University of Calgary, Calgary, Alberta, T2N 1N4, Canada

JUDY D. WALL
Department of Biochemistry, University of Missouri-Columbia, Columbia, MO 65211, USA

JUERGEN WIEGEL
Department of Microbiology, Center for Biological Resource Recovery, University of Georgia, Athens, GA 30602, USA

1
Anaerobes in the Recycling of Elements in the Biosphere

Howard Gest

Microorganisms are responsible for the natural recycling of a number of chemical elements in the biosphere. The recycling obviously occurs on a massive scale and is particularly important in regard to nitrogen, carbon, sulfur, oxygen, and hydrogen. These elements are used, in one form or another, in the biosynthetic and bioenergetic processes of both aerobic and anaerobic microorganisms. Global cyclic transformations of the elements requires the participation of various kinds of organisms, primarily bacteria, and each "metabolic type" specializes in catalysis of a specific portion of the overall cycle. An example in point is the anaerobic reduction of sulfate to sulfide. Anaerobes are found in environments where dioxygen has been displaced by gaseous products of anaerobic metabolism, such as CH_4, CO_2, hydrogen, and H_2S. Despite sensitivity to oxygen, anaerobic bacteria also persist in circumstances usually thought to be aerobic in character. Thus they commonly occur in microenvironments where oxygen is constantly removed by the respiration of aerobes, as in small soil particles.

Stephenson (1947) pointed out that the number and variety of chemical reactions known to be catalyzed by bacteria far exceeded those attributable to other living organisms. Moreover, she noted, "Amongst heterotrophs it is as anaerobes that bacteria specially excel. . . . It is in the use of hydrogen acceptors that bacteria are specially developed as compared with animals and plants." This is another way of saying that anaerobes are redox specialists, which have special systems for oxidizing energy-rich substrates without recourse to molecular oxygen.

Who First Observed Anaerobic Life?

The conventional wisdom is that the first observation of anaerobic microbial life was made by Louis Pasteur. In fact, Pasteur *re*discovered the anaerobic lifestyle. The first person actually to see anaerobic microorganisms was Antony van Leeuwenhoek, who did a remarkable experiment in 1680,

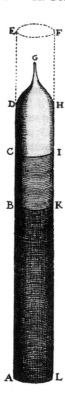

FIGURE 1.1. Diagram illustrating Leeuwenhoek's pepper tube experiment. (From Leeuwenhoek's letter no. 32 to the Royal Society of London, June 14, 1680.)

described in detail in one of his famous letters to the Royal Society of London (Dobell 1960).

Leeuwenhoek used two identical glass tubes, each filled about halfway with crushed pepper powder (to line *BK* in Fig. 1.1). As shown in Figure 1.1, clean rain water was added to line *CI*. Using a flame, he sealed one of the tubes at point *G*; the aperture of the other tube was left open. Leeuwenhoek said [see Dobell 1960, pp. 197–198]. that, after several days, "I took a little water out of the second glass, through the small opening at G; and I discovered in it a great many very little animalcules, of divers sort having its own particular motion." After 5 days, he opened the sealed tube in which some pressure had developed, forcing liquid out. He expected not to see "living creatures in this water." But, in fact, he observed "a kind of living animalcules that were round and bigger than the biggest sort that I have said were in the other water." Clearly, in the sealed tube, the conditions had become quite anaerobic owing to consumption of oxygen by aerobes. In 1913, the great microbiologist Martinus Beijerinck repeated Leeuwenhoek's experiment exactly and identified *Clostridium butyricum* as a prominent organism in the sealed pepper infusion tube fluid. Beijerinck (1913) commented:

We thus come to the remarkable conclusion that, beyond doubt, Leeuwenhoek in his experiment with the fully closed tube had cultivated and seen genuine anaerobic bacteria, which would happen again only after 200 years, namely, about 1862 by Pasteur. That Leeuwenhoek, one hundred years before the discovery of oxygen and the composition of air, was not aware of the meaning of his observations is understandable. But the fact that in the closed tube he observed an increased gas pressure caused by fermentative bacteria and in addition saw the bacteria, prove in any case that he not only was a good observer, but also was able to design an experiment from which a conclusion could be drawn.

Two Important Element Cycles

The Nitrogen Cycle

The most noteworthy multistage element cycles in which bacteria play important roles are the nitrogen and sulfur redox cycles. The fixation of nitrogen is a reductive process that provides organisms with nitrogen in a form usable for the synthesis of amino acids, nucleic acids, and other cell constituents. In essence, the overall conversion to the key intermediate, ammonia, can be represented as:

$$N_2 + 8H \rightarrow 2\,NH_3 + H_2 \qquad (1.1)$$

This way of summarizing nitrogen fixation implies that all nitrogenases have the capacity to produce hydrogen under certain conditions. The nitrogenase-catalyzed production of hydrogen is a major physiologic process in the metabolism of photosynthetic bacteria during anaerobic phototrophic growth, when ammonia and nitrogen are absent and cells depend on certain amino acids as nitrogen sources (see later).

Table 1.1 lists free-living anaerobic bacteria that fix nitrogen and have been used for experimental studies during recent decades. Note that the list

TABLE 1.1. Free-living nitrogen-fixing anaerobes.

Chemoorganotrophs	Phototrophs	Chemolithotrophs
Clostridium spp.	*Chromatium*	*Methanococcus*
Desulfotomaculum	*Chlorobium*	*Methanosarcina*
Desulfovibrio	*Heliobacillus*	
	Heliobacterium	
	Heliophilum	
	Rhodobacter	
	Rhodomicrobium	
	Rhodopila	
	Rhodopseudomonas	
	Rhodospirillum	
	Thiocapsa	

Source: Madigan and co-workers (2000).

includes methanogens, anaerobes that are of special interest in the carbon cycle. Methanogens produce CH_4 primarily by reducing CO_2 with hydrogen, and this process is clearly of huge magnitude in the biosphere (Ehhalt 1976). It occurs copiously in lake sediments, swamps, marshes, and paddy fields. The methanogens are also abundant in the anaerobic digestion chambers of many ruminant animals and termites (Madigan, et al. 2000).

Nitrogen in organic combination in living organisms is recycled to inorganic nitrogen after their death through the activities of various microorganisms. In brief, organic nitrogen is converted to ammonia (ammonification), which is then nitrified in two successive aerobic stages: (1) oxidation of ammonia to nitrite by *Nitrosomonas* and (2) oxidation of nitrite to nitrate by *Nitrobacter*. Completion of the cycle requires anaerobic reduction of nitrate to nitrogen, referred to as denitrification. The latter is accomplished mainly by bacteria capable of growing either as aerobes or anaerobes, typically species in the genera *Pseudomonas, Paracoccus,* and *Bacillus*. Historically, such metabolic types have been referred to by clumsy terms such as *facultative aerobe* and *facultative anaerobe*. A more sensible term is *amphiaerobe,* meaning "on both sides of oxygen." Amphiaerobe is defined as an organism that can use either oxygen (like an aerobe) or, as an alternative, some other energy-conversion process that is independent of oxygen (Chapman and Gest 1983).

The Sulfur Cycle

Anaerobes are particularly prominent in the cyclic interconversions of inorganic sulfur compounds. Reduction of sulfate to hydrogen sulfide by species of *Desulfovibrio* and *Desulfobacter* is of widespread occurrence and of economic significance, because of the corrosive properties of H_2S. The sulfide is also produced from S^0 by related organisms of the genus *Desulfuromonas*. Beijerinck was the first to establish that sulfide in the biosphere is produced mainly by bacterial reduction of sulfate. In 1894, he and his assistant, van Delden, isolated and described *Spirillum desulfuricans*, later renamed *Desulfovibrio desulfuricans*, providing the first pure cultures of a sulfate reducer. *Unculturable*, a favorite word of some contemporary molecular biologists, was not in Beijerinck's vocabulary (see later).

Anaerobic recycling of sulfide to sulfate ($H_2S \rightarrow S^0 \rightarrow SO_4^{2-}$) is a specialty of anoxygenic purple and green photosynthetic bacteria (*Chromatium, Chlorobium,* etc.) that can use sulfide as an electron donor for CO_2 reduction. The coordinated cross-feeding of the sulfate reducers and the sulfide-using photosynthetic bacteria frequently results in massive blooms of *Chromatium* spp. For example, this is commonly seen on shores of the Baltic Sea when sea grass and other plants become covered by drifting sand. Decomposition of the organic matter is coupled with bacterial sulfate reduction, yielding large quantities of H_2S; the conditions become ideal for growth of purple photosynthetic bacteria.

By interesting coincidence, the old dogma that cytochromes are not present in anaerobes was demolished by discovery, at about the same time, of c-type cytochromes in *Desulfovibrio* and anoxygenic photosynthetic bacteria (Kamen and Vernon 1955).

The Meaning of Diversity

During the last two decades of the twentieth century, biodiversity became a focus of discussion by many biologists and environmentalists. Inevitably, this led to more interest in the diversity of microorganisms. Unfortunately, the word *diversity* can have several meanings, and the one in mind is frequently not specified. Molecular biologists interested in evolution have championed differences in 16S RNA sequences as the primary indicators of the diversity of genera and species of prokaryotes. This has led to the view that "molecular phylogenetic techniques have provided methods for characterizing natural microbial communities without the need to cultivate organisms" (Hugenholtz and Pace 1996). Moreover, it has been said that "the types and numbers of organisms in natural communities can be surveyed by sequencing rRNA genes obtained from DNA isolated directly from cells in their ordinary environments. Analyzing microbial communities in this way is more than a taxonomic exercise because the sequences can be used to develop insights about organisms" (Pace 1996). Pace also made the assertion that the use of sequences allows us to infer properties of uncultivated organisms and "survey biodiversity rapidly and comprehensively."

Associated with such claims, the myth of unculturability of most prokaryotes has been promoted by statements such as "only a small fraction of less than 1% of the cells observed by microscopy (i.e., in *natural sources*) can be recovered as colonies on standard laboratory media" (Amann 2000). Applying this vague criterion is obviously misleading. How many well-known organisms—anaerobes, autotrophs, nutritionally fastidious bacteria, etc.—described in *Bergey's Manual* will grow in so-called standard media? Obviously, very few. Casual acceptance of some molecular biologist's views has led ecologist Wilson (1999) to some further, essentially unverifiable, extrapolations:

How many species of bacteria are there in the world? Bergey's *Manual of Systematic Bacteriology*, the official guide updated to 1989, lists about 4000. There has always been a feeling among microbiologists that the true number, including the undiagnosed species, is much greater, but no one could even guess by how much. Ten times more? A hundred? Recent research suggests that the answer might be at least a thousand times greater, with the total number ranging into the millions.

Some remarks by Amann (2000) are relevant to the question of the number of bacterial species extant:

Another methodological artifact are chimeric sequences which can be formed during PCR amplification of mixed template at a frequency of several percent.... The assumption that each rRNA sequence is equivalent to a species is as shaky as the still wide spread assumption that from the frequency of an rRNA clone in a library the relative abundance of the respective organism in the environment can be estimated.

For those who are impatient, I note Amann's estimate that "If there are *just* [emphasis added] one million species that ultimately can be cultured and if their complete taxonomic description proceeds at a rate of 1,000 species/year it would take roughly the next millenium to get a fairly complete overview on microbial diversity." My own experience tells me that if there are, say, 50,000 truly distinctive bacterial species still unknown, their isolation and characterization will be a long time in coming.

Our understanding of the roles of bacteria in the cycles of nature is based on characterization of the biochemical activities of isolated species of anaerobes and aerobes—in other words, on phenotypic patterns, which define *biochemical diversity* or, one might say, *metabolic diversity*. There is, of course, no way that biochemical diversity can be reconstructed simply by processing information contained in rRNA genes. A detailed analysis of the meaning of diversity in the prokaryotic world was provided by Palleroni (1997), and his conclusions are worthy of attention:

Modern approaches based on the use of molecular techniques presumed to circumvent the need for culturing prokaryotes, fail to provide sufficient and reliable information for estimation of prokaryote diversity. Many properties that make these organisms important members of the living world are amenable to observation only through the study of living cultures. Since current culture techniques do not always satisfy the need of providing a balanced picture of the microflora composition, future developments in the study of bacterial diversity should include improvements in the culture methods to approach as closely as possible the conditions of natural habitats. Molecular methods of microflora analysis have an important role as guides for the isolation of new prokaryotic taxa.

Since the 1980s, there has been a great escalation of research on prokaryotes; but, as far as I can tell, our knowledge of the principal reactions in the element cycles of nature has not changed appreciably. No doubt there are still unknown ancillary chemical cycles catalyzed by bacteria. One likely possibility is indicated by a recent report that a lithoautotrophic bacterium isolated from a marine sediment can obtain energy for anaerobic growth by oxidation of phosphite(P^{+3}) to phosphate(P^{+5}) while simultaneously reducing sulfate to H_2S (Schink and Friedrich, 2000). Evidently, establishment of such cycles will require the time-honored approach of isolation and characterization of pure cultures or well-defined consortia. We can expect that as new details emerge, we will learn that anaerobes are as biochemically diverse as other kinds of prokaryotes, perhaps more so. Another example in point is given by the description of new genera of sulfate reducers isolated from permanently cold Arctic marine sediments. Isolates of the new genera *Desulfofrigus, Desulfofaba*, and *Desulfotalea* grew at the in situ temperature of −1.7°C (Knoblauch et al. 1999).

The Historical Role of Anaerobes on Earth

More than 50 years of geochemical research has established that the atmosphere of the early Earth was essentially anaerobic. It is estimated that 2 billion years before the present, there was still virtually no molecular oxygen in the atmosphere. Since fossils of microorganisms ~3.5 billion years old have been found, it follows that for ~1.5 billion years the Earth must have been populated by anaerobic prokaryotes. It is reasonable to believe that anaerobic green and purple photosynthetic bacteria were the precursors of the first organisms capable of oxygenic photosynthesis, the cyanobacteria. When oxygen began to accumulate in the biosphere, anaerobes presumably faced a crisis of oxygen toxicity. Undoubtedly, many anaerobes died while others retreated to anaerobic locales, where we find their descendents today. Still others apparently evolved protective devices, such as superoxide dismutase or the rudiments of oxygen respiration. Another mechanism for avoiding oxygen toxicity, recently discovered in the hyperthermophilic anaerobe *Pyrococcus furiosus*, depends on the enzyme superoxide reductase, which reduces superoxide to H_2O_2; the latter is then reduced to water by peroxidases (Jenney et al. 1999).

Connected with the kinetics of oxygen evolution in the early atmosphere is the question of the origin of sulfate, required by the anaerobic sulfate reducers. Did the latter organisms evolve only after oxygen accumulation led to oxidation of reduced sulfur to sulfate? This notion was challenged by Peck (1974), who concluded:

Sulfate reducing bacteria were not antecedents of photosynthetic bacteria, but rather evolved from ancestral types which were photosynthetic bacteria. Although initially surprising, this evolutionary relationship is consistent with the idea that the accumulation of sulfate, the obligatory terminal electron acceptor for the sulfate reducing bacteria, was the result of bacterial photosynthesis.

As noted earlier, sulfate is generated when sulfide is the electron donor for anaerobic growth of purple and green photosynthetic bacteria.

We still have only foggy notions of early prokaryotic evolution. If anything, the picture has recently become more obscure because new evidence indicating extensive "horizontal" gene transfer among bacterial species casts doubt on current prokaryotic phylogenetic trees that branch from a main trunk, as in an actual tree. With this in mind, Doolittle (2000) proposed a more complex pattern of interconnecting prokaryotic evolutionary lines that strikes me as resembling the ramifications of a dollop of spaghetti.

Molecular Hydrogen: Electron Currency in Anaerobic Metabolism

Molecular hydrogen is encountered in the metabolic patterns of a variety of prokaryotes, either as an electron donor or as an end product (Gest 1954). The ability to produce hydrogen by reduction of protons with metabolic electrons was probably an ancient mechanism for achieving redox balance in energy-yielding bacterial fermentations. Gray and Gest (1965) referred to hydrogenase as a "delicate control valve for regulating electron flow" and concluded that "the hydrogen-evolving system of strict anaerobes represents a primitive form of cytochrome oxidase, which in aerobes effects the terminal step of respiration, namely, the disposal of electrons by combination with molecular oxygen."

The nitrogenase-catalyzed energy-dependent production of hydrogen by photosynthetic bacteria noted earlier appears to represent another kind of regulatory function. When the bacteria grow on organic acids (e.g., malate), using certain amino acids (e.g., glutamate) as nitrogen sources, nitrogenase is derepressed and functions as a hydrogen-evolving catalyst. Under such conditions, the supplies of ATP produced by photophosphorylation and of electrons generated from organic substrates evidently are in excess relative to the demands of the biosynthetic machinery. Nitrogenase then performs as a "hydrogenase safety valve," catalyzing hydrogen formation by energy-dependent reduction of protons. If molecular nitrogen becomes available, hyrogen evolution stops because ATP and the electron supply are used for production of ammonia, which is rapidly consumed for synthesis of amino acids and other nitrogenous compounds. Thus light-dependent hydrogen formation via nitrogenase has been interpreted to reflect "energy idling" when this is required for integration of energy conversion and biosynthetic metabolism (Gest 1972; Hillmer and Gest 1977; Gest 1999). Of interest in connection with the several functions of nitrogenase, is the suggestion of Broda and Peschek (1980) that nitrogenase evolved from an early ATP-requiring hydrogenase "that supported fermentations by ensuring the release of H_2."

Conclusion

It is likely that comparative structural and other studies of hydrogenases and nitrogenases will eventually illuminate events in the early evolution of energy-yielding mechanisms. We are indebted to the anaerobes for their necessary roles as recycling agents in Earth's element cycles.

Acknowledgments. My research on photosynthetic bacteria is supported by National Institutes of Health grant GM 58050. I also thank Dr. Hans van

Gemerden, University of Groningen (Netherlands) for translation of Beijerinck's 1913 paper, written in Dutch.

References

Amann R. 2000. Who is out there? Microbial aspects of biodiversity. Syst Appl Microbiol 23:1–8.

Beijerinck MW. 1913. De infusies en de ontdekking der bakteriën Jaarb Kon Akad Wetensch 1913, p 1–28

Broda E, Peschek GA. 1980. Evolutionary considerations on the thermodynamics of nitrogen fixation. Biosystems 13:47–56.

Chapman DJ, Gest H. 1983. Terms used to describe biological energy conversions, interactions of cellular systems with molecular oxygen, and carbon nutrition. In: Schopf JW, editor. Earth's earliest biosphere; its origin and evolution. Princeton, NJ: Princeton University Press; p 459–63.

Dobell C. 1960. Antony van Leeuwenhoek and his "little animals." New York: Dover; pp. 197–8.

Doolittle WF. 2000. Uprooting the tree of life. Sci Am 282:90–5.

Ehhalt DH. 1976. The atmospheric cycle of methane. In: Schlegel HG, Gottschalk G, Pfennig N, editors. Microbial prodution and utilization of gases. Göttingen, Germany: Goltze; p 13–22.

Gest H. 1954. Oxidation and evolution of molecular hydrogen by microorganisms. Bact Rev 18:43–73.

Gest H. 1972. Energy conversion and generation of reducing power in bacterial photosynthesis. Adv Microb Physiol 7: 243–82.

Gest H. 1999. Bioenergetic and metabolic process patterns in anoxyphototrophs. In: Peschek GA, Löffelhardt W, Scmetterer G, editors. The phototrophic prokaryotes. New York: Kluwer Academic/Plenum. P 11–9.

Gray CT, Gest H. 1965. Biological formation of molecular hydrogen. Science 148:186–92.

Hillmer P, Gest H. 1977. H_2 metabolism in the photosynthetic bacterium *Rhodopseudomonas capsulata*: H_2 production by growing cultures. J Bacteriol 129:724–31.

Hugenholtz P, Pace NR. 1996. Identifying microbial diversity in the natural environment: a molecular phylogenetic approach. Trends Biotechnol 14:190–7.

Jenney FE Jr, Verhagen MFJM, Cui X, Adams MWW. 1999. Anaerobic microbes: oxygen detoxification without superoxide dismutase. Science 286:306–9.

Kamen MD, Vernon LP. 1955. Comparative studies on bacterial cytochromes. Biochim Biophys Acta 17:10—22.

Knoblauch C, Sahm K, Jorgensen, BB. 1999. Psychrophilic sulfate-reducing bacteria isolated from permanently cold Arctic marine sediments: description of *Desulfofrigus oceanense* gen. nov., sp. nov., *Desulfofrigus fragile* sp. nov., *Desulfofaba. gelida* gen. nov., sp. nov., *Desulfotalea psychrophila* gen. nov., sp. nov. and *Desulfotalea arctica* sp. nov. Int J Syst Bacteriol 49:1631–43.

Madigan MT, Martinko JM, Parker J. 2000. Biology of microorganisms. Upper Saddle River, NJ: Prentice Hall.

Pace NR. 1996. New perspective on the natural microbial world: molecular microbial ecology. ASM News 62:463–70.

Palleroni NJ. 1997. Prokaryotic diversity and the importance of culturing. Ant V Leeuwenhoek 72:3–19.

Peck HD Jr. 1974. The evolutionary significance of inorganic sulfur metabolism. In: Carlile MJ, Skehel JJ, editors. Evolution in the microbial world. 24th symposium of the Society for General Microbiology. Cambridge: Cambridge University Press. P 241–62.

Schink B, Friedrich M. 2000. Phosphite oxidation by sulphate reduction. Nature 406:37.

Stephenson M. 1947. Some aspects of hydrogen transfer. Ant V Leeuwenhoek J Microbiol Serol 12: 33–48; see p 34.

Wilson EO. 1999. The diversity of life. New York: Norton; p 142–143.

2
The Diversity of Energy Sources of Microorganisms

Hans Günter Schlegel

This book is occupied with the recent progress that has been achieved in the area of the biochemistry and physiology of anaerobic bacteria. The width of the theme requires special knowledge and survey. To facilitate the survey, I should like to direct a glance on the collateral sciences of microbiology and deal with the question when the knowledge was obtained on which our present research is based. The formulation of the biological questions is old; the answering, however, requires knowledge on the properties of the substances that surround us, that means physics and chemistry. In this short contribution, I pose the question of how the exploration of materials, with which physiology and biochemistry deal, came about. In essence it is a chapter on analytical chemistry and physics as well as on the modes of biological energy conversions.

Toward the Exploration of the Constituents of Living Organisms

The fundamentals of knowledge on the composition of materials used by humans were already collected in ancient times. The known methods—such as preparation of bread, wine, beer, vinegar, soap, and cosmetics; tissue staining; and tanning of hides—today belongs to chemical technology. Various methods, like pressing of fatty oils and destillation of etherized oils, were used, too. Seven metals were known: gold, silver, copper, iron, lead, tin, and mercury. Among the acids only acetic acid and among the alkaline compounds only soda and potash were known. Simple chemical operations, such as weighing, filtration, evaporation, crystallization, and destillation were also known. Further exploration with respect to the composition of materials started in the thirteenth century. Three epochs can be differentiated, and all three saw contributions to the understanding of the metabolism of organisms.

In the first epoch, the properties of metals were explored. New metals (zinc, arsenic, antimony, and bismuth) were discovered. The specific weights of some metals were determined, with only 5% deviation. The metals were dissolved in "oleum," or sulfuric acid. Some metals started to play a role in the medication of diseases. The outstanding representative of the use of metals in medicine, in iatrochemistry, was Paracelsus (1493–1541). Much of the technological knowledge on metals, metallurgy, was summarized by Georg Agricola (1496–1555) in *De re metallica* (1556) and other books.

When the metals were studied, the release of gases and vapors was observed. For example, Paracelsus reported that when iron was dissolved in sulfuric acid "air comes out like a wind." The differentiation of these kinds of "air" took more than two centuries. Research on the analysis of gases was extremely productive in developing the basic methods to handle and study gases and even paved the way to design techniques for elementary analysis of the constituents of organisms. Thus the study of gases became the second epoch of analytical chemistry; it is called "the pneumatic age."

To keep the time scale in mind, the willow tree experiment performed by Jan Baptist van Helmont (1577–1644) is worth mentioning. Van Helmont was born in Brussels, studied in Leuwen, and spent the major part of his lifetime in the neighborhood of Brussels. He was a chemist, physiologist, and physician. In his own person curious contradictions were combined. He represents the transition from the Scholastic Age to the Age of Enlightenment. On the one hand, he was highly impressed by the Copernican view, by Harvey's discovery of the circulation of blood, and by Bacon's essays; and he was a careful observer and was able to undertake simple experiments. He coined the word *gas*. He regarded the gases that were formed during wine fermentation and during combustion of charcoal as identical. His scientific observations and experiences were published by his son in 1648 in Amsterdam under the title *Ortus medicinae*. On page 109 of that work we find the concise description of a great experiment. In English translation it says:

But that all plants directly and materially are produced solely from the element of water, I have learnt from this experiment. I took an earthenware pot, placed in it 200 lb of soil dried in an oven, moistened it with rainwater and planted in it a willow shoot weighing 5 lb. Finally, after five years, a tree hat grown and weighed 169 lb and about 3 oz. But the earthenware pot was constantly wet only with rainwater or distilled water, if it was necessary; and it was ample and imbedded in the ground: and to prevent dust from flying around and mixing with the soil, I covered the pot with an iron plate coated with tin and pierced with many holes. I did not add the weight of the fallen leaves of four autumns. Finally, I again dried the soil in the pot, and there were the same 200 lb minus about 2 oz. Therefore, 164 lb of wood, bark, and root had arisen alone from water.

Thus for answering a stinging question van Helmont performed a noteworthy adequate experiment, but he was unable to draw the correct conclusions from it because the prerequisites were lacking. It took more than 150 years of research in chemistry to explore the composition of air and the basic constituents of plants.

TABLE 2.1. The pneumatic chemists.

Gaseous compound	Discoverer	Year of discovery
Hydrogen	Henry Cavendish (1731–1810)	1766
Oxygen	Carl W. Scheele (1749–1819)	1771
Oxygen	Joseph Priestley (1733–1804)	1772
Nitrogen	David Rutherford (1749–1819)	1772
Ammonia (NH_3)	Joseph Priestley	1774
HCl	Joseph Priestley	1774
Nitrous oxide (N_2O)	Joseph Priestley	1774
Carbon dioxide (CO_2)	Torbern Bergman (1735–1785)	1774
	Antoine L. Lavoisier (1743–1794)	1775
Hydrogen sulfide (H_2S)	Carl W. Scheele	1776
Methane (CH_4)	Alexander Volta (1745–1827)	1776
Air analysis	Henry Cavendish	1783

Step by step, the tools used for quantitative determinations were improved (Table 2.1). The balance was known since ancient times. Robert Boyle introduced the pneumatic vessel and Joseph Priestley used mercury as barrier fluid. Stephen Hales invented the gasometer; Henry Cavendish, the eudiometer. The great masters of gas analysis, Cavendish, Priestley, Carl W. Scheele, and Torbern Bergman, worked at almost the same time and some discoveries were made by them independently from the others. Cavendish was the most versatile and successful among them. When he studied the composition of air and measured the content of oxygen and nitrogen he found that about 1% was missing. This difference was due to the noble gases discovered about 100 years later by John William Rayleigh. Cavendish's research indicated the precision of the methods developed and used by the pneumatic chemists. The work of the pneumatic chemists provided the knowledge of those gases involved in the gas metabolism of microorganisms.

Gas analysis prepared the tools for the elementary analysis of natural substances. And it was Scheele who prepared from plants the first organic compounds, such as tartaric acid, malic acid, oxalic acid, gallic acid, citric acid, lactic acid, and glycerol. It is justified to regard him as the founder of organic chemistry. The other great chemist who started the series of discoveries in gas analysis was Antoine Laurent Lavoisier. He understood that combustion depends on the presence of oxygen and thus replaced the phlogiston theory of G.E. Stahl with a new combustion theory. Lavoisier made the first experiments to determine the composition of organic compounds and found carbon, hydrogen, and oxygen as their constituents. Lavoisier's work and theory were soon accepted, and he is considered to be the first to design the basic methods of the elementary analysis of organic compounds. After his early death, the work was not continued until Joseph Louis Gay-Lussac (1778–1850) and Jöns Jakob Berzelius (1779–1848) and, later, Justus Liebig (1803–1873), Jean Baptiste Dumas (1800–1884), and their

TABLE 2.2. Discovery and first elementary analysis of substrates and products of microorganisms.

Substance	Year of first mention (author)	Year of first analysis (author)
Oxalic acid	1776 (Scheele)	1811 (Gay-Lussac)
Acetic acid	1783 (Berthollet)	1814 (Berzelius)
Succinic acid	~1600	1815 (Berzelius)
Tartaric acid	1770 (Scheele)	1830 (Berzelius)
Malic acid	1785 (Scheele)	1830 (Liebig)
Benzoic acid	1780 (Scheele)	1832 (Wöhler, Liebig)
Formic acid	1761 (Marggraf)	1832 (Pélouze)
Glucose	1792 (Lowitz)	1834 (Liebig)
Sucrose	1747 (Marggraf)	1834 (Liebig)
Mannitol	1806 (Proust)	1834 (Liebig, Oppermann)
Gallic acid	1785 (Scheele)	1834 (Pélouze)
Pyruvic acid	1830 (Berzelius)	1835 (Berzelius)
Methanol		1835 (Dumas, Péligot)
Glycerol	1779 (Scheele)	1836 (Pélouze)
Phenol	1834 (Runge)	1842 (Laurent)
Fumaric acid	1833 (Winckler)	1843 (J. Pélouze)
Propionic acid	from plants	1844 (Gottlieb)
Glycolic acid		1851 (Strecker)
Citric acid	1784 (Scheele)	1851 (Rochleder, Willigk)
Catechol	1825 (Faraday)	1851 (Wagner)
Glyoxylic acid		1856 (Debus)
Lactic acid	1780 (Scheele)	1858 (Wurtz)
Isopropanol		1862 (Friedel)
Hexadecane		1869 (Zincke)
Ethanol	1796 (Lowitz)	1875 (Gutzeit)
Lactose	17th century	1879 (Demole)
Ribose		1881 (Fischer, Piloty)
Fructose	~1800 in fruits	1881 (Jungfleisch, Lefram)
Isocitric acid		1887 (Fittig)
Mannose		1888 (Fischer, Hirschberger)
Oxaloacetic acid		1895 (Michael, Bucher)
α-Ketoglutaric acid		1908 (Blaise, Gault)
Deoxyribose		1927 (Meisenheimer, Jung)

collaborators and students succeeded in the analysis of many compounds (Table 2.2).

Liebig completed the methods to determine the carbon and hydrogen content of organic compounds. Table 2.2 shows the most important compounds involved in basic biochemistry. The table was composed by consulting Gmelin's (1829) handbook and the pertinent original literature and contains the author and year of the original isolation and nomination of some organic compounds, alcohols, and sugars and in addition the author and year of the first publication of the first correct analysis of the composition of the compound. The table indicates that the most important compounds involved in basic metabolism were analyzed by Berzelius, Liebig,

Pélouze, and Dumas. Knowledge of the compounds enabled the biologists to speculate about the metabolism of plants, animals, and microorganisms on a scientific basis; to use some pure organic compounds for comparison and as substrates; and to design the corresponding experiments. The commercial availability enabled chemists to study the organic compounds more closely. One of the outstanding developments was started with the studies of Eilhard Mitscherlich on isomorphism of crystals, which was continued by Louis Pasteur (1822–1895) on the tartaric acids. The discoveries made early in his scientific career motivated Pasteur to spend his life performing research in the chemistry of microorganisms. By in 1858–1861, he had studied alcoholic fermentation by yeast and the production of lactic, acetic, and butyric acids by bacteria; thus he discovered anaerobic energy conversion by fermentation, "la vie sans l'air."

The methods for elementary analysis allowed researchers to study several fermentations in more detail. There was enough evidence indicating that fermentation and putrefaction occurred under anoxic conditions and that fermentation and putrefaction are different from each other only by the products. As Pasteur was a chemist himself, some chemists and physiologists of the time did not find it below their dignity to study putrefaction and to choose the dirtiest among the anoxic natural ecosystems, sewer sludge (Kloakenschlamm), as a model system. The outstanding pioneer of the analysis of the products of putrefaction was Felix Hoppe-Seyler (1825–1895). He added sugar or organic acids to sewer sludge. For example, formic acid was fermented to carbon dioxide and hydrogen and acetic acid, to carbon dioxide and methane. In the presence of sulfate, acetic acid was converted to carbon dioxide and hydrogen sulfide. With some goodwill we can assign to Hoppe-Seyler the discovery of the anaerobic food chain, in which gaseous hydrogen plays a prominent role. These studies, published in 1876 and 1886, were done with crude cultures. Hoppe-Seyler (1876) concluded: "The number of reductions of organic substances in putritive processes is obviously extraordinary great, . . . the formation of mannitol from galactose or glucose, of propionic acid from lactic acid, and succinic acid from tartaric or malic acids are such . . . reductions." The products of putrefaction could at this time, however, not yet be attributed to the activities of specific bacteria.

The data obtained from studying crude cultures were not as bad as one would assume. For example, the stoichiometric relationships between the consumption of sugar and the production of alcohol and carbon dioxide were determined even before yeast became known as the causative agent of alcoholic fermentation (Gay Lussac 1810). Furthermore, the general equation for the formation of propionic acid was calculated by A. Fitz in Strassburg in 1878 on the basis of studies with crude cultures inoculated with cow excrements:

$$3 \text{ lactate} \rightarrow 2 \text{ propionate} + \text{acetate} + CO_2 + H_2O. \tag{2.1}$$

It did not deviate from the values found with pure cultures (Freudenreich and Orla-Jensen 1906; van Niel 1928). The analytical methods allowed the determination of the products of glucose fermentation by *Escherichia coli*, and the general equation calculated (Harden 1901) does not differ from that accepted today. Harden even correctly deduced that hydrogen arises from formic acid; the *Aerobacter* modification was recognized by Harden in 1906. Thus the methods for the analysis of fermentation products were completed within the nineteenth century.

Physics and chemistry provided further methods useful for investigating and describing bacteria and understanding the biochemical background. Each progress in physics contributed to chemical analysis. The important optical methods of analysis were spectroscopy, spectrography, flame photometry, colorimetry, and spectrophotometry. They helped differentiate among pigments of green plants, phototrophic bacteria, blood, cells, muscle cells, and cytochromes as well as accessory pigments. Little inventions improved the practical work in the laboratory, such as the Bunsen burner (1857) and the water jet pump (1868). Physicochemistry added many principles and methods of determinations, such as hydrogen ion concentration, the pH term, a variety of titration methods, and polarimetry. Chromatography methods, also started in the middle of the nineteenth century, were developed further by Michail Tswett, a botanist, but not introduced as a general analytical tool until 1941 by Martin and Synge. It took a long time to develop the method of electrophoresis on paper or in gels to its present simplicity and routine application.

Toward Understanding Modes of Biological Energy Conversion

At least from the time of van Helmont on, the chemists when separating and describing gases usually examined the effect of the gases on animals and plants. Lavoisier (1777) understood that aerobic respiration is the process in which oxygen is consumed and carbon dioxide is produced, shortly after the discovery of oxygen (1771). However, it took about half a century to make aerobic respiration more comprehensible from a physical point of view (Joule 1843; Mayer 1845; and many others).

Oxygenic photosynthesis was the second mode of energy conversion whose principles were understood. Experiments showing that oxygen is involved, carbon dioxide is the source of carbon, and light is the energy source was contributed by J. Ingenhouse (1730–1799), J. Senebier (1742–1809), and T. de Saussure (1767–1845). J.R. Mayer (1814–1878) provided the most comprehensible enlightening explanations.

The third mode of energy conversion, fermentation, was discovered by Pasteur. After examining alcoholic fermentation by yeast, he studied several bacterial fermentations, including butyric acid fermentation and its causative bacterium *Vibrio butyrique*, which evidently obtained energy under anoxic conditons.

In chronological order, anoxygenic photosynthesis was the fourth mode of energy conversion. Although several species of purple bacteria had been found in nature and their green and red pigments had been described, their dependence on light for growth was not recognized before the photophysiological experiments of Theodor Wilhelm Engelmann (1843–1909) led to the conclusions that purple bacteria are phototrophs (1888). The reason for the absence of oxygen evolution was, however, explained later by H. Molisch (1907), J. Buder (1919), and C.B. van Niel (1931).

The fifth mode of energy generation, chemosynthesis or lithotrophy, was discovered by the Russian botanist Sergius N. Winogradsky (1856–1953), when he was working in the laboratory of Anton de Bary (1831–1888) in Strassburg in 1887. Originally, he intented to reevaluate the monomorphism–pleomorphism controversy and chose *Beggiatoa* as a model organism. He collected the black mud surface of ponds and observed *Beggiatoa* filaments under the microscope. He saw the filaments accumulate sulfur droplets intracellulary, when he added hydrogen sulfide, and saw the droplets disappear when hydrogen sulfide was absent. He saw the filaments grow and the cells divide. The bacteria seemed to prefer the absence of organic substrates. Thus Winogradsky concluded that *Beggiatoa* gains metabolic energy from the oxidation of hydrogen sulfide and the accumulated sulfur, a type of respiration with inorganic hydrogen donors that he called inorgoxidation (chemosynthesis, today lithotrophy). From where the inorgoxidizers gain the cellular carbon Winogradsky discovered when studying the nitrifyers (1891). He succeeded in isolating nitrifyers from soil and growing them in a purely mineral medium. By determining the amount of nitrite and nitrate produced from ammonia as well as the cell carbon formed, he showed a constant stoichiometric ratio to exist between the products of ammonia oxidation and assimilation of carbon. His conclusion—that in these bacteria the process of inorgoxidation is linked to autotrophic CO_2 fixation—was fully justified. Thus Winogradsky discovered a new mode of living which we call today chemolithoautotrophy. It enables a large metabolic group of bacteria to grow in mineral solutions with inorganic hydrogen donors, such as ammonia, nitrite, sulfur, hydrogen sulfide, thiosulfate, ferrous iron, molecular hydrogen, and carbon monoxide and with carbon dioxide as the sole carbon source.

A sixth type of bacterial energy conversion is anaerobic respiration. The reduction of nitrate to nitrogen and N_2O had already been shown in the 1880s. The first sulfate-reducing bacterium, the strict anaerobic *Spirillum desulfuricans*, had been grown in pure culture by M.W. Beijerinck (1851–1931) by 1895 and shown to be able to grow on simple organic acids as hydrogen donors and sulfate as hydrogen acceptors. After the discovery of a cytochrome (C_3) by Postgate (1954), the energy-conversion process, earlier called "dissimilatory sulfate reduction," was renamed sulfate respiration. "Dissimilatory nitrate reduction" by facultative anaerobic bacteria was renamed nitrate respiration. And later it was discovered that methane formation from CO_2 and hydrogen and ferric iron reduction are anaerobic respiration processes, too.

Harry Peck's Scientific Career

Having reviewed the development of the knowledge about the important gases and organic compounds involved in metabolism and about the modes of energy conversion in bacteria that were known in 1950, when Harry Peck had to decide about his way into science, I would like to add a few remarks on his career and a word of thanks. Peck chose microbiology for his bachelors and masters degree work. With Cochrane at Wesleyan University (Connecticut) Peck studied the basic metabolism of *Streptomyces coelicolor*, using resting cells and cell-free extracts and employing ^{14}C-labeled acetate and glucose, to find out whether the tricarboxylic acid cycle and the pentose phosphate pathway were present in this genus. Then worked with Howard Gest to get his Ph.D. (1955). Gest was familiar with gaseous hydrogen and, with Martin Kamen, had discovered the photoproduction of molecular hydrogen by *Rhodospirillum rubrum* (done in Washington University, 1949), and had become interested in the formic hydrogen lyase system of *E. coli*. Thus Peck became familiar with hydrogenases in facultative anaerobic bacteria and clostridia, discovered NAD reduction with hydrogen, encountered the diversity of hydrogenases, worked with resting cells and cell-free extracts, and compared various assays (deuterium-hydrogen exchange included).

Peck then became interested in sulfate-reducing bacteria, which he had got to know in Gest's laboratory. To study the reduction of sulfate, Peck worked in Fritz Lipmann's laboratory in Massachussetts General Hospital (1956) and with Lipmann at Rockefeller University (1957). Lipmann started work on active sulfate in 1954 with Helmut Hilz as a postdoctoral fellow and studied the activation of sulfate to APS and PAPS. Lipmann had left the active sulfate projects by 1957 and started, at Rockefeller University, the studies on protein synthesis. Peck published one paper on the reduction of sulfate with hydrogen in extracts of *Desulfovibrio desulfuricans* (1959) and one on APS as an intermediate on the oxidation of thiosulfate by *Thiobacillus thioparus* (1960).

In 1958, Peck joined the staff of the enzymology group of the Oak Ridge National Laboratory and continued there to work on sulfur metabolism of chemolithoautotrophic bacteria. In 1965, he was called to Athens. Before moving there, he spent a year in the laboratory of Jacques Senez and Jean LeGall. And then he explored the research niche of hydrogenase and sulfur metabolism, which is a gold mine still today, in various directions.

Acknowledgment. I am very grateful to Harry Peck and Howard Gest for having made me acquainted with their work by sending their reprints to me. In the DDR, where I worked, we did not have access to the American journals published during and after World War II. The first of Peck reprints that I ordered were then accompanied by more, and subsequent publications followed. I also thank Dr. Günther Beer, Chemistry Department, Georg-August-Universtität Göttingen, for help with composing the tables.

Reference

Gmelin L. 1829. Handbuch der theoretischen Chemie. 3. Aufl. 2. Band 1. Abtheilung. Frankfurt: F. Varrentrapp.

Suggested Reading

Lieben F. 1935. Geschichte der physiologischen Chemie. Leipzig and Wien: Deuticke.

Schlegel HG. 1999a. Bacteriology paved the way to cell biology: a historical account. In: Lengeler JW, Drews G, Schlegel HG, editors. Biology of the prokaryotes. Stuttgart: Thieme, Blackwell.

Schlegel HG. 1999b. Geschichte der Mikrobiologie. Acta historica Leopoldina 28. Halle/Saale: Deutsche Akademie der Naturforscher Leopoldina.

Szabadváry F. 1966. Geschichte der analytischen Chemie. Braunschweig: Vieweg und Sohn.

Wehefritz V, Kováts Z, editors. 1994. Bibliography on the history of chemistry and chemical technology, 17th to the 19th century. Munich and Paris: Saur.

3
Mechanism of Hydrogen Activation

Simon P.J. Albracht

In this report a few recent contributions of my group to some exciting developments in the field of hydrogenases are described. These new findings have added to our understanding of how hydrogenases work. Hydrogenases (reaction: $H_2 \Leftrightarrow 2H^+ + 2e^-$) are found in many microorganisms. They are functional in the anaerobic degradation of organic substrates, enabling an organism to dispose of excess reducing equivalents in the form of hydrogen. Other organisms can use this released hydrogen as electron and energy sources for growth.

There are two classes of metal-containing hydrogenases. The [NiFe]-hydrogenases have a $NiFe(CN)_2(CO)$ group as active site (Fig. 3.1), which is attached to the protein via four Cys thiols. These enzymes are usually involved in the uptake of hydrogen. For the [NiFe]-hydrogenase from *Desulfovibrio vulgaris* Miyazaki, it has been proposed that one of the two CN ligands is replaced by SO (Higuchi et al. 1997). The [Fe]-hydrogenases, functional in the production of dihydrogen, contain an Fe-Fe site, where the iron atoms are also coordinated by CO, CN, and thiols. Here the direct connection to the protein involves only one Cys residue. The two bridging thiols are not provided by the protein but presumably by a 1,3-dithiopropanol molecule (Nicolet et al. 1999, 2000). All the enzymes contain at least one [4Fe-4S] close to the bimetallic site. In [Fe]-hydrogenases this proximal cluster is directly attached to an iron atom (called Fe1) of the Fe-Fe site via a Cys thiol bridge. In [NiFe]-hydrogenases, the conserved proximal cluster is about 12 Å from Ni-Fe site and is located in a different subunit. As discussed later, I assume that the bimetallic site plus the conserved, proximal cluster function as an effective two-electron-accepting unit in nearly all hydrogenases. Most enzymes have additional Fe-S cluster, which will not be discussed here. Some relevant Fourier Transform Infrared (FTIR) spectra, which provide essential complementary information for the active-site structures, are shown in Figure 3.1 and are discussed hereafter.

20

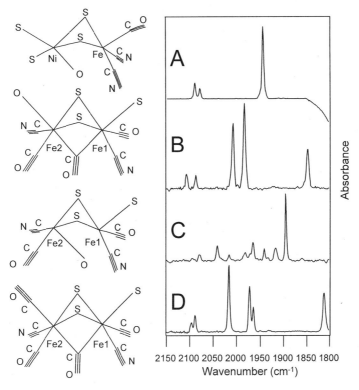

FIGURE 3.1. Structures and FTIR spectra of the bimetallic sites in metal-containing hydrogenases. **A**, Active site of the inactive *Desulforibrio gigas* [NiFe]-hydrogenase (2frv) (Volbeda Garcin, Piras et al. 1996) and FTIR spectrum in the 2150–1800 cm^{-1} spectral region from the *Allochromatium vinosum* enzyme in the aerobic, inactive state. **B**, Fe–Fe site from *Clostridium pasteurianum* ([Fe]-hydrogenase I crystallized in the presence of 2 mM dithionite (1feh) (Peters et al. 1998) and the FTIR spectrum of the *D. vulgaris* [Fe]-hydrogenase in the aerobic inactive state. **C**, Fe–Fe site of the active *Desulfovibrio desulfuricans* [Fe]-hydrogenase crystallized under 10% hydrogen (1hfe) (Nicolet et al. 1999) and the FTIR spectrum of active *D. vulgaris* [Fe]-hydrogenase treated with 100% hydrogen. **D**, Fe–Fe site from the *C. pasteuri-anum* enzyme treated with CO (1c4a and 1c4c) (Lemon and Peters 1999) and the FTIR spectrum of active *D. vulgaris* [Fe]-hydrogenase treated with CO. The pictures were deduced from the Protein Data Bank (PDB) crystal structure files, invoking the information from FTIR studies on the *A. vinosum* [NiFe]-hydrogenase (Happe et al. 1997; Pierik et al. 1999) and the *D. vulgaris* Hildenborough [Fe]-hydrogenase (Pierik et al. 1998a).

FTIR Spectra and Structure

The structure of the active sites in both classes of hydrogenases has emerged only from a combination of the three-dimensional data and the results from FTIR studies on closely related enzymes. An end-on bound CO group to the iron atom in the Ni-Fe site of the *D. gigas* enzyme could explain the band at $1944\,cm^{-1}$ (Fig. 3.1A) in the enzyme from *A. vinosum* (formerly called *Chromatium vinosum*). The symmetrical and antisymmetrical stretch vibrations from the two CN groups in the latter enzyme (at 2090 and $2079\,cm^{-1}$, respectively), could explain two of the diatomic ligands in the former enzyme. Likewise, the diatomic ligands in the Fe-Fe sites of the *C. pasteurianum* and *D. desulfuricans* enzymes can now be recognized in the FTIR spectra from the *D. vulgaris* Hildenborough enzyme (Fig. 3.1B–D). The assumption of three CO groups, two end-on ones and one bridging one, in the *C. pasteurianum* enzyme crystals, prepared under nitrogen in the presence of 2mM dithionite (Fig. 3.1B), can explain the bands at 2007, 1983, and $1847\,cm^{-1}$, respectively, in the aerobic inactive *D. vulgaris* enzyme. The two CN bands (2106 and $2087\,cm^{-1}$) found in the latter enzyme can explain one of the diatomic ligands on each iron atom in the former enzyme. In the *C. pasteurianum* enzyme crystals prepared with dithionite under nitrogen an oxygen species is bound to Fe2. This makes each of the iron atoms six coordinate, and so the enzyme may not be active in this state.

The FTIR spectrum of the hydrogen-activated *D. vulgaris* enzyme, subsequently treated with CO, showed two CN bands, three bands from end-on bound CO and one from a bridging CO (Fig. 3.1D). Lemon and Peters (1999) determined the structure of the *C. pasteurianum* enzyme with added CO, which is bound in a light-sensitive way (Bennett et al. 2000). They found that the oxygen species on the Fe2 atom (Fig. 3.1B) was replaced by a diatomic molecule, probably CO (Fig. 3.1D). This fits perfectly with the FTIR spectrum of the CO-inhibited *D. vulgaris* enzyme. Two end-on bound CO molecules on Fe2 are expected to have a vibrational interaction, whereby the symmetrical and antisymmetrical stretch vibrations can differ considerably in frequency. This probably can explain two of the three bands in the $2050–1950\,cm^{-1}$ spectral region (Fig. 3.1D). Studies with ^{13}CO will provide further insight into the interpretation of the spectrum. Crystals of the *D. desulfuricans* [Fe]-hydrogenase prepared under 10% hydrogen (Nicolet et al. 1999) did not show a bridging ligand (Fig. 3.1C). The interpretation of the diatomic ligands was aided by the FTIR spectra of the *D. vulgaris* enzyme. The latter enzyme did not show a band from a bridging CO when reduced with hydrogen. The FTIR spectrum is, however, difficult to interpret and it is not yet clear what happens to the bridging CO ligand upon reaction of the active site with hydrogen.

As the enzymatic reaction involves H_2, H^+, and electrons and the active sites are deeply buried in the protein, the X-ray crystallographers have deduced possible pathways for these substrates to enter and leave the enzyme. The Fe-S clusters are no doubt involved in the transfer of electrons to and from the active bimetallic site. One or more hydrophobic channels, leading directly to one of the metal atoms in the bimetallic site, have been found in the protein structures (Montet et al. 1997; Frey 1998; Nicolet et al. 2000). In [NiFe]-hydrogenases, the channel points to the nickel atom. In [Fe]-hydrogenases the channel points to Fe2. Hence, it is likely that these metal atoms are directly involved in hydrogen activation. In the following, several aspects concerning the possible mechanism of action of both classes of hydrogenases are discussed.

Heterolytic Splitting and Hydride Oxidation

The splitting of hydrogen is presumably a heterolytic process (Krasna 1979). The conversion of *para*-H_2 to *ortho*-H_2, a reaction apparently not involving any redox reactions, is inhibited in ^2H_2O. This was interpreted as:

$$E + p\text{-}H_2 \leftrightarrow EH^- + H_a^+ \tag{3.1}$$

$$EH^- + H_b^+ \leftrightarrow E + o\text{-}H_2 \tag{3.2}$$

In D_2O, HD was found instead of o-H_2. It is presently assumed that binding of hydrogen to a metal ion in the bimetallic active site weakens the H-H bond sufficiently to enable this reaction. Oxidation of the hydride is expected to be a two-electron process, and hydrogenases should, therefore, contain a redox unit capable of accepting these two electrons simultaneously. I assume here that the bimetallic center plus the conserved proximal Fe-S cluster perform this task.

[NiFe]-Hydrogenases

Electron paramagnetic resonance (EPR) studies on [NiFe]-hydrogenases indicated that the active site can exist in at least seven different states, four inactive states and three active ones (Albracht 1994). Inspection with FTIR confirmed this (Bagley et al. 1995; De Lacey et al. 1997). The inactive states (called ready and unready) have an extra oxygen species (Van der Zwaan et al. 1990) near to the nickel atom, presumably spaced between the nickel and iron atoms (Volbeda et al. 1995, 1996) (Fig. 3.1A). For the *D. vulgaris* Miyazaki enzyme, this species has been proposed to be sulfur rather than oxygen (Higuchi et al. 1997). The O/S species blocks the rapid activation of hydrogen. Upon reductive activation, this species leaves the active site, presumably as H_2O (or H_2S). In this report, only the three active states are considered (Ni_a-S, Ni_a-C*, and Ni_a-SR) (Fig. 3.2). The states are observed in many [NiFe]-hydrogenases, like the ones from *D. gigas* and *A. vinosum*;

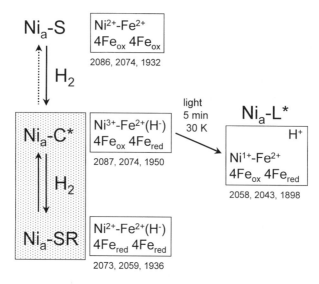

FIGURE 3.2. Overview of the three states of the active standard [NiFe]-hydrogenase from *A. vinosum*. The wavelengths indicate the infrared frequencies for the two CN groups and the CO group, respectively. The reactions with hydrogen are fast (*thick arrows*) or extremely slow (*dotted arrow*). Protons are not shown. *a*, active; *C*, C state; *L*, light-induced state; *R*, reduced; *S*, EPR silent; *, the active site in this state is a $S = {}^1/_2$ system (detectable by EPR); *4Fe*, [4Fe-4S] cluster.

these will be referred to as standard [NiFe]-hydrogenases. As discussed below, there are also [NiFe]-hydrogenases that can adopt only one or two of these states.

The activity of active enzyme is usually assayed with artificial electron acceptors or donors (usually dyes). It has been shown that when the *A. vinosum* enzyme is directly attached to an electrode, its hydrogen-oxidizing activity is much higher than that obtained with dyes (Pershad et al. 1999). Even under 10% hydrogen, the diffusion of hydrogen to the active site was shown to be the rate-limiting step. This means that in normal assays, the reaction with dyes is probably rate limiting. It also indicates that electron transfer and the ejection of H^+ by the enzyme are fast processes.

Competition Between Hydrogen and Carbon Monoxide

It has long been known that carbon monoxide acts as a competitive inhibitor of most hydrogenases. This indicates that CO and hydrogen compete for the same binding site in the enzyme. EPR studies showed that under certain conditions, CO can directly bind to nickel (Van der Zwaan et al. 1986, 1990) in the Ni_a-C* state. Both, the Ni_a-C* state and the induced,

EPR-detectable CO state converted to one and the same EPR state, the Ni_a-L* state, when illuminated at temperatures <50K. For the Ni_a-C* state, but not for the CO-induced state, the rate of this conversion showed a strong isotope (H/D) effect (Van der Zwaan et al. 1985, 1986). At the time, when only nickel was known to be in the active site, this was interpreted as an argument for the binding of hydrogen (H_2 or H⁻) or CO to the same coordination site on nickel. Although ^{13}CO (the nuclear spin of ^{13}C is $\frac{1}{2}$) gave a nearly isotropic hyperfine interaction of 30 G with the nickel-based unpaired spin, the hyperfine interaction with hydrogen (a much stronger I $=\frac{1}{2}$ magnet) was only about 5 G. This indicated that the early interpretation of hydrogen binding to nickel was not so straightforward.

The initial FTIR studies (Bagley et al. 1994) also demonstrated binding of CO to the active site, at that time still believed to consist of nickel only. At the present level of understanding of the active Ni-Fe site, these data indicate that the $2060 cm^{-1}$ band of the externally added CO is best interpreted as CO binding (end-on) to nickel and not to iron. A higher frequency is expected for binding to iron (like that of the internal CO bound to iron), and vibrational interaction with the internal CO would be expected as well. None of the inactive states can bind CO; activation is absolutely required (Bagley et al. 1995). Both EPR and FTIR studies indicate that it is the Ni_a-S state that binds CO best. Under 1 bar of CO, active enzyme is completely in the Ni_a-S·CO state; the Fe-S clusters in this active (but inhibited) enzyme can be reduced or oxidized, without any effect on the status of the active site (Surerus et al. 1994).

Recent rapid-mixing rapid-freezing studies have further elucidated the binding of CO and hydrogen. Figure 3.3 summarizes these studies (Happe et al. 1999). The experiment demonstrated that (in the absence of light) CO does not react with the Ni_a-SR state or with the Ni_a-C* state. It is only on the level of the Ni_a-S state that CO competes with hydrogen for binding to the active site. With hydrogen, the active site in the Ni_a-S state is converted to Ni_a-C* (and Ni_a-SR):

$$Ni_a\text{-}S + H_2 \rightarrow Ni_a\text{-}C* \tag{3.3}$$

As indicated in Figure 3.2, this is probably just the binding of hydrogen (presumably as H⁻ + H⁺) to the active site at some distance from nickel. This binding induces the oxidation of the nickel center ($Ni^{2+} \rightarrow Ni^{3+}$), whereby the electron initially enters the proximal Fe-S cluster. CO can bind to the divalent nickel in the Ni_a-S state and thereby fixes the active site in the Ni_a-S·CO state:

$$Ni_a\text{-}S + CO \rightarrow Ni_a\text{-}S\text{·}CO \tag{3.4}$$

So, although the binding sites for CO and hydrogen are not expected to involve the same coordination site, the Ni_a-S state cannot bind hydrogen and CO at the same time: The two gasses compete for binding.

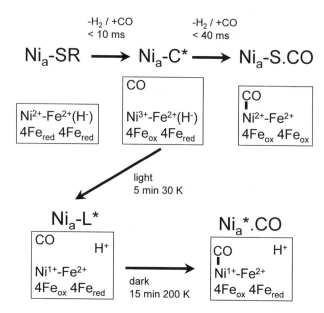

FIGURE 3.3. Reactions of CO with hydrogen-reduced *A. vinosum* hydrogenase (Happe et al. 1999). Starting with enzyme plus 0.8mM hydrogen (equivalent to 1 bar; Ni_a-SR state), a transient Ni_a-C* state was detected within 10ms when the solution was mixed with CO-saturated buffer (in the dark). Thereafter, a rapid decline of the Ni_a-C* state was noticed (conversion into Ni_a-S·CO). The sample obtained at 10ms could be converted to the Ni_a-L* state by illumination at 30K. Raising the temperature to 200K did not reverse process; instead a state was detected in which CO was directly bound to nickel (Ni_a*·CO). Protons are not shown.

We could show that 10ms after mixing hydrogen-reduced enzyme with CO-saturated buffer, CO was present very close to nickel, although it did not bind to the active site. A sample obtained 10ms after mixing in the dark (frozen in isopentane at 130K) showed the same amount of Ni_a-C* signal in EPR at 30K as a sample obtained by mixing with argon-saturated buffer. Upon illumination, both samples showed the well-known conversion into the Ni_a-L* state. Illumination removes an exchangeable hydrogen species from the active site (Van der Zwaan et al. 1985; Chapman et al. 1988; Fan et al. 1991; Whitehead et al. 1993). At the same time, the Ni-EPR signal drastically changes, and the $v(CO)$ of the intrinsic CO group bound to iron shifts $52\,cm^{-1}$ to a lower frequency (Fig. 3.2), indicating a large increase in charge density on the iron atom. In the absence of CO, this light-induced process is reversed by warming to 200K. Upon warming of the CO-treated sample to 200K in the dark, the original state did not return, however.

Instead, an EPR signal characteristic for the binding of CO to Ni_a-C* (Van der Zwaan et al. 1986, 1990) was observed (here called Ni_a*·CO). The results were interpreted as follows.

Within 10 ms after mixing, a CO molecule is located very close to the nickel atom. It does not bind, presumably because the low charge density at the nickel site (the nickel atom and its surrounding ligands) does not encourage this. This suggested to my group that the formal oxidation of nickel in the Ni_a-C* state is trivalent. Light induces the dissociation of a hydride, bound at or close to iron, such that the two electrons remain on the active site. Most of this charge density flows to the positive nickel plus its thiol ligands, drastically changing the EPR signal. Due to a limited mobility at 30K, the nearby CO molecule cannot yet bind, but if we raise the temperature to 200K, it can bind to the nickel atom with its attractive large-charge density. The CO thereby fixes this charge density on the nickel site. An FITR spectrum of this state is not yet known. This experiment also demonstrated the competition of hydrogen and CO for binding to the active site: If H_2/H^- is bound (presumably at or close to iron), CO cannot bind. Only after photodissociation of the hydrogen species, can the CO (present close to nickel) compete with the hydrogen species for binding, even at 200K. In view of the much weaker hyperfine interaction of ^1H versus ^{13}CO, the competition probably does not involve the same coordination site. As the competition on the level of the Ni_a-C* state is not observed in the dark, it is presumably not relevant for the inhibitory action of CO during turnover.

A [NiFe]-Hydrogenase Resistant to Carbon Monoxide and Oxygen

The soluble, NAD-reducing [NiFe]-hydrogenase (SH) from *Ralstonia eutropha* is capable of activating and oxidizing hydrogen in the presence of oxygen. In fact, this bacterium can thrive on Knallgas. The NADH produced by the SH is oxidized by a respiratory chain, including a terminal oxidase. The enzyme is also insensitive to carbon monoxide. The sequence information on this enzyme (Tran-Betcke et al. 1990) predicts the presence of a normal Ni-Fe site in the large hydrogenase subunit (HoxH). The small hydrogenase subunit (HoxY) is smaller than the one from the *D. gigas* enzyme and misses the C-terminal part with the Cys residues for two more Fe-S clusters. HoxY holds only the conserved Cys residues for the proximal cluster. In addition, the intact enzyme contains two more subunits (HoxF and HoxU), which form an NADH-dehydrogenase (or diaphorase) module.

What makes this enzyme insensitive to oxygen and CO? Our studies on this enzyme (Happe et al. 2000) taught us three things: (1) we could not

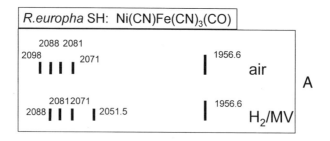

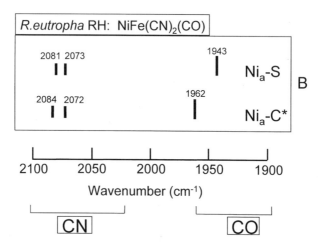

FIGURE 3.4. The infrared bands in the 2120–1900 cm^{-1} spectral region of the soluble NAD-reducing hydrogenase (SH) and the regulatory hydrogenase (RH) from *Ralstonia eutropha*. **A**, Aerobic inactive SH as isolated and SH after reduction by hydrogen in the presence of 5 mM methyl viologen (MV) for 60 min at 30°C (Happe et al. 2000). **B**, Aerobic RH (Ni$_a$-S state) and upon reaction with H$_2$ (Ni$_a$-C* state) (Pierik et al. 1998b).

find any significant EPR signals from nickel under any redox condition, (2) the FTIR spectrum of SH is similar to that of standard [NiFe]-hydrogenases but in addition shows two extra bands in the CN region (Fig. 3.4), and (3) only one of the four bands ascribed to metal-bound CN shifted upon redox changes. As the v(CO) at 1956 cm^{-1} is not shifting, it was concluded that the CN responsible for the 2098 cm^{-1} band is not bound to iron but to nickel (Happe et al. 2000). This is in striking contrast to the EPR and FTIR behavior of the *A. vinosum* or *D. gigas* enzymes. Hence a Ni(CN)Fe(CN)$_3$(CO) prosthetic group was proposed in which the iron site is six coordinate (two bridging thiolates, three CN, one CO) and the nickel site, five coordinate. The present idea is that this introduces so much steric hindrance, that the nickel site is available only for hydrogen and not for the larger CO or oxygen molecules. The resting state of the aerobically purified SH is not active. Instantaneous full activation under hydrogen occurs only upon the addition of a small amount (5μM) of NADH (Schneider and Schlegel 1976). With hydrogen alone, it takes at least 30 min at 30°C (Happe et al. 2000). We currently assume that reducing equivalents from NADH, which can rapidly reach the Ni-Fe site via the diaphorase module, removes a species (possible oxygen), blocking the sixth ligand position of the nickel atom. As a result, the infrared band from the nickel-bound CN group shifts from 2098 to 2052 cm^{-1} (Fig. 3.4).

The lack of any EPR signal of the Ni-Fe site and the lack of shifts in the bands from the $Fe(CN)_3(CO)$ moiety suggest that the Ni-Fe site is not involved in redox changes during the oxidation of hydrogen. The presence of nickel is essential for activity, however; and so it is proposed that the heterolytic cleavage of hydrogen occurs upon binding of hydrogen to Ni^{2+} in the Ni-Fe site. Such a cleavage was reported for a Ni^{2+} thiolate complex (Sellmann et al. 2000). Thereafter, the two reducing equivalents of the hydride are captured by another redox factor. It was found that the flavodoxin fold formed by the first 170 amino acids of the small hydrogenase subunit of the *D. gigas* enzyme is probably conserved in all hydrogenases (Albracht and Hedderich 2000). This flavodoxin fold forms the main part of HoxY. Schneider and Schlegel (1978) reported that the SH contains more than one FMN group and that optimal activity was obtained when two FMN molecules were bound to the SH. It was, therefore, proposed (Albracht and Hedderich 2000) that the HoxY subunit contains this second FMN molecule and that this FMN accepts the two electrons from the hydride. No FMN has been found in purified standard [NiFe]-hydrogenase, possibly because there is not enough space to accommodate this molecule. The experience with SH indicates that only one vacant coordination site on nickel is sufficient for hydrogen activation. This is what I suspect for standard [NiFe]-hydrogenases as well.

Hydrogen Sensor

Ralstonia eutropha also contains a protein called the regulatory hydrogenase (RH), which is required for the regulation of the biosynthesis of SH and the membrane-bound hydrogenase in this bacterium (Lenz and Friedrich 1998). Only when hydrogen is sensed by RH is biosynthesis of hydrogenases started. EPR and FTIR experiments with overexpressed crude RH (Pierik et al. 1998b) indicate that it contains a $NiFe(CN)_2(CO)$ site, like in normal [NiFe]-hydrogenases. This is in agreement with predictions from sequence analyses (Lenz and Friedrich 1998). The RH has some very special properties, however: (1) it can exist in only two redox states (Ni_a-S and Ni_a-C*, Fig. 3.4); (2) it does not bind CO, and the reaction with hydrogen is not perturbed by oxygen; and (3) its activity, as measured with

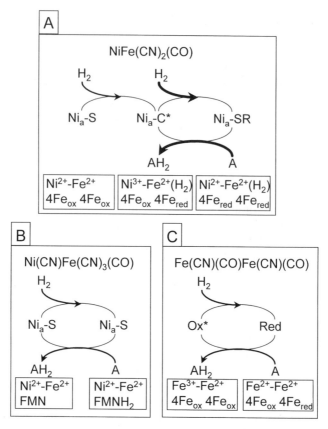

FIGURE 3.5. The reactions of hydrogen with the bimetallic sites in hydrogenases. **A**, Enzymes with a standard Ni-Fe active site. The binding of hydrogen to the Ni_a-S state leads to the Ni_a-C* state. It is assumed that the nickel ion is thereby oxidized and that the electron moves to the Fe-S clusters. For unknown reasons, in RH no further reaction with hydrogen is possible. In standard [NiFe]-hydrogenases only the Ni_a-C* and Ni_a-SR states are in rapid equilibrium with hydrogen (Coremans et al. 1990; Happe et al. 1999) (See Fig. 3.1). I propose here that the enzyme shuttles between these two states during rapid turnover of hydrogen (*thick arrows*). This is in line with the lack of such a high activity in the RH. Thus the binding of hydrogen to the Ni_a-S state brings the enzyme in the highly active Ni_a-C* state. The subsequent reaction with a second hydrogen molecule reduces the Ni_a-C* state to the Ni_a-SR state: One electron goes to nickel and the other goes to the proximal cluster. It cannot be excluded that the two electrons from hydride are transiently stored in the Ni-Fe site (whereby Ni^{3+} might be formally reduced to Ni^+) **B**, In the active site of the *R. eutropha* SH, no hydrogen binding is required to activate the Ni-Fe site. Starting with aerobic enzyme, the only requirement is to once remove a blocking ligand on nickel by reduction via the NADH-dehydrogenase module by a small amount of NADH. Thereafter, hydrogen is activated at the nickel atom, and the two electrons from H^- are transferred to the nearby FMN.

artificial dyes, is two orders of magnitude lower than that of normal [NiFe]-hydrogenases.

[Fe]-Hydrogenases

Early studies on [Fe]-hydrogenase I from *C. pasteurianum* (Erbes et al. 1975) already noted that the active enzyme is in rapid equilibrium with hydrogen and shows two different EPR-detectable states. These observations are still valid. Without hydrogen, a sharp, rhombic EPR signal (g_{123} = 2.099, 2.041, 2.001) is detected. In the presence of hydrogen, this signal disappears and a rather broad spectrum reminiscent to interacting [4Fe-4S]$^+$ clusters appears. The reaction of active enzyme with hydrogen is complete within 10 ms (Erbes et al. 1975).

Conclusion

The information described in this report leads me to the representation of the reactions of hydrogen with the [NiFe]- and [Fe]-hydrogenases shown in Figure 3.5.

Acknowledgments. I wish to acknowledge my present co-workers, Winfried Roseboom, Boris Bleijlevens, Sergei Kurkin, Eddy van der Linden, and Bart Faber, for their contributions to the ongoing research in my group and for stimulating discussions.

Figure 3.5. *Continued.* The H_2-NAD$^+$ reaction is inhibited neither in air nor in the presence of CO. **C,** The possible reactions of hydrogen with the Fe-Fe site of active [Fe]-hydrogenases. In the oxidized state, the bimetallic center shows a S = 1/2 EPR signal, presumably due to an Fe^{3+}-Fe^{2+} pair (an Fe^{2+}-Fe$^+$ pair cannot be excluded). Whether the unpaired spin is localized on iron (Pierik et al. 1998a) or elsewhere (Popescu and Münck 1999) is not known. Hydrogen is presumably reacting at the vacant coordination site on Fe2 (Fig. 3.1C). After the heterolytic splitting, the two reducing equivalents from the hydride are rapidly taken up by the Fe-Fe site (one electron) and the attached proximal cluster (one electron). Subsequently, the electron is transferred from the proximal cluster to the other Fe-S clusters in the enzyme. Under equilibrium conditions, the proximal cluster in the active enzyme appears to be always in the oxidized [4Fe-4S]$^{2+}$ state (Popescu and Münck 1999). Protons are not shown.

References

Albracht SPJ. 1994. Nickel hydrogenases: in search of the active site. Biochim Biophys Acta 1188:167–204.

Albracht SPJ, Hedderich R. 2000. Learning from hydrogenases: location of a proton pump and of a second FMN in bovine NADH: ubiquinone oxidoreductase (complex I) FEBS Lett 485:1–6.

Bagley KA, Duin EC, Roseboom W, et al. 1995. Infrared-detectable groups sense changes in charge density on the nickel site in hydrogenase from *Chromatium vinosum*. Biochemistry 34:5527–35.

Bagley KA, van Garderen CJ, Chen M, et al. 1994. Infrared studies on the interaction of carbon monoxide with divalent nickel in hydrogenase from *Chromatium vinosum*. Biochemistry 33:9229–36.

Bennett B, Lemon BJ, Peters JW. 2000. Reversible carbon monoxide binding and inhibition at the active site of the Fe-only hydrogenase. Biochemistry 39:7455–60.

Chapman A, Cammack R, Hatchikian EC, et al. 1988. A pulsed EPR study of redox-dependent hyperfine interactions for the nickel centre of *Desulfovibrio gigas* hydrogenase. FEBS Lett 242:134–8.

Coremans JMCC, van Garderen CJ, Albracht SPJ. 1992. On the redox equilibrium between H_2 and hydrogenase. Biochim Biophys Acta 1119:148–56.

De Lacey AL, Hatchikian EC, Volbeda A, et al. 1997. Infrared-spectroelectrochemical characterization of the [NiFe] hydrogenase of *Desulfovibrio gigas*. J Am Chem Soc 119:7181–9.

Erbes DL, Burris RH, Orme-Johnson WH. 1975. On the iron-sulfur cluster in hydrogenase from *Clostridium pasteurianum* W5. Proc Natl Acad Sci USA 72:4795–9.

Fan C, Teixeira M, Moura J, et al. 1991. Detection and characterization of exchangeable protons bound to the hydrogen-activating nickel site of *Desulfovibrio gigas* hydrogenase: a ^1H and ^2H Q-band ENDOR study. J Am Chem Soc 113:20–4.

Frey M. 1998. Nickel-iron hydrogenases: structural and functional properties. Struct Bonding 90:98–126.

Happe RP, Roseboom W, Albracht SPJ. 1999. Pre-steady-state kinetics of the reactions of [NiFe]-hydrogenase from *Chromatium vinosum* with H_2 and CO. Eur J Biochem 259:602–9.

Happe RP, Roseboom W, Egert G, et al. 2000. Unusual FTIR and EPR properties of the H_2-activating site of the cytoplasmic NAD-reducing hydrogenase from *Ralstonia eutropha*. FEBS Lett 446:259–63.

Happe RP, Roseboom W, Pierik AJ, et al. 1997. Biological activation of hydrogen. Nature 385:126.

Higuchi Y, Yagi T, Noritake Y. 1997. Unusual ligand structure in Ni-Fe active center and an additional Mg site in hydrogenase revealed by high resolution X-ray structure analysis. Structure 5:1671–80.

Krasna AI. 1979. Hydrogenase: properties and applications. Enzyme Microb Technol 1:165–72.

Lemon BJ, Peters JW. 1999. Binding of exogenously added carbon monoxide at the active site of the iron-only hydrogenase (CpI) from *Clostridium pasteurianum*. Biochemistry 38:12969–73.

Lenz O, Friedrich B. 1998. A novel multicomponent regulatory system mediates H_2 sensing in *Alcaligenes eutrophus*. Proc Natl Acad Sci USA 95:12474–9.

Montet Y, Amara P, Volbeda A, et al. 1997. Gas access to the active site of Ni-Fe hydrogenases probed by X-ray crystallography and molecular dynamics. Nat Struct Biol 4:523–6.

Nicolet Y, Piras C, Legrand P, et al. 1999. The structure of *Desulfovibrio desulfuricans* Fe-hydrogenase shows unusual coordination to active site Fe. Structure 7:12–23.

Nicolet Y, Lemon BL, Fontecilla-Camps JC, Peters JW. 2000. A novel FeS cluster in Fe-only hydrogenases. Trends Biochem Sci 25:138–43.

Pershad HR, Duff JL, Heering HA, et al. 1999. Catalytic electron transport in *Chromatium vinosum* [NiFe]-hydrogenase: application of voltammetry in detecting redox-active centers and establishing that hydrogen oxidation is very fast even at potentials close to the reversible H^+/H_2 value. Biochemistry 38: 8992–9.

Peters JW, Lanzilotta WN, Lemon BJ, Seefeldt LC. 1998. X-ray crystal structure of the Fe-only hydrogenase (CpI) from *Clostridium pasteurianum* to 1.8 Å resolution. Science 282:1853–8.

Pierik AJ, Hulstein M, Hagen WR, Albracht SPJ. 1998a. A low-spin iron with CN and CO as intrinsic ligands forms the core of the active site in [Fe]-hydrogenases. Eur J Biochem 258:572–8.

Pierik AJ, Schmelz M, Lenz O, et al. 1998b. Characterization of the active site of a hydrogen sensor from *Alcaligenes eutrophus*. FEBS Lett 438:231–5.

Pierik AJ, Roseboom W, Happe RP, et al. 1999. Carbon monoxide and cyanide as intrinsic ligands to iron in the active site of [NiFe]-hydrogenases. NiFe(CN)$_2$CO, biology's way to activate H_2. J Biol Chem 274:3331–7.

Popescu CV, Münck E. 1999. Electronic structure of the H cluster in [Fe]-hydrogenases. J Am Chem Soc 121:7877–84.

Schneider K, Schlegel HG. 1976. Purification and properties of the soluble hydrogenase from *Alcaligenes eutrophus* H16, Biochim Biophys Acta 452:66–80.

Schneider K, Schlegel HG. 1978. Identification and quantitative determination of the flavin component of soluble hydrogenase from *Alcaligenes eutrophus*. Biochem Biophys Res Commun 84:564–71.

Sellmann D, Geipel F, Moll M. 2000. [Ni(NHPnPr(3))('S(3)')], the first nickel thiolate complex modeling the nickel cysteinate site and reactivity of [NiFe] hydrogenase. Angew Chem Int Educ Engl 39:561–3.

Surerus KK, Chen M, van der Zwaan JW, et al. 1994. Further characterization of the spin coupling observed in oxidized hydrogenase from *Chromatium vinosum*. A Mössbauer and multifrequency EPR study. Biochemistry 33:4980–93.

Tran-Betcke A, Warnecke N, Böcker C, et al. 1990. Cloning and nucleotide sequence of the genes for the subunits of NAD-reducing hydrogenase of *Alcaligenes eutrophus* H16. J Bacteriol 172:2920–9.

Van der Zwaan JW, Albracht SPJ, Fontijn RD, Roelofs YBM. 1986. EPR evidence for direct interaction of carbon monoxide with nickel in hydrogenase from *Chromatium vinosum*. Biochim Biophys Acta 872:208–15.

Van der Zwaan JW, Albracht SPJ, Fontijn RD, Slater EC. 1985. Monovalent nickel in hydrogenase from *Chromatium vinosum*: light sensitivity and evidence for direct interaction with hydrogen. FEBS Lett 179:271–7.

Van der Zwaan JW, Coremans JMCC, Bouwens ECM, Albracht SPJ. 1990. Effect of $^{17}O_2$ and ^{13}CO on EPR spectra of nickel in hydrogenase from *Chromatium vinosum*. Biochim Biophys Acta 1041:101–10.

Volbeda A, Charon M-H, Piras C, et al. 1995. Crystal structure of the nickel-iron hydrogenase from *Desulfovibrio gigas*. Nature 373:580–7.

Volbeda A, Garcin E, Piras C, et al. 1996. Structure of the [NiFe] hydrogenase active site: evidence for biologically uncommon Fe ligands. J Am Chem Soc 118: 12989–96.

Whitehead JP, Gurbiel RJ, Bagyinka C, et al. 1993. The hydrogen binding site in hydrogenase: 35 GHz ENDOR and XAS studies of the Ni-C (reduced and active form) and the Ni-L photoproduct. J Am Chem Soc 115:5629–35.

4
Reductive Activation of Aerobically Purified *Desulfovibrio vulgaris* Hydrogenase: Mössbauer Characterization of the Catalytic H Cluster

BOI HANH HUYNH, PEDRO TAVARES, ALICE S. PEREIRA, ISABEL MOURA, AND JOSÉ J.G. MOURA

Hydrogenases are a class of iron–sulfur enzymes that are used by a variety of microorganisms to catalyze the oxidation of molecular hydrogen and/or the reduction of protons to yield hydrogen during metabolism. According to the metals contained in their prosthetic groups, hydrogenases are grouped into two categories: the nickel-iron-containing hydrogenases and the all-iron-containing hydrogenases (Adams 1990; Przybyla et al. 1992). The periplasmic hydrogenase of *Desulfovibrio vulgaris* (Hilldenbourough) is an iron hydrogenase (Huynh et al. 1984) composed of a large and a small subunit of molecular mass of approximately 46 and 10 kDa, respectively (Voordouw and Brenner 1985; Prickril et al. 1986). Nucleotide-derived amino acid sequence analysis (Voordouw and Brenner 1985) and spectroscopic investigations (Grande et al. 1983; Huynh et al. 1984; Patil et al. 1988; Pierik et al. 1992) suggested that *D. vulgaris* hydrogenase contains two ferredoxin type [4Fe-4S] clusters (F clusters) at the N-terminal end of the large subunit and a catalytic H cluster, which exhibits unusual spectroscopic properties. X-ray crystallographic structures of two iron hydrogenases, one isolated from *Desulfovibrio desulfuricans* (Nicolet et al. 2000) and the other from *Clostridium pasteurianum* (Peters et al. 1999), were solved to high resolution, and the structure of the H cluster was revealed. The H cluster was shown to consist of two iron clusters, a binuclear iron cluster ($[2Fe]_H$) and a ferredoxin type [4Fe-4S] cluster ($[4Fe-4S]_H$), bridged by a cysteinyl sulfur (Fig. 4.1).

The binuclear $[2Fe]_H$ cluster is believed to be the substrate binding and activation site. It has a most unusual structure with the two iron atoms (Fe1 and Fe2) bridged by two thiolates of a 1,3-propanedithiol group and a diatomic molecule, probably a CO (Peters et al. 1999; Nicolet et al. 2000).

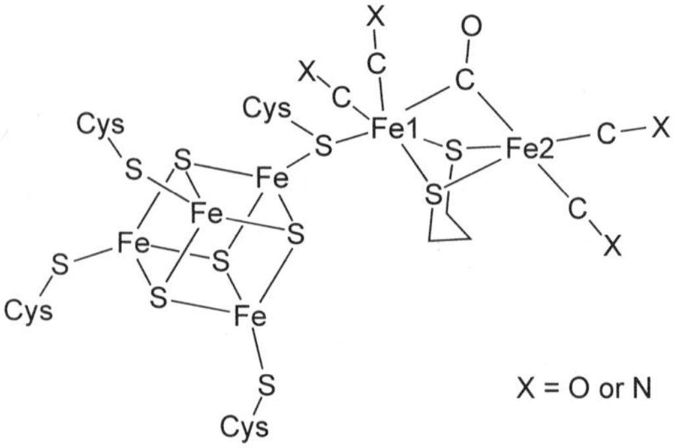

FIGURE 4.1. H cluster based on the X-ray crystallographic structures of *D. desulfuricans* hydrogenase (Nicolet et al. 1999) and *C. pasteurianum* hydrogenase I (Peters et al. 1999). Peters et al. (1999) showed a terminal water molecule to coordinate to Fe2.

Each iron atom is terminally coordinated by one CO and one CN^-. The coordination of Fe1 is completed by the additional bridging cysteinyl sulfur. For Fe2, the sixth coordination site may be empty, as in the *D. desulfuricans* enzyme (Nicolet et al. 1999), or occupied by a solvent molecule, as observed in the *C. pasteurianum* enzyme (Peters et al. 1999). The assignment for the diatomic ligands is supported by infrared spectroscopic evidence (Pierik et al. 1998), and similar diatomic ligands have also been found for the corresponding binuclear [Ni-Fe] cluster of the nickel–iron hydrogenases (Volbeda et al. 1995).

In contrast to other iron hydrogenases, the purification of which requires strict anaerobicity owing to their sensitivity toward oxygen, the *D. vulgaris* hydrogenase can be purified aerobically. The aerobically purified *D. vulgaris* enzyme is inactive, however, and requires a reductive activation. Similar to other iron hydrogenases, the activated *D. vulgaris* enzyme is extremely air sensitive. This reductive activation process is irreversible and has been studied by electron paramagnetic resonance (EPR) spectroscopy (Patil et al. 1988; Pierik et al. 1992). The H cluster of the aerobically purified *D. vulgaris* hydrogenase was found to be EPR silent. Upon reduction, the H cluster displayed a rhombic EPR signal with *g* values at 2.06, 1.96, and 1.89 (the rhombic 2.06 signal). The intensity of this signal maximized at a redox potential of approximately −110 mV and reached a maximum concentration of 0.7 spin/molecule. Upon further lowering of the redox potential, the rhombic 2.06 signal disappeared and was replaced by another rhombic signal with *g* values at 2.10, 2.04, and 2.00 (the rhombic 2.10 signal). The maximum accumulation of this signal was about 0.4 spin/molecule, which occurred at −300 mV. The rhombic 2.10 signal is common to all the iron hydrogenases and has been assigned to the oxidized form of active H cluster ($H_{OX-2.10}$) (Adams 1990). The rhombic 2.10 signal disappears below

−320 mV and reappears upon reoxidation of the enzyme under anaerobic conditions. The rhombic 2.06 signal, however, did not reappear upon reoxidation (Pierik et al. 1992). This observation is consistent with the irreversibility of the reductive activation process and suggests that the rhombic 2.06 signal represents an inactive form of the H cluster ($H_{OX-2.06}$). Because both the rhombic 2.10 and the rhombic 2.06 signals are originating from an $S = 1/2$ system, if the conversion of $H_{OX-2.06}$ to $H_{OX-2.10}$ were to represent a reduction at the H cluster, it would have to be a two-electron reduction step. This was considered to be unlikely, and the conversion of $H_{OX-2.06}$ to $H_{OX-2.10}$ was suggested to represent a redox-dependent conformational change at the H cluster (Patil et al. 1988).

In an effort to obtain more information on the electronic properties of this unusual H cluster and to gain further insights into the reductive activation process of *D. vulgaris* hydrogenase, we used Mössbauer spectroscopy to characterize in detail the Fe-S clusters in samples of *D. vulgaris* enzyme equilibrated at different redox potentials during a reductive titration. A total of five samples (as-purified, −110 mV, −310 mV, −350 mV, and CO reacted) were studied. Details of the study are published elsewhere (Pereira et al. 2001). Here, we present the major results obtained for the H cluster in various oxidation states related to the reductive activation.

It is important to point out that *D. vulgaris* hydrogenase contains three multinuclear iron clusters and each cluster may exist in equilibrium between two different oxidation states in each sample. Consequently, the raw Mössbauer spectra are complex, consisting of overlapping spectra originating from different iron sites of these various clusters. For clarity, we present only the deconvoluted spectra of the H cluster. These spectra were prepared by removing the contributions of other iron species from the raw spectra. Details of the analysis are available (Pereira et al. 2001).

The H_{OX+1} State

In the as-purified *D. vulgaris* enzyme, the two F clusters are in the [4Fe-4S]$^{2+}$ oxidation state. The H cluster is in a state that is one oxidizing equivalent above the $H_{OX-2.06}$ or $H_{OX-2.10}$ state and is thus designated H_{OX+1}. The Mössbauer spectra of H_{OX+1} can be prepared from the spectra of the as-purified enzyme by removing the contributions of the F clusters from the raw data. Figure 4.2 shows such a prepared spectrum for H_{OX+1}. This spectrum can be least-squares fitted with three equal-intensity quadrupole doublets, each doublet corresponding to a pair of equivalent iron atoms of the H cluster. The parameters obtained for the three quadrupole doublets are $\Delta E_Q = 1.24 \pm 0.05$ mm/s and $\delta = 0.47 \pm 0.03$ mm/s for doublet 1, $\Delta E_Q = 0.82 \pm 0.05$ mm/s and $\delta = 0.44 \pm 0.03$ mm/s for doublet 2, and $\Delta E_Q = 1.09 \pm 0.05$ mm/s and $\delta = 0.16 \pm 0.04$ mm/s for doublet 3. Spectra recorded at strong applied fields indicate that H_{OX+1} has a diamagnetic ground state.

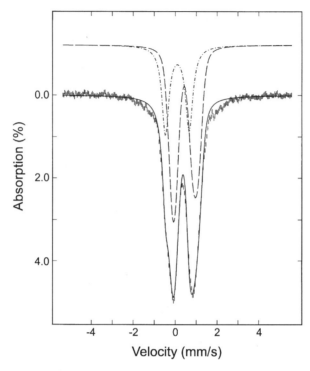

FIGURE 4.2. Mössbauer spectrum of H_{OX+1} (*hatched marks*) prepared from the spectra of the aerobically purified *D. vulgaris* hydrogenase recorded at 4.2 K in a magnetic field of 0.05 T applied parallel to the γ-rays. *Solid line*, a least-squares fit to the data, assuming three equal-intensity quadrupole doublets; *dotted-and-dashed line*, doublet corresponding to the $[2Fe]_H$ cluster; *dashed line*, superposition of two doublets (doublets 1 and 2) corresponding to the $[4Fe-4S]_H$ cluster.

On the basis of the Mössbauer parameters, doublets 1 and 2 are attributed to the $[4Fe-4S]_H$ cluster. The parameters determined for doublets 1 and 2, particularly the isomer shift values, indicate that the $[4Fe-4S]_H$ cluster is in the 2+ oxidation state (Middleton et al. 1978; Trautwein et al. 1991). In general, a $[4Fe-4S]^{2+}$ cluster can be viewed as composed of two antiferromagnetically coupled Fe(III)Fe(II) pairs forming a diamagnetic ground state. Doublets 1 and 2 therefore represent the valence delocalized Fe(III)Fe(II) pairs of the $[4Fe-4S]_H^{2+}$ cluster. Doublet 3 is assigned to the binuclear $[2Fe]_H$ cluster. The observation of one single doublet for two iron atoms indicates that the two iron atoms in $[2Fe]_H$ are indistinguishable by Mössbauer spectroscopy and suggests they have an equivalent spin and oxidation state. With the CO and CN^- ligands, the iron atoms in $[2Fe]_H$ are expected to be low spin. The small isomer shift (0.16 mm/s) observed for doublet 3 indicates that the two iron atoms are indeed low spin. The oxidation state of the iron atoms, however, cannot be determined unambiguously from the Mössbauer data because the Mössbauer parameters ΔE_Q and δ are relatively insensitive to the oxidation state for low-spin iron ions, particularly in cases involving strong ligands, such as CO and CN^- (Greenwood and Gibb 1971; Debrunner 1989). Currently, only a limited number of iron model complexes with mixed $S/CO/CN^-$ ligands have been synthesized, and Mössbauer investigations have been reported for only three of them (Hsu et al. 1997; LeCloirec et al. 1999). Compared to these model complexes, the isomer shift of doublet 3, 0.16 mm/s, is similar to the value 0.15 mm/s reported for a mononuclear low-spin Fe(II) complex (Hsu et al. 1997) and is considerably larger than the values, 0.05 and 0.03 mm/s (at 77 K), reported for two low-spin diFe(I) complexes (LeCloirec et al. 1999). On the basis of this comparison, the oxidation state of the $[2Fe]_H$ in H_{OX+1} is either diFe(II) or diFe(III). For the latter assignment, a diamagnetic ground state can be achieved through antiferromagnetic coupling between the two low-spin Fe(III) ions. In addition, a diFe(I) assignment for the $[2Fe]_H$ in H_{OX+1} is not favored because, if the oxidation state of the cluster were diFe(I), further reduction of the $[2Fe]_H$ cluster (which is observed experimentally) would be difficult, if not impossible.

The $H_{OX-2.06}$ State

Analysis of the Mössbauer spectra of the *D. vulgaris* hydrogenase sample poised at -110 mV reveals that the F clusters are in the $[4Fe-4S]^{2+}$ state, whereas the H cluster is present in an equilibrium between H_{OX+1} and $H_{OX-2.06}$. The sample contains 0.37 equivalent of H_{OX+1} and 0.63 equivalent of $H_{OX-2.06}$. Consequently, the spectra of $H_{OX-2.06}$ can be obtained from the spectra of the -110 mV sample by removing the contributions of other species from the raw data using the parameters listed in the previous section. Two such prepared Mössbauer spectra of $H_{OX-2.06}$ are shown in Figure 4.3. These spectra can be decomposed into three spectral components of equal contributions: A diamagnetic quadrupole doublet and two paramagnetic components. The diamagnetism and parameters obtained for the quadrupole doublet ($\Delta E_Q = 1.08 \pm 0.05$ mm/s and $\delta = 0.17 \pm 0.04$ mm/s) are practically identical to those of the doublet 3 of H_{OX+1}, indicating that this doublet is associated with the $[2Fe]_H$ cluster and that conversion of the diamagnetic H_{OX+1} to the $S = 1/2$ $H_{OX-2.06}$ state does not involve changes at the $[2Fe]_H$ cluster. Consequently, changes must occur at the $[4Fe-4S]_H$ cluster.

With the quadrupole doublet assigned to the $[2Fe]_H$ cluster, the two remaining paramagnetic components have to be arising from the $[4Fe-4S]_H$ cluster. Detailed analysis of the magnetic components recorded under various magnetic field supports this assignment. The parameters obtained from the analysis are $\Delta E_Q = 1.12 \pm 0.05$ mm/s, $\delta = 0.49 \pm 0.03$ mm/s, $\eta = 1.4$,

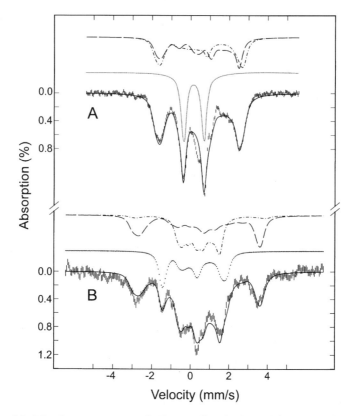

FIGURE 4.3. Mössbauer spectra of $H_{OX-2.06}$ (*hatched marks*) prepared from the spectra of *D. vulgaris* hydrogenase poised at −110 mV. The data were recorded at 4.2 K in a magnetic field of 0.05 T **A**, or 8 T **B** applied parallel to the γ-rays. Theoretical simulations for the diamagnetic $[2Fe]_H$ cluster (*dotted lines*), the Fe(II)Fe(III) pair (*dotted-and-dashed lines*), and the diferrous pair (*dashed lines*), of the $[4Fe-4S]_H^{+1}$ cluster are shown. The solid lines are superpositions of these three simulated spectral components.

and $A/g_n\beta_n = -(19.9 \pm 1.0, 26.8 \pm 1.5, 22.3 \pm 1.0)$ T for component I and $\Delta E_Q = -2.30 \pm 0.05$ mm/s, $\delta = 0.57 \pm 0.03$ mm/s, $\eta = -0.3$, and $A/g_n\beta_n = +(15.7 \pm 1.0, 9.5 \pm 1.5, 17.5 \pm 1.0)$ T for component II. These parameters are typical for $[4Fe-4S]^+$ clusters. Without exception, the spectra of $[4Fe-4S]^+$ clusters are composed of two spectral components representing two antiferromagnetically coupled pairs of iron atoms, a valence delocalized Fe(II)Fe(III) pair and a diferrous pair (Middleton et al. 1978; Trautwein et al. 1991). The opposite signs of the **A** tensors determined for components I and II indicate that the two iron pairs in the $[4Fe-4S]_H^+$ cluster are indeed antiferromagnetically coupled. Similar to other $[4Fe-4S]^+$ clusters, component I—which has smaller magnitudes of ΔE_Q and δ and thus represents the mixed valence pair—displays negative A values, whereas component II—which has larger magnitudes of ΔE_Q and δ and thus represents the diferrous pair—shows positive A values. Consequently, on the basis of the Mössbauer data, it can be concluded that the $H_{OX-2.06}$ state is one electron further reduced than the H_{OX+1} state. The reducing equivalent is localized on the $[4Fe-4S]_H$ cluster, and the reduction of H_{OX+1} to $H_{OX-2.06}$ has no effect on the electronic properties of the $[2Fe]_H$ cluster.

The $H_{OX-2.10}$ State

Figure 4.4 shows the Mössbauer spectra of the $H_{OX-2.10}$ species. Analysis of the Mössbauer spectra indicates that the $-310\,mV$ sample contains 0.5 equivalent of $H_{OX-2.10}$ state. Compared to the $H_{OX-2.06}$ spectra (Fig. 4.3), the $H_{OX-2.10}$ exhibits much reduced magnetic splittings (i.e., reduced magnetic hyperfine interactions). Also, the quadrupole doublet representing $[2Fe]_H$ in H_{OX+1} and $H_{OX-2.06}$ has disappeared, indicating that the state of the $[2Fe]_H$ cluster has been changed. Detailed analysis of the data shows that the spectra of $H_{OX-2.10}$ can be decomposed into four spectral components with the following parameters: $\Delta E_Q = 0.85 \pm 0.06\,mm/s$, $\delta = 0.13 \pm 0.04\,mm/s$, $\eta = 0$, and diamagnetic for component A; $\Delta E_Q = 0.67 \pm 0.06\,mm/s$, $\delta = 0.14 \pm 0.04\,mm/s$, $\eta = 0$, and $A/g_n\beta_n = -12.0 \pm 0.5\,T$ for component B; $\Delta E_Q = 1.14 \pm 0.10\,mm/s$, $\delta = 0.44 \pm 0.03\,mm/s$, $\eta = 0$, and $A/g_n\beta_n = -6.2 \pm 0.7\,T$ for component C; and $\Delta E_Q = 1.34 \pm 0.10\,mm/s$, $\delta = 0.44 \pm 0.03\,mm/s$, $\eta = -0.8$, and $A/g_n\beta_n = +6.2 \pm 0.7\,T$ component D. Components A and B have an intensity that is half of that of components C and D and are attributed to the $[2Fe]_H$ cluster, with each component representing one iron atom. Components C and D are attributed to the $[4Fe-4S]_H$ cluster, with each component representing a pair of iron atoms. The small isomer shift values observed for components A and B indicate that the iron atoms in the $[2Fe]_H$ cluster remain low spin (as expected). The diamagnetism detected for component A indicates that it represents a low-spin Fe(II) site, whereas the paramagnetism observed for component B indicates either a low-spin Fe(III) or low-spin Fe(I) site. In other words, the Mössbauer data show that the $[2Fe]_H$ cluster in $H_{OX-2.10}$ is a valence localized Fe(I)Fe(II) or Fe(II)Fe(III) pair.

For [4Fe-4S] clusters, the isomer shift parameter is a good indication of its oxidation state (Trautwein et al. 1991; Popescu and Münck 1999). The values 0.44 and 0.43 mm/s determined for components C and D indicate that the $[4Fe-4S]_H$ cluster in $H_{OX-2.10}$ is in the 2+ oxidation state. As mentioned above, a $[4Fe-4S]^{2+}$ cluster generally has a diamagnetic ground electronic state, which is a consequence of the antiferromagentic coupling of the two valence delocalized Fe(II)Fe(III) pairs. It is, therefore, intriguing to observe that components C and D display magnetic hyperfine interactions, albeit small in magnitude. This phenomenon indicates that the $[4Fe-4S]_H^{2+}$ cluster is not magnetically isolated and thus must be interacting with a

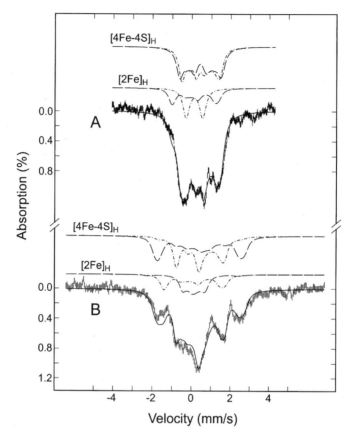

FIGURE 4.4. Mössbauer spectra of $H_{OX-2.10}$ (*hatched marks*) prepared from the spectra of *D. vulgaris* hydrogenase poised at −310 mV. The data were recorded at 4.2 K in a magnetic field of 0.05 T **A**, or 8 T **B** applied parallel to the γ-rays. Theoretical simulations for the individual iron sites of the $[2Fe]_H$ cluster (*dotted-and-dashed lines*, component A; *dashed lines*, component B) and of the $[4Fe-4S]_H^{2+}$ cluster (*dotted-and-dashed lines*, component C; *dashed lines*, component D) are shown above the experimental data. *Solid lines*, superpositions of these four simulated spectral components.

nearby paramagnetic species. In fact, the observed A values, both in signs and in magnitudes, can be explained by a theoretical model that assumes a $[4Fe-4S]^{2+}$ cluster exchange coupled via one of its corner iron atom to a $S = 1/2$ center, such as a low-spin Fe(III) or Fe(I) (Belinsky 1996; Xia et al. 1997; Popescu and Münck 1999; Pereira et al. 2000). Consequently, the Mössbauer data demonstrated that $H_{OX-2.10}$ is composed of a $[4Fe-4S]_H^{2+}$ cluster exchange coupled to an $S = 1/2$ mixed valence $[2Fe]_H$ cluster. In other words, the reducing equivalent that localized in the $[4Fe-4S]_H^{1+}$ cluster in $H_{OX-2.06}$ is removed from the cluster in $H_{OX-2.10}$ and must be, therefore, relocate in the $[2Fe]_H$ cluster. Because there are two alternative assignments—diFe(III) or diFe(II)—for the $[2Fe]_H$ cluster in $H_{OX-2.06}$, there are also two possible corresponding assignments for the $[2Fe]_H$ in $H_{OX-2.10}$; Fe(II)Fe(III) or Fe(I)Fe(II), respectively.

Conclusion

The aerobically purified iron hydrogenase from *D. vulgaris* is inactive and requires a reductive activation process. Previous EPR investigations indicate that this process is irreversible and involves the conversion of the catalytic H cluster from an EPR silent state (H_{OX+1}) to an $S = 1/2$ $H_{OX-2.10}$ state via a transient $S = 1/2$ $H_{OX-2.06}$ state (Patil et al. 1988; Pierik et al. 1992). The $H_{OX-2.10}$ state is observed in the as-purified forms of other iron hydrogenases that are purified anaerobically (Adams 1990) and represents the oxidized form of the active H cluster. To provide further insights into the reductive activation process, Mössbauer spectroscopy was used to characterize the Fe-S clusters in samples of *D. vulgaris* hydrogenase prepared at various stages during a reductive activation (Pereira et al. 2001). Characteristic parameters describing the electronic structures of several key redox states of the H cluster were obtained from this study and the Mössbauer results for the H_{OX+1}, $H_{OX-2.06}$, and $H_{OX-2.10}$ states were presented.

Consistent with the X-ray crystallographic structure of the H cluster, the Mössbauer spectra can be understood as originating from an exchange coupled $[4Fe-4S]_H$-$[2Fe]_H$ unit. In H_{OX+1}, the system ground state is diamagnetic. The $[4Fe-4S]_H$ cluster is in the diamagnetic 2+ oxidation state. The two iron atoms in $[2Fe]_H$ are low spin and have the same oxidation and spin state. Either a diFe(II) or an antiferromagnetically coupled diFe(III) assignment is possible. The data further indicate that both $H_{OX-2.06}$ and $H_{OX-2.10}$ are isoelectronic. Both are one electron further reduced than H_{OX+1}. Consequently, the Mössbauer data provided direct evidence to support our previous suggestion based on the EPR studies (Patil et al. 1988) that conversion of $H_{OX-2.06}$ to $H_{OX-2.10}$ does not involve reduction of the H cluster but rather reflects a conformational change that affects the electronic properties of the H cluster. The particulars of this alteration in electronic properties were also revealed by the Mössbauer data. Compared to the states of the two subclusters in H_{OX+1}, the $[4Fe-4S]_H$ cluster in $H_{OX-2.06}$ is reduced to the paramagnetic 1+ oxidation state, whereas the state of the $[2Fe]_H$ cluster remains the same. In $H_{OX-2.10}$, the $[4Fe-4S]_H$ cluster is reoxidized to the 2+ state and the reducing equivalent is transferred to the $[2Fe]_H$ cluster, resulting in a mixed-valence binuclear cluster (Fig. 4.5). Thus the Mössbauer data revealed that the reductive activation of the aerobically purified *D. vulgaris* hydrogenase begins with reduction of the $[4Fe-4S]_H$ cluster from the 2+ oxidation state (in H_{OX+1}) to the 1+ oxidation state (in $H_{OX-2.06}$) followed by transfer of the reducing equivalent from the $[4Fe-4S]_H^{1+}$ cluster to the

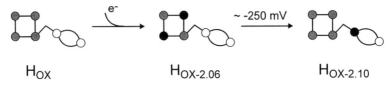

$$H_{OX} \qquad H_{OX-2.06} \qquad H_{OX-2.10}$$

FIGURE 4.5. Changes of oxidation states of the $[2Fe]_H$ (*ellipse*) and $[4Fe-4S]_H$ (*square*) clusters during reductive activation of *D. vulgaris* hydrogenase. The $[2Fe]_H$ cluster is assumed to begin with a diFe(III) state in H_{OX+1}, although a diFe(II) state is an equally probable starting state (see text). In the latter assignment, the oxidation state of the $[2Fe]_H$ cluster in $H_{OX-2.10}$ would be Fe(I)Fe(II). *Black circles*, valence-localized Fe(II) sites; *open circles*, valence-localized Fe(III) sites; *gray circles*, valence-delocalized Fe(II)Fe(III) pairs. The iron atoms in the $[4Fe-4S]_H$ cluster are high spin, and the iron atoms in the $[2Fe]_H$ cluster are low spin.

binuclear $[2Fe]_H$ cluster (Fig. 4.5). On the basis of this realization, it is reasonable to assume that the irreversible conformational change that activates the H cluster must alter the redox properties of either one or both subclusters of H to promote the relocation of the reducing equivalent from the $[4Fe-4S]_H$ cluster to the binuclear $[2Fe]_H$ cluster. Specific conformational changes involved in this activation step, however, is not know.

Acknowledgments. We thank Jean LeGall, Daulat S. Patil, and Shao H. He for the growth of *D. vulgaris* and for purification and biochemical characterization of the ^{57}Fe-enriched hydrogenase. This work is supported in part by grants from the National Institutes of Health (GM 47295 and GM 58778 to B.H.H.) and from PRAXIS (to J.J.G.M. and I.M.).

References

Adams MWW. 1990. The structure and mechanism of iron-hydrogenases. Biochim Biophys Acta 1020:115–45.

Belinsky MI. 1996. Hyperfine evidence of strong double exchange in multimetallic {[Fe$_4$S$_4$]-Fe} active center of *Escherichia coli* sulfite reductase. J Biol Inorg Chem 1:186–8.

Debrunner PG. 1989. Mössbauer spectroscopy of iron porphyrins. Phys Bioinorg Chem Ser Iron Porphyrins 4:137–234.

Grande HJ, Dunham WR, Averill B, et al. 1983. Electron paramagnetic resonance and other properties of hydrogenases isolated from *Desulfovibrio vulgaris* (strain Hildenborough) and *Megasphaera elsdenii*. Eur J Biochem 136:201–7.

Greenwood NN, Gibb TC. 1971. Mössbauer spectroscopy. London: Chapman & Hall.

Hsu H-F, Koch SA, Popescu CV, Münck E. 1997. Chemistry of iron thiolate complexes with CN and CO. Models for the [Fe(CO)(CN)$_2$] structural unit in Ni-Fe hydrogenase enzymes. J Am Chem Soc 119:8371–2.

Huynh BH, Czechowski MH, Krüger HJ, et al. 1984. *Desulfovibrio vulgaris* hydrogenase: a nonheme iron enzyme lacking nickel that exhibits anomalous EPR and Mössbauer spectra. Proc Natl Acad Sci USA 81:3728–32.

LeCloirec A, Best SP, Borg S, et al. 1999. A di-iron dithiolate possesing structural elements of the carbonyl/cyanide sub-site of the H center of Fe-only hydrogenase. Chem Commun 22:2285–6.

Middleton P, Dickson DPE, Johnson CE, Rush JD. 1978. Interpretation of the Mössbauer spectra of the four-iron ferredoxin from *Bacillus stearothermophilus*. Eur J Biochem 88:135–41.

Nicolet YY, Lemon BJ, Fontecilla-Camps JC, Peters JW. 2000. A novel FeS cluster in Fe-only hydrogenases. Trends Biochem Sci 25:138–43.

Nicolet Y, Piras C, Legrand P, et al. 1999. *Desulfovibrio desulfuricans* iron hydrogenase: the structure shows unusual coordination to an active site Fe binuclear center. Structure (London) 7:13–23.

Patil DS, Moura JJG, He SH, et al. 1988. EPR-detectable redox centers of the periplasmic hydrogenase from *Desulfovibrio vulgaris*. J Biol Chem 263: 18732–8.

Pereira AS, Tavares P, Moura I, et al. 2001. Mössbauer characterization of the iron-sulfur clusters in *Desulfovibrio vulgaris* hydrogenase. J Am Chem Soc 123:2771–82.

Peters JW, Lanzilotta WN, Lemon BJ, Seefeldt LC. 1999. X-ray crystal structure of the Fe-only hydrogenase (CpI) from *Clostridium pasteurianum* to 1.8 Ångstrom resolution. Science (Washington, DC) 282:1853–8.

Pierik AJ, Hagen WR, Redeker JS, et al. 1992. Redox properties of the iron-sulfur clusters in activated iron-hydrogenase from *Desulfovibrio vulgaris* (Hildenborough). Eur J Biochem 209:63–72.

Pierik AJ, Hulstein M, Hagen WR, Albracht SPJ. 1998. A low-spin iron with CN and CO as intrinsic ligands forms the core of the active site in [Fe]-hydrogenases. Eur J Biochem 258:572–8.

Popescu CV, Münck E. 1999. Electronic structure of the H cluster in [Fe]-hydrogenases. J Am Chem Soc 121:7877–84.

Prickril BC, Czechowski MH, Przybyla AE, et al. 1986. Putative signal peptide on the small subunit of the periplasmic hydrogenase from *Desulfovibrio vulgaris*. J Bacteriol 167:722–5.

Przybyla AE, Robbins J, Menon N, Peck HD Jr. 1992. Structure-function relationships among the nickel-containing hydrogenases. FEMS Microbiol Rev 88:109–35.

Trautwein AX, Bill E, Bominaar EL, Winkler H. 1991. Iron-containing proteins and related analogs. Complementary Mössbauer, EPR and magnetic susceptibility studies. Struct Bonding (Berlin) 78:1–95.

Volbeda A, Charon M-H, Piras C, et al. 1995. Crystal structure of the nickel-iron hydrogenase from *Desulfovibrio gigas*. Nature 373:580–7.

Voordouw G, Brenner S. 1985. Nucleotide sequence of the gene encoding the hydrogenase from *Desulfovibrio vulgaris* (Hildenborough). Eur J Biochem 148:515–20.

Xia J, Hu Z, Popescu CV, et al. 1997. Mössbauer and EPR study of the Ni-activated α-subunit of carbon monoxide dehydrogenase from *Clostridium thermoaceticum*. J Am Chem Soc 119:8301–12.

5
Iron-Sulfur Cluster Biosynthesis

Jeffrey N. Agar, Dennis R. Dean, and Michael K. Johnson

Iron-sulfur clusters constitute one of the most ancient, ubiquitous, and structurally and functionally diverse classes of biological prosthetic groups. For reviews see Cammack (1992), Johnson (1994, 1998), Beinert et al. (1997), Beinert and Kiley (1999), and Beinert (2000). Indeed there are now known to be in excess of 120 distinct types of Fe-S cluster-containing enzymes and proteins, distributed over all three kingdoms of life, and the list is growing rapidly.

The crystallographically defined structures of homometallic biological Fe-S clusters, together with the known core oxidation states and corresponding ground state spin states, are shown in Figure 5.1. Of particular relevance with respect to the process of cluster biosynthesis, is the observation that $[Fe_2(\mu_2\text{-}S)_2]$ rhombs can be considered the basic building block for assembly, based on both structural and electronic considerations. Cubane-type $[Fe_4S_4]$ clusters can be assembled from two $[Fe_2S_2]$ units, whereas $[Fe_3S_4]$ and $[Fe_8S_7]$ clusters can be assembled from $[Fe_4S_4]$ units via loss of one iron and cluster fusion, respectively. With the exception of the anti-ferromagnetically coupled, valence-localized $S = 1/2$ $[Fe_2S_2]^+$ centers, Mössbauer studies indicate that valence-delocalized $[Fe_2S_2]^+$ fragments are integral components of all higher nuclearity clusters. The first examples of valence-delocalized $[Fe_2S_2]^+$ centers were identified and characterized in a variant form of a 2Fe ferredoxin and shown to have ferromagnetically coupled $S = 9/2$ ground states (Crouse et al. 1995; Achim et al. 1996; Johnson et al. 1998a). Antiferromagnetic interaction between $S = 9/2$ $[Fe_2S_2]^+$ fragments and $S = 2$ Fe^{2+}, $S = 5/2$ Fe^{3+}, $S = 5$ $[Fe_2S_2]^{2+}$, $S = 9/2$ $[Fe_2S_2]^+$, or $S = 4$ $[Fe_2S_2]^0$ fragments, can then be invoked to rationalize the ground state electronic properties of $[Fe_3S_4]^{0,-}$ and $[Fe_4S_4]^{3+,2+,+}$ clusters.

The primary role of Fe-S clusters lies in mediating biological electron transport. Accordingly, the ubiquitous $[Fe_2S_2]^{2+,+}$, $[Fe_3S_4]^{+,0}$, and $[Fe_4S_4]^{2+,+}$ clusters are integral components of the respiratory and photosynthetic electron-transport chains, as well as a wide range of enzymes and proteins involved with the metabolism of carbon, oxygen, hydrogen, sulfur, and nitrogen. In addition, Fe-S clusters are also known to constitute, in whole

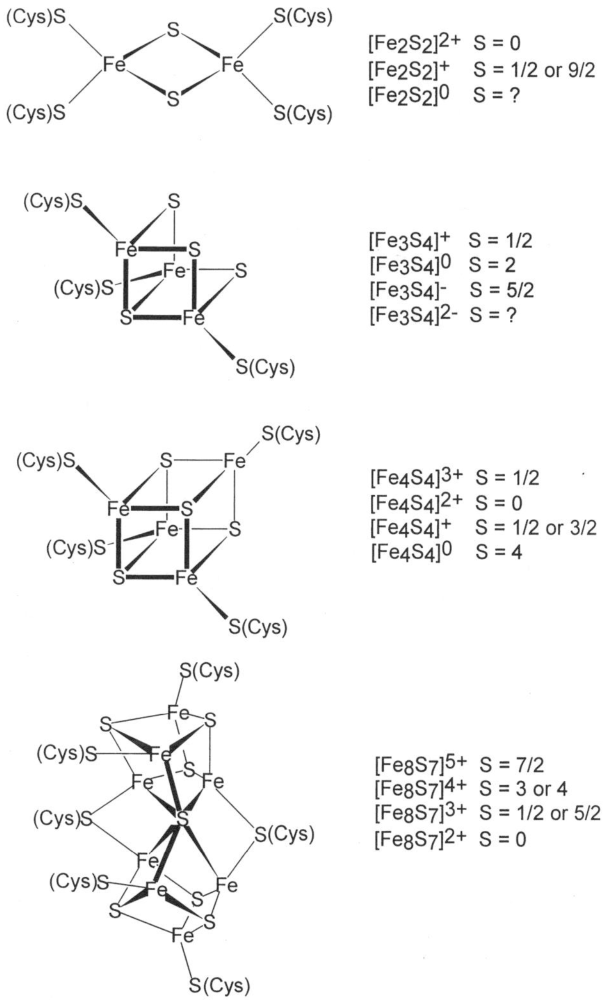

$[Fe_2S_2]^{2+}$ S = 0
$[Fe_2S_2]^{+}$ S = 1/2 or 9/2
$[Fe_2S_2]^{0}$ S = ?

$[Fe_3S_4]^{+}$ S = 1/2
$[Fe_3S_4]^{0}$ S = 2
$[Fe_3S_4]^{-}$ S = 5/2
$[Fe_3S_4]^{2-}$ S = ?

$[Fe_4S_4]^{3+}$ S = 1/2
$[Fe_4S_4]^{2+}$ S = 0
$[Fe_4S_4]^{+}$ S = 1/2 or 3/2
$[Fe_4S_4]^{0}$ S = 4

$[Fe_8S_7]^{5+}$ S = 7/2
$[Fe_8S_7]^{4+}$ S = 3 or 4
$[Fe_8S_7]^{3+}$ S = 1/2 or 5/2
$[Fe_8S_7]^{2+}$ S = 0

FIGURE 5.1. Structures, oxidation states and spin states of crystallographically defined Fe-S clusters. Complete cysteinyl ligand (as shown) is most common, although a limited number of examples are known in which histidine, serine, and aspartate residues can act as cluster ligands. The spin states denoted by a *question mark* have yet to be determined, and the $[Fe_3S_4]^{-}$ cluster has been observed only as a fragment in heterometallic $[MFe_3S_4]^{+}$ clusters in which M is a divalent transition metal ion.

or in part, the active sites of a diverse range of redox and nonredox enzymes, and there is strong evidence for roles in a variety of sensing and regulatory processes. New roles are also beginning to emerge in mediating disulfide reduction in two one-electron steps in a class of ferredoxin-dependent disulfide reductases (Staples et al. 1998; Dai et al. 2000), generating a

5′-deoxyadenosyl radical intermediate in S-adenosylmethionine-dependent Fe-S enzymes (Johnson et al. 1998b; Cosper et al. 2000; Sofia et al. 2001), and as the sulfur donor in the final step of biotin biosynthesis (Gibson et al. 1999; Bui et al. 1998). However, while the functions of biological Fe-S centers continue to proliferate, the mechanism of cluster biosynthesis, until recently, was poorly understood. This process is central to understanding iron metabolism in both aerobic and anaerobic organisms, as well as regulatory and biosynthetic processes involving cluster assembly/disassembly.

Analogs of $[Fe_2S_2]$ and $[Fe_4S_4]$ clusters have been synthesized from $Fe^{3+,2+}$, S^{2-}, and thiolates by spontaneous self-assembly in aprotic media (Berg and Holm 1982). Moreover, the $[Fe_2S_2]$ and $[Fe_4S_4]$ centers in many ferredoxins and simple Fe-S proteins can be reassembled in vitro via spontaneous self assembly by anaerobically incubating apo-proteins with $Fe^{3+/2+}$ and S^{2-} in the presence of dithiothreitol or β-mercaptoethanol (Malkin and Rabinowitz 1966). However, both the transport and the internal metabolism of iron in the cell are tightly regulated, owing to the insolubility of Fe^{3+} ion in aqueous solution at neutral pH and the ability of Fe^{2+} ion to interact with oxygen and generate hydroxyl radicals, resulting in damage to a variety of cellular components. In addition, S^{2-} would be extremely toxic to the cell at the levels required for efficient spontaneous self-assembly of biological Fe-S centers. Hence the low levels and high toxicity of free $Fe^{3+/2+}$ and S^{2-} within the cell argue strongly against spontaneous self-assembly as a viable mechanism for the biosynthesis of Fe-S centers. However, although the need for specific biochemical pathways for mediating Fe-S cluster biosynthesis has long been recognized, significant progress in understanding this important biosynthetic process has occurred only over past few years. Much of the current understanding of Fe-S cluster biosynthesis comes from studies of the organization and function of genes in the aerobic nitrogen-fixing bacterium *Azotobacter vinelandii*.

This review is a summary of our recent genetic, biochemical, and biophysical, studies of the Fe-S cluster assembly proteins of *A. vinelandii*, with particular emphasis on the role of the NifU and IscU proteins. The results reveal insight into the mechanism of both general and nitrogenase-specific Fe-S cluster biosynthesis in *A. vinelandii* and indicate a common mechanism for Fe-S cluster assembly that is used throughout nature.

Fe-S Cluster Biosynthesis Genes

Studies of the organization and function of the *nif* genes involved with nitrogen fixation in *A. vinelandii*, revealed two gene products—NifS and NifU—that are essential for full activation of both nitrogenase component proteins (Dean et al. 1993) (Fig. 5.2). Knockout mutants of either gene resulted in slow diazatrophic growth and incomplete incorporation of the Fe-S centers in both the Fe-protein and MoFe-protein, suggesting that both

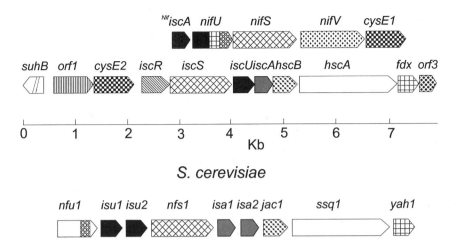

FIGURE 5.2. Organization of *nif*-specific and general Fe-S cluster assembly (*isc*) genes in *A. vinelandii*. The genes encoding for equivalent proteins or domains in *Saccharomyces cerevisiae* are shown for comparison.

were involved in Fe-S cluster biosynthesis. This was substantiated by the discovery that NifS is a homodimeric, pyridoxal phosphate-dependent L-cysteine desulfurase that catalyzes the reductive conversion of cysteine to alanine and sulfide via an enzyme-bound persulfide intermediate (Zheng et al. 1993, 1994). The ubiquitous and essential role of NifS-like enzymes in sulfur trafficking, in general (Kambampati and Lauhon 2000), and in Fe-S cluster biosynthesis, in particular (Kispal et al. 1999; Li et al. 1999; Schwartz et al. 2000), is now firmly established. X-ray crystal structures have recently appeared for three members of this general class of enzyme (Clausen et al. 2000; Fujii et al. 2000; Kaiser et al. 2000). These structures have revealed some interesting variations and provided additional mechanistic details, while confirming the essential elements of the original mechanistic proposal for cysteine desulfurization. For example, *Synechocystis* has a cystine C-S lyase in which the cysteine persulfide is formed on the substrate cystine rather than on a conserved active site cysteine residue (Clausen et al. 2000).

Because NifS provides sulfur, NifU is clearly a good candidate for mobilizing iron or providing a scaffold for NifS-mediated Fe-S cluster biosynthesis. Heterologous overexpression and purification of *A. vinelandii* NifU yielded a homodimer of 33 kDa subunits, each containing a redox-active $[Fe_2S_2]^{2+,+}$ cluster, $E_m = -254 \, mV$ vs. NHE (Fu et al. 1994). The primary sequence of NifU suggests a modular protein, with three distinct domains (Fig. 5.3). The N-terminal domain NifU-1 is homologous with the ubiqui-

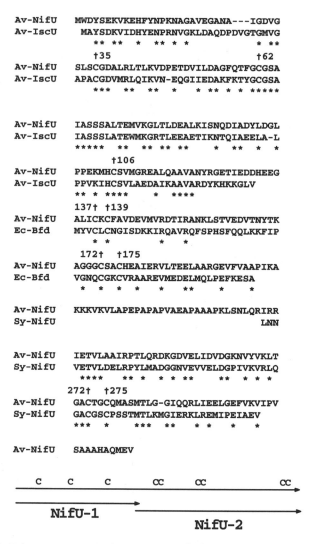

FIGURE 5.3. Primary sequence of *A. vinelandii* (*Ar*) NifU showing the homology of discrete domains with *A. vinelandii* IscU, *E. coli* (*Ec*) Bfd and *Synechocystis* (*Sy*) NifU. †, conserved cysteines in *A. vinelandii* NifU; *, identical residues. A representation of NifU, showing the nine cysteine residues and the segments corresponding to NifU-1 and NifU-2, is also shown.

tous IscU proteins that are involved with general Fe-S cluster assembly in almost all organisms, including *A. vinelandii* (Zheng et al. 1998). Each of the three cysteines (residues 35, 62, and 106) is conserved in all NifU and IscU proteins identified thus far. The middle domain has sequence homology with the $[Fe_2S_2]$-containing bacterioferritin-associated ferredoxin (Bfd) from *Escherichia coli*, which has a potential, albeit unproven, role in reductive release of iron from bacterioferritin (Garg et al. 1996; Quail et al. 1996). The four conserved cysteines in this domain (residues 137, 139, 172, and 175), are therefore good candidates for the ligands to the redox active $[Fe_2S_2]^{2+,+}$ cluster in NifU. The C-terminal domain contains two cysteines (residues 272 and 275) and is homologous with a small (68 residues) NifU-like protein from *Synechocystis* and a domain of the Nfu1 protein from *Saccharomyces cerevisiae* (Fig. 5.2). Both of these proteins have been implicated in mediating Fe-S cluster assembly in their respective organisms (Garland et al. 1999; Schilke et al. 1999; Nishio and Nakai 2000).

Genes encoding homologs to NifU and NifS are located within the genomes of a wide variety of non-nitrogen-fixing bacteria (Zheng et al. 1998). In bacteria they have been designated *isc* genes to indicate the proposed role of their products in general *i*ron-*s*ulfur *c*luster assembly or repair. In accord with this hypothesis, *isc* genes have been shown to be essential for *A. vinelandii* viability (Zheng et al. 1998) and a comparison of the organization of the *nif*-specific and *isc*-gene cluster regions in this organism is shown in (Fig. 5.2). The NifS and IscS proteins are highly homologous, and the proposal that these pyridoxal phosphate-dependent L-cysteine desulfurases are involved with the mobilization of sulfur for iron–sulfur cluster assembly is further supported by the presence of a *cysE*-like homologs in *A. vinelandii*. O-Acetylserine synthase, the product of the *cysE* gene, catalyzes the rate-limiting step in cysteine biosynthesis. *CysE2* is located in another gene cluster immediately upstream from the *iscSUA* genes, and *cysE1* is located downstream of *nifU* and *nifS* in the *nif* gene cluster (Fig. 5.3). Hence the role of the *cysE1* and *cysE2* gene products is likely to be boosting the cysteine pool in response to an increased demand for the mobilization of sulfur for Fe-S cluster formation.

IscU is one of the most conserved proteins in nature (Hwang et al. 1996) and plays a crucial role in Fe-S cluster biosynthesis in almost all forms of life (Zheng et al. 1998; Garland et al. 1999; Schilke et al. 1999). The in vitro studies of *A. vinelandii* IscU and NifU discussed herein indicate that the function of IscU and the N-terminal domain of NifU are to provide a scaffold for NifS/IscS-mediated Fe-S cluster biosynthesis. Although IscU is truncated, compared to NifU, and does not contain a sequence corresponding to the redox-active $[Fe_2S_2]^{2+,+}$ cluster-binding region, there is another gene within the *isc* gene cluster, *fdx*, that encodes for a $[Fe_2S_2]^{2+,+}$ ferredoxin (Ta and Vickery 1992; Jung et al. 1999). The primary sequence of ligating cysteine residues ($C-X_5-C-X_2-C-X_{35}-C$) and the spectroscopic/redox properties of this large class of $[Fe_2S_2]$-containing ferredoxins are similar to those of mammalian and bacterial hydroxylase-type $[Fe_2S_2]$ ferredoxins that were previously characterized as electron donors to P450s. While it seems likely that this $[Fe_2S_2]$ ferredoxin and the $[Fe_2S_2]$-domain of NifU both have redox functions, their specific roles in Fe-S cluster biosynthesis have yet to be determined.

A third gene, *iscA*, is common to both the *nif* and *isc* operons and corresponds to the region previously termed *orf6* in the *A. vinelandii nif* gene cluster. The IscA gene products all contain three conserved cysteine residues (35, 99, and 101 in both *A. vinelandii* IscA proteins) (Zheng et al. 1998). Although gene knockout or depletion experiments in eukaryotic cells implicate a role for the IscA family of proteins in Fe-S cluster biosynthesis (Jensen and Culotta 2000; Kaut et al. 2000; Pelzer et al. 2000), a specific function has yet to be identified. The conserved cysteines make them candidates for iron or sulfur transport and/or for Fe-S cluster assembly or transport. Despite the similarities between the Nif and Isc proteins involved with Fe-S cluster biosynthesis, the process may be considerably more complex in the Isc system. For example, there are heat-shock cognate (Hsc) proteins and a putative regulatory protein, IscR, encoded in the *isc* gene cluster, which are not present in the *nif*-specific genes. The *hscA* and *hscB* genes encode proteins that have a high degree of sequence identity compared to the molecular chaperone proteins encoded by *dnaJ* and *dnaK*, respectively (Zheng et al. 1998). A specific role for these proteins as molecular chaperones that assist in the maturation of Fe-S proteins is implied by the recent the finding that IscU is a substrate for HscA (Hoff et al. 2000; Silberg et al. 2001). This suggests a specialized function for HscB in the assembly, stabilization, or transfer of Fe-S clusters formed on IscU. The IscR protein contains a DNA-binding motif and a potential Fe-S cluster-binding region involving three conserved cysteines and one conserved histidine in a $C–X_5–C–X_5–C–X_2–H$ arrangement. Hence IscR may be involved with regulating Fe-S cluster biosynthesis in response to cluster incorporation or transformation at a cluster-binding domain.

In accord with the notion that Fe-S clusters are one of the most ancient types of prosthetic group, the Isc proteins have been widely conserved in prokarya, eukarya, and archaea. In eukaryotic cells, the Fe-S cluster assembly proteins are located in the mitochondrial matrix, and the available evidence indicates that preformed clusters are likely to be exported into the cytosol via an ATP-binding cassette (ABC) transporter for the assembly of cytosolic Fe-S proteins (Kispal et al. 1999; Lill and Kispal 2000). *Saccharomyces cerevisiae* is the best characterized eukaryotic system investigated thus far, and homologs of many of the bacterial *isc* genes have been identified and shown to be involved in Fe-S biosynthesis on the basis of both biochemical and genetic evidence (Strain et al. 1998; Garland et al. 1999; Kispal et al. 1999; Schilke et al. 1999; Jensen and Culotta 2000; Kaut et al. 2000; Lange et al. 2000; Lill and Kispal 2000; Pelzer et al. 2000); e.g., Nfu1p contains a domain corresponding to the C-terminal domain of NifU, Isu1 and Isu2 correspond to IscU, Nfs1p corresponds to NifS and IscS, Isa1p and Isa2p correspond to IscA, Jac1p corresponds to HscB, Ssq1p corresponds to HscA, and Yah1p corresponds to Fdx (Fig. 5.2). Hence understanding the function of the Isc system in prokaryotes is likely to provide insight into iron homeostasis and Fe-S cluster assembly in mitochondria, with potential relevance to iron-storage diseases and the control of cellular iron uptake.

NifU as a Scaffold for Cluster Biosynthesis

Our initial attempts to investigate the role of the NifU and IscU proteins in cluster biosynthesis focused on *A. vinelandii* NifU. The Nif-specific Fe-S cluster assembly system is likely to be simpler than the general Isc system for two reasons: (1) it does not appear to require molecular chaperones and (2) it is likely to be optimized specifically for the assembly of the cubane-type [Fe$_4$S$_4$] clusters that are present in the Fe-protein and in the double-cubane P-clusters. Size-exclusion chromatography demonstrated that the homodimeric NifS and NifU proteins form a transient 1:1 complex (Yuvaniyama et al. 2000), suggesting that these two proteins function in concert to facilitate cluster biosynthesis. Mutational studies involving individual Cys-to-Ala substitutions demonstrated that each of the first seven cysteines (residues 35, 62, 106, 137, 139, 172, and 175) are required for normal rates of diazotrophic growth and that two pairs of closely spaced cysteine residues (residues 137, 139, 172, and 175) provide the ligands to the redox-active [Fe$_2$S$_2$]$^{2+,+}$ cluster that is present in NifU as purified (Agar et al. 2000b). In contrast, mutational studies indicate that the two remaining cysteine residues (residues 272 and 275) are not required for full physiological function of NifU. Although the C–X–X–C arrangement and the surrounding residues suggest a redox-active disulfide, a specific role for these two cysteines remains to be elucidated.

The modular nature of the NifU protein has been demonstrated by constructing plasmids encoding for the N-terminal domain (NifU-1) corresponding to the IscU protein and the C-terminal domain containing the ligands to the intrinsic or permanent [Fe$_2$S$_2$]$^{2+,+}$ cluster (NifU-2) (Fig. 5.3). NifU-2 was purified and found to be a homodimer containing one [Fe$_2$S$_2$]$^{2+,+}$ per subunit (Agar et al. 2000b). Both the redox potential and the spectroscopic properties of the [Fe$_2$S$_2$]$^{2+,+}$ cluster were not significantly perturbed compared to holo-NifU (Agar et al. 2000b). The diamagnetic ($S = 0$) oxidized cluster was characterized by absorption and resonance Raman spectroscopies, and the paramagnetic ($S = 1/2$) reduced cluster was investigated by absorption, variable-temperature magnetic circular dichroism (MCD) and electron paramagnetic resonance (EPR) spectroscopies. Hence the N-terminal domain of NifU can be removed without any significant effect on the properties of the [Fe$_2$S$_2$]$^{2+,+}$ cluster in the C-terminal domain. As discussed above, NifU-2 is itself likely to be made up of two distinct domains, and studies of purified fragment containing only the two C-terminal cysteines (residues 272 and 275) are currently in progress to address the role of these cysteine residues.

The possibility that the N-terminal domain of NifU is involved with mobilizing iron for cluster biosynthesis was assessed by monitoring the ability of both NifU-1 and NifU to bind either Fe^{2+} or Fe^{3+} ion (Agar et al. 2000b). No spectroscopic evidence for Fe^{2+} ion binding was observed, but both absorption and resonance Raman studies afforded evidence for binding of up to one Fe^{3+} ion per NifU-1 or NifU monomer. In the case of NifU, the spectroscopic signature of bound mononuclear Fe^{3+} was superimposed on that of the indigenous $[Fe_2S_2]^{2+}$ center. After subtracting the contributions from the $[Fe_2S_2]^{2+}$ center, the spectroscopic properties of the mononuclear Fe^{3+} sites in NifU and NifU-1 were found to be similar, in accord with a modular structure for NifU and Fe^{3+} binding exclusively within the NifU-1 domain. On the base of the frequencies and relative intensities of the Fe-S stretching modes observed in the resonance Raman spectra ($\nu_{sym} = 314\,cm^{-1}$ and $\nu_{asym} = 368\,cm^{-1}$) and the intensities of the $S \rightarrow Fe^{3+}$ charge transfer bands in the visible absorption spectra, the coordination environment of the mononuclear Fe^{3+} site was interpreted as tetrahedral, with three cysteinate and one oxygenic ligand. Cys-to-Ala mutational studies of NifU-1 indicated that all three of the N-terminal cysteine residues (residues 35, 62, and 106) are involved with ligating the mononuclear Fe^{3+} center, but the nature of the oxygenic ligand has yet to be determined (i.e., water or a protein side chain).

However, near-stoichiometric Fe^{3+} ion binding to NifU-1 or NifU was observable only in experiments conducted at 2°C in anaerobic samples that had been pretreated with dithiothreitol to ensure reduction of any intrasubunit or intersubunit disulfides. At room temperature, <10% of the NifU-1 or NifU was in a Fe^{3+} bound form, and colorimetric analysis indicates that the remainder of the Fe is in solution was in the form of free Fe^{2+} ion. Hence this mononuclear Fe^{3+}-bound species is more likely to be an intermediate in the reduction of Fe^{3+} ion by NifU or NifU-1 rather than an initial step in cluster assembly on the NifU-1 domain of NifU. In this connection, it is important to note that Fe^{3+} is rapidly reduced to Fe^{2+} by cysteine in aqueous solution (Schubert, 1932). The physiological significance (if any) of the apparent ferric reductase activity associated with the NifU-1 domain of NifU remains to be established.

The presence of permanent $[Fe_2S_2]^{2+,+}$ clusters in holo-NifU impeded spectroscopic characterization of transient clusters assembled in a NifS-mediated process. However, unambiguous evidence for NifS-directed assembly of oxidatively and reductively labile $[Fe_2S_2]^{2+}$ clusters on NifU-1 was obtained using the combination of UV-visible absorption and resonance Raman spectroscopies (Yuvaniyama et al. 2000). The anaerobic reaction mixture involved NifU : ferric ammonium citrate : β-mercaptoethanol : L-cysteine : NifS in a 50 : 100 : 5000 : 1000 : 1 ratio. The presence of catalytic amounts of NifS facilitated meaningful UV-visible absorption results and monitoring the time course of cluster assembly. Although ferric ammonium citrate was used for these cluster assembly studies, control experiments showed that Fe^{3+} is rapidly reduced by a 10-fold excess of cysteine and β-mercaptoethanol under anaerobic conditions and that ferric ammonium citrate can be replaced by ferrous ammonium sulfate in the reaction mixture without altering the rate or product of the NifS-directed cluster assembly. Hence Fe^{2+} ion rather than Fe^{3+} ion is required for assembly of a transient $[Fe_2S_2]^{2+}$ cluster on NifU-1. Mutational studies demonstrated that the presence of each of the cysteine residues in NifU-1 is required for assembly of this transient cluster, and the rate of assembly was found to depend linearly on NifS. Accurate iron and sulfide determinations were impeded by the extreme lability of cluster during purification procedures, but the UV-visible extinction coefficients before purification indicated that no more than one $[Fe_2S_2]^{2+}$ cluster was assembled per NifU-1 homodimer under these conditions.

Several variants of NifU-1 have been investigated, and one, involving a D37A substitution, was found to contain substoichiometric amounts of a bound $[Fe_2S_2]^{2+}$ cluster as purified (Yuvaniyama et al. 2000). However, the spectroscopic properties of the cluster in this variant are distinct from those observed in wild-type NifU-1. This is best illustrated by comparison of the resonance Raman spectra in the Fe-S stretching region (Fig. 5.4). $[Fe_2S_2]^{2+}$ clusters exhibit readily identifiable resonance Raman spectra using excitation into the intense $S \rightarrow Fe^{3+}$ charge transfer bands (Spiro et al. 1988). Moreover, the spectra provide a detailed monitor of the cluster environment, since the Fe-S stretching frequencies and enhancement profiles are sensitive to cluster ligation, hydrogen bonding interactions, and the conformation of bound cysteine residues (Han et al. 1989a, 1989b). A comparison of the resonance Raman spectra of the permanent $[Fe_2S_2]^{2+}$ clusters in holo-NifU with the transient $[Fe_2S_2]^{2+}$ clusters assembled in NifU-1 and IscU reveals a similar pattern of bands; the major difference involves a progressive increase in the corresponding frequencies. The increased Fe-S stretching frequencies in NifU-1 and IscU can be interpreted in terms of complete cysteinyl ligation with changes in Fe–S–C–C dihedral angles compared to the permanent clusters in NifU or in terms of one or two non-cysteinyl cluster ligands. As discussed below, Mössbauer studies of the $[Fe_2S_2]^{2+}$ clusters assembled in IscU strongly support the latter explanation.

Although the resonance Raman spectrum of the D37A NifU-1 as purified is characteristic of a $[Fe_2S_2]^{2+}$ cluster, the vibrational frequencies and enhancements of discrete bands indicate major differences in the cluster environment compared to that assembled in NifU-1. Hence D37 is an important residue in determining the stability and environment of the cluster in NifU-1. It is also a potential candidate for a cluster ligand, because it is conserved in all NifU and IscU sequences. The ability of the D37A variant to stabilize a cluster in the NifU-1 domain of A. vinelandii NifU (Yuvaniyama et al. 2000) prompted Cowan and co-workers to investigate the equivalent site-specific modification in human IscU (Foster et al. 2000). Although these authors were not successful in assembling a cluster in wild-

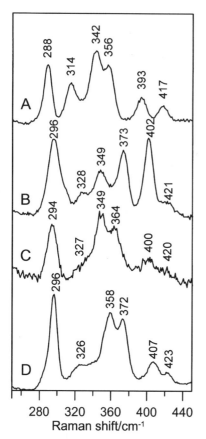

FIGURE 5.4. Resonance Raman spectra of $[Fe_2S_2]^{2+}$ centers in *A. vinelandii*. **A,** NifU as isolated. **B,** D37A NifU-1 as isolated. **C,** NifU-1 repurified after NifS-mediated cluster assembly. **D,** IscU containing two $[Fe_2S_2]^{2+}$ clusters per dimer; purified fraction after IscS-mediated cluster assembly. All spectra were recorded using 457-nm excitation at 17 K and with 6 cm^{-1} resolution. Vibrational modes resulting from lattice modes of ice have been subtracted.

type human IscU, the D37A variant was also found to contain a $[Fe_2S_2]^{2+}$ center as purified.

IscU as a Scaffold for Fe-S Cluster Biosynthesis

The lability of transient clusters assembled in NifU and spectroscopic inter-ference from the permanent $[Fe_2S_2]^{2+,+}$clusters impeded detailed character-ization of the assembled clusters and the time course of NifS-mediated cluster assembly in holo-NifU. Hence we turned our attention to *A. vinelandii* IscU. As with NifU and NifS, 1:1 complex formation between homodimeric IscU and IscS was demonstrated by both size-exclusion chro-matography and chemical cross-linking (Agar et al. 2000c). Furthermore, preliminary absorption, resonance Raman, and analytical studies provided convincing evidence for IscS-mediated assembly of reductively labile $[Fe_2S_2]^{2+}$ clusters on IscU (Agar et al. 2000c). Provided the sample was manipulated under strictly anaerobic conditions, the cluster was stable and resistant to chelating agents. Indeed samples treated with EDTA yielded analytical data indicative of one $[Fe_2S_2]^{2+}$ cluster per IscU dimer. *A. vinelandii* IscU was therefore used for a detailed investigation of the time course of cluster assembly (Agar et al. 2000a).

Experiments were carried out using a reaction mixture containing IscU with a 5-fold excess of ferric ammonium citrate and 10-fold excesses of β-mercaptoethanol and L-cysteine and IscU:IscS ratios in the range 460:1 and 28:1 and 1:1 in order to vary the rate of reaction. Preparative fast protein liquid chromatography (FPLC) was used to separate the fractions contained in samples taken at discrete time intervals. In each case, the results indicated a sequential time course of events involving three discrete and FPLC-resolvable cluster-bound forms of IscU, termed fractions 2, 3 and 4 (Agar et al. 2000a): Apo-iscU → Fraction 2 → Fraction 3 → Fraction 4. The entire reaction takes several hours with a 460:1 IscU:IscS ratio but is completed in minutes with the more physiological 1:1 stoichiometry. Cysteine and β-mercaptoethanol would be expected to reduce Fe^{3+} to Fe^{2+} under anaerobic conditions, and replacing ferric ammonium citrate with ferrous ammonium sulfate had no significant effect on the rate or sequence of reactions, in accord with Fe^{2+} rather than Fe^{3+} being used for cluster assembly.

The combination of analytical, absorption, resonance Raman, and Mössbauer was used to establish the number, type, and properties of clusters in each resolved fraction (Fig. 5.5) (Agar et al. 2000a). Fraction 2 contained 2Fe atoms per IscU dimer, and the spectroscopic data indicated one $[Fe_2S_2]^{2+}$ cluster per dimer. Fraction 3 contained 4Fe atoms per IscU dimer, and the spectroscopic data indicated two spectroscopically identical $[Fe_2S_2]^{2+}$ clusters per IscU dimer. Only one of the two $[Fe_2S_2]^{2+}$ clusters in Fraction 3 was removed by treatment with iron chelators, such as EDTA, to yield a sample identical to Fraction 2. Fraction 4 contained 4Fe atoms per IscU dimer, and the spectroscopic data indicated one $[Fe_4S_4]^{2+}$ cluster per IscU dimer. No apo-IscU was observed during formation of the $[Fe_4S_4]^{2+}$ cluster, and no fractions containing either one $[Fe_2S_2]^{2+}$ cluster and one $[Fe_4S_4]^{2+}$ cluster or two $[Fe_4S_4]^{2+}$ clusters were observed. Taken together with the sequential nature of the reaction, this suggests that the $[Fe_4S_4]^{2+}$ cluster is formed by direct reductive coupling of the two $[Fe_2S_2]^{2+}$ clusters assembled on IscU rather than by degradation of $[Fe_2S_2]^{2+}$ clusters followed reassembly as a $[Fe_4S_4]^{2+}$ cluster or by building onto each individual $[Fe_2S_2]^{2+}$ cluster.

The discovery that $[Fe_4S_4]^{2+}$ clusters can be produced by IscS-mediated cluster assembly on IscU (Agar et al. 2000a) was unexpected. In particular, it appears to be inconsistent with the preliminary studies on IscU (Agar et al. 2000c) and the results described above for NifU-1 (Yuvaniyama et al. 2000), both of which were interpreted in terms of $[Fe_2S_2]^{2+}$ clusters as the sole product of IscS- or NifS-mediated clusters assembly. However, under

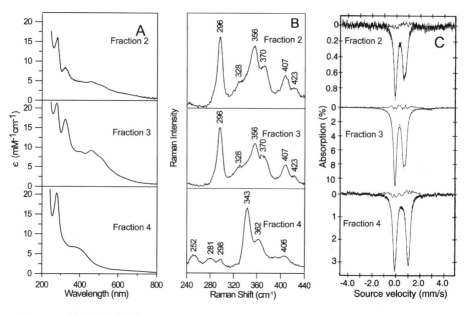

FIGURE 5.5. UV-visible absorption, resonance Raman, and Mössbauer spectra of Fe-S cluster-containing fractions of *A. vinelandii* IscU purified by FPLC from a reaction mixture containing IscU, L-cysteine, ferric ammonium citrate, β-mercaptoethanol, and catalytic amounts of IscS (see text). **A,** UV-visible absorption spectra. **B,** Raman spectra recorded at 17 K, using 457-nm excitation with 6 cm^{-1} resolution. The vibrational modes from ice have been subtracted. **C,** Mössbauer spectra at 4.2 K with an applied field of 50 mT for samples prepared using ^{57}Fe ferric ammonium citrate. *Solid line through datum points,* simulated data, *solid line near baseline,* the residual (experimental minus simulated data). Simulation parameters: δ = 0.32 mm/s and ΔE_Q = 0.90 mm/s for site 1 and δ = 0.25 mm/s and ΔE_Q = 0.64 mm/s for site 2 in fraction 2; δ = 0.33 mm/s and ΔE_Q = 0.92 mm/s for site 1 and δ = 0.26 mm/s and ΔE_Q = 0.67 mm/s for site 2 in fraction 3; two valence delocalized pairs with δ = 0.44 mm/s and ΔE_Q = 0.98 mm/s for pair 1 and δ = 0.44 mm/s and ΔE_Q = 1.23 mm/s for pair 2 in fraction 4.

the conditions used for preliminary IscS-mediated cluster assembly on IscU, the resulting sample would be expected to be a mixture of fractions 2 and 3, and EDTA treatment would result in a homogeneous sample containing one $[Fe_2S_2]^{2+}$ cluster per IscU dimer, as observed. Furthermore recent Mössbauer studies of NifS-mediated cluster assembly on ^{56}Fe holo-NifU, using ^{57}Fe ferric ammonium citrate, show that a $[Fe_4S_4]^{2+}$ cluster is the final product of cluster assembly at the transient cluster-binding site. Hence the observation of only $[Fe_2S_2]^{2+}$ clusters in NifU-1 is likely to be a consequence of incomplete assembly and/or truncation of the NifU-2 domain.

The spectroscopic properties of homogeneous samples of IscU containing $[Fe_2S_2]^{2+}$ or $[Fe_4S_4]^{2+}$ clusters provide additional insight into the properties and ligation of both types of cluster (Agar et al. 2000a). Both clusters were rapidly and irreversibly lost on exposure to air or reduction with dithionite. Nevertheless the Mössbauer spectrum of the $S = 0$ $[Fe_2S_2]^{2+}$ clusters provides strong evidence for one or two noncysteinyl ligands at one iron site. The two antiferromagnetically coupled high-spin Fe^{3+} sites have distinct isomer shifts and quadrupole splittings, indicating different ligand fields. Indeed, the observed parameters are similar to those observed for Rieske proteins, which have two histidyl ligands on one iron site (Fig. 5.5). Although the Fe-S stretching frequencies observed in the resonance Raman spectra argue against histidyl ligation, they are consistent with partial noncysteinyl (N or O) ligation for the $[Fe_2S_2]^{2+}$ clusters in IscU (see above). Since the $[Fe_2S_2]^{2+}$ clusters have no more than three cysteine ligands, they each can be subunit bridging or located within individual subunits. The Mössbauer parameters and resonance Raman spectrum of fraction 4 are characteristic of a $[Fe_4S_4]^{2+}$ cluster (Fig. 5.5). Although the Mössbauer parameters for $[Fe_4S_4]^{2+}$ clusters are relatively insensitive to the nature of the ligands, the anomalously high frequency of the totally symmetric breathing mode of the $[Fe_4S_4]$ core (343 cm^{-1}, compared to 333–339 cm^{-1} for clusters with complete cysteinyl ligation) is best interpreted in terms of a $[Fe_4S_4]^{2+}$ cluster with one noncysteinyl ligand (Agar et al. 2000a). Because the $[Fe_4S_4]^{2+}$ cluster is likely to have three cysteinyl ligands, it can also be subunit bridging or located exclusively within one subunit.

Conclusion

In vitro studies have demonstrated that the IscU and the homologous N-terminal domain of NifU provide a scaffold for IscS- or NifS-mediated assembly of both $[Fe_2S_2]^{2+}$ and $[Fe_4S_4]^{2+}$ clusters using Fe^{2+} ion as source of iron and cysteine as the source of sulfur. Moreover, cluster assembly on IscU has been shown to be a sequential process that proceeds via well-defined intermediate species involving one $[Fe_2S_2]^{2+}$ cluster per homodimer and two $[Fe_2S_2]^{2+}$ per homodimer. The final product, containing one $[Fe_4S_4]^{2+}$ cluster per homodimer, is formed by reductive coupling of the two $[Fe_2S_2]^{2+}$ clusters. Both the $[Fe_2S_2]^{2+}$ and the $[Fe_4S_4]^{2+}$ clusters are likely to have partial noncysteinyl ligation, and we speculate that this is important in mediating cluster release and/or conversion chemistry. The nature and extent of the noncysteinyl ligation has yet to be determined, but spectroscopic studies on the D37A variant of NifU-1 raise the possibility of D37 as a ligand to the $[Fe_2S_2]^{2+}$ cluster in wild-type NifU.

These observations form the basis for the working hypothesis for the mechanism of cluster assembly on the IscU scaffold (Fig. 5.6). The proposal that the reductive coupling of two $[Fe_2S_2]^{2+}$ clusters to yield a $[Fe_4S_4]^{2+}$ cluster involves disulfide formation between two of the released cysteine residues is based on model chemistry. Holm and co-workers showed that

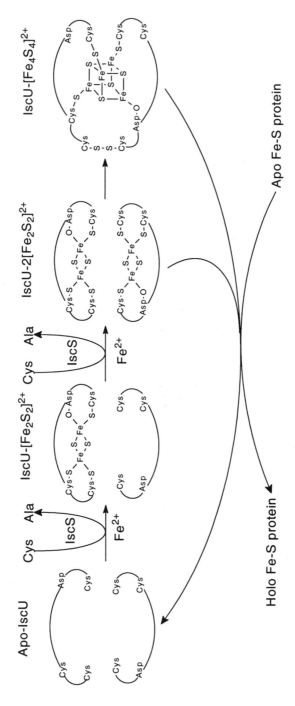

FIGURE 5.6. Working hypothesis for the mechanism of IscS-mediated cluster assembly on IscU.

reductive coupling of two $[Fe_2S_2]^{2+}$ clusters to form a $[Fe_4S_4]^{2+}$ cluster and a dithiolate is a viable pathway for the synthesis of $[Fe_4S_4]^{2+}$ clusters (Hagen et al. 1981). Reductive disulfide cleavage is clearly an attractive proposal for mediating release of the $[Fe_4S_4]^{2+}$ core. Since the first and second $[Fe_2S_2]^{2+}$ clusters assembled on IscU differ in terms of lability, it is possible that either a $[Fe_2S_2]^{2+}$ or $[Fe_4S_4]^{2+}$ core can be released and transferred intact to an apo Fe-S protein. This scheme is presented in the spirit of a working model that provides a basis for future experimentation. In particular, it should be emphasized that there is currently no direct evidence for the subunit location of either the $[Fe_2S_2]^{2+}$ or $[Fe_4S_4]^{2+}$ clusters (i.e., bridging or within one subunit of the homodimer), the extent and type of the non-cysteinyl cluster ligation, and the proposed intact transfer of either $[Fe_2S_2]^{2+}$ or $[Fe_4S_4]^{2+}$ clusters to apo iron–sulfur proteins. These aspects, together with the detailed mechanism of assembly of the initial $[Fe_2S_2]^{2+}$ cluster and the mechanism of cluster release, are under active investigation in our laboratories. The role of IscA, the molecular chaperones HscA and HscB, and the putative regulatory protein IscR are also fascinating areas for further investigation.

Addendum

Several notable advances have been made in understanding the biogenesis of Fe-S clusters since this chapter was submitted for publication early in 2001. A succinct summary of the most important advances in relation to the results presented in this chapter is presented below.

Intact transfer of a $[Fe_2S_2]^{2+}$ cluster assembled on an IscU protein (ISU1 from *Schizosaccharoyces pombe*) to an apo 2Fe ferredoxin has been demonstrated (Wu et al. 2002b). In addition, the factors determining the rate of cluster transfer from human and *S. pombe* IscU-type proteins to apo 2Fe ferredoxins have been investigated, and the critical role of the conserved aspartate residue in effecting efficient cluster transfer has been established (Wu et al. 2002c). Progress has also been made in understanding the mechanism of IscS-mediated assembly of $[Fe_2S_2]^{2+}$ clusters on IscU proteins. Direct transfer of S from IscS to IscU in bacterial systems has been demonstrated using [^{35}S]cysteine radiotracer studies (Urbina et al. 2001) and mass spectrometry (Smith et al. 2001). Futhermore, the IscU protein from *Thermatoga maritima* has been shown to bind ferrous and ferric ions using fluorescence and isothermal calorimetry and these iron sites are proposed to provide the nucleation sites for the assembly of $[2Fe-2S]^{2+}$ clusters (Nuth et al. 2002). HscA has been shown to interact with specific region of *E. coli* IscU with a LPPVK motif that is invariant among all members of the IscU family of proteins (Hoff et al. 2002). Clearly it will be of great interest to investigate the effect of binding HscA on the scaffolding and cluster transfer properties of IscU proteins.

IscA has been shown to provide an alternative scaffold to IscU for the assembly of $[Fe_2S_2]^{2+}$ cluster cores in *E. coli* (Ollagnier-de-Choudens et al. 2001) and *S. pombe* (Wu et al. 2002a) and the the *nif*-specific IscA (^{Nif}IscA) has been shown to provide an alternative scaffold to NifU for the assembly of both $[Fe_2S_2]^{2+}$ and $[Fe_4S_4]^{2+}$ cluster cores (Krebs et al. 2001). However, the need for alternative scaffold proteins for Fe-S cluster assembly has yet to be specifically addressed. IscR has been purified from *E. coli* and shown to be a $[Fe_2S_2]$ cluster-containing transcription repressor of the genes encoding for Fe-S cluster assembly proteins (Schwartz et al. 2001). This raises the intriguing possibility of transcription of the *isc* operon being regulated by transfer of a $[Fe_2S_2]$ cluster assembled on an IscU or IscA scaffold to IscR. Perhaps the most exciting development has been the evidence for a third bacterial system for Fe-S clusters assembly, typified by the *sufABCDSE* operon in *E. coli* (Takahashi and Tokumoto 2002). Although this system appears to provide a back up for the Isc system in *E. coli*, it is likely to be the only system for Fe-S cluster assembly in many archaea. The Suf system promises to be the evolutionary ancestor of the Isc and Nif systems, but none of the constituent proteins have been purified thus far. SufS and SufA appear to be related to NifS/IscS and IscA, respectively, based on sequence homology, but the roles of SufBCDE remain to be elucidated. It will clearly be of great interest to investigate the Fe-S cluster assembly mechanism in the Suf system for comparison of the pathways that are now emerging for the Isc and Nif systems.

Acknowledgments. This research was supported by grants from the National Institutes of Health (GM62524 to M.K.J.) and the National Science Foundation (MCB9630127 to D.R.D.).

References

Achim C, Golinelli M-P, Bominaar EL, et al. 1996. Mössbauer study of Cys56Ser mutant 2Fe ferredoxin from *Clostridium pasteurianum*: evidence for double exchange in an $[Fe_2S_2]^+$ cluster. J Am Chem Soc 118:8168–9.

Agar JN, Krebs B, Frazzon J, et al. 2000a. IscU as a scaffold for iron-sulfur cluster biosynthesis: sequential assembly of [2Fe-2S] and [4Fe-4S] clusters in IscU. Biochemistry 39:7856–62.

Agar JN, Yuvaniyama P, Jack RF, et al. 2000b. Modular organization and identification of a mononuclear iron-binding site within the NifU protein. J Biol Inorg Chem 5:167–77.

Agar JN, Zheng L, Cash VL, et al. 2000c. Role of the IscU protein in iron-sulfur cluster biosynthesis: IscS-mediated assembly of a $[Fe_2S_2]$ cluster in IscU. J Am Chem Soc 122:2136–7.

Beinert H. 2000. Iron-sulfur proteins: ancient structures, still full of surprises. J Biol Inorg Chem 5:2–15.

Beinert H, Kiley PJ. 1999. Fe-S proteins in sensing and regulatory functions. Curr Opin Chem Biol 3:152–7.

Beinert H, Holm RH, Münck E. 1997. Iron-sulfur clusters; nature's modular, multipurpose structures. Science 277:653–9.

Berg JM, Holm RH. 1982. Structures and reactions of iron-sulfur protein clusters and their synthetic analogs. In: Spiro TG, editor. Iron-sulfur proteins. New York: Wiley-Interscience. p 1–66.

Bui BTS, Florentin D, Fournier F, et al. 1998. Biotin synthase mechanism: on the origin of sulphur. FEBS Lett 440:226–30.

Cammack R. 1992. Iron-sulfur clusters in enzymes: themes and variations. Adv Inorg Chem 38:281–322.

Clausen T, Kaiser JT, Steegborn C, et al. 2000. Crystal structure of the cystine C-S lyase from *Synechocystis*: stalibization of cysteine persulfide for FeS cluster biosynthesis. Proc Natl Acad Sci USA 97:3856–61.

Cosper NJ, Booker SJ, Ruzicka FJ, et al. 2000. Direct FeS cluster involvement in generation of a radical in lysine 2,3-aminomutase. Biochemistry 39:15668–73.

Crouse BR, Meyer J, Johnson MJ. 1995. Spectroscopic evidence for a reduced Fe_2S_2 cluster with a $S = 9/2$ ground state in mutant forms of *Clostridium pasteurianum* 2Fe ferredoxin. J Am Chem Soc 117:9612–13.

Dai S, Schwendtmayer C, Schürmann P, et al. 2000. Redox signaling in chloroplasts: cleavage of disulfides by an iron-sulfur cluster. Science 287:655–8.

Dean DR, Bolin JT, Zheng L. 1993. Nitrogenase metalloclusters: structures, organization, and synthesis. J Bacteriol 1756:6737–44.

Foster MW, Mansy SS, Hwang J, et al. 2000. A mutant human IscU protein contains a stable $[2Fe-2S]^{2+}$ center of possible functional significance. J Am Chem Soc 122:6805–6.

Fu W, Jack RF, Morgan TV, et al. 1994. *nifU* gene product from *Azotobacter vinelandii* is a homodimer that contains two identical [2Fe-2S] clusters. Biochemistry 33:13455–63.

Fujii T, Maeda M, Mihara H, et al. 2000. Structure of a NifS homologue: X-ray structure analysis of CsdB, an *Escherichia coli* counterpart of mammalian selenocysteine lyase. Biochemistry 39:1263–73.

Garg RP, Vargo CJ, Cui X, Kurtz DM Jr. 1996. A [2Fe-2S] protein encoded by an open reading frame upstream of the *E. coli* bacterioferritin gene. Biochemistry 35:6297–301.

Garland SA, Hoff K, Vickery LE, Culotta VC. 1999. *Saccharomyces cerevisiae ISU1* and *ISU2*: members of a well-conserved gene family for iron-sulfur cluster assembly. J Mol Biol 294:897–907.

Gibson KJ, Pelletier DA, Turner IM Sr. 1999. Transfer of sulfur to biotin from biotin synthase (bioB protein). Biochem Biophys Res Commun 254:632–5.

Hagen KS, Reynolds JG, Holm RH. 1981. Definition of reaction sequences resulting in self-assembly of $[Fe_4S_4(SR)_4]^{2-}$ clusters from simple reactants. J Am Chem Soc 103:4054–63.

Han S, Czernuszewicz RS, Kimura T, et al. 1989a. Fe_2S_2 protein resonance Raman revisited: structural variations among adrenodoxin, ferredoxin, and red paramagnetic protein. J Am Chem Soc 111:3505–11.

Han S, Czernuszewicz RS, Spiro TG. 1989b. Vibrational spectra and normal mode analysis for [2Fe-2S] protein analogues using ^{34}S, ^{54}Fe, and ^2H substitution: coupling of Fe-S stretching and S-C-C bending modes. J Am Chem Soc 111:3496–504.

Hoff KG, Silberg JJ, Vickery LE. 2000. Interaction of the iron-sulfur cluster assembly protein IscU with the Hsc66/Hsc20 molecular chaperone system of *Escherichia coli*. Proc Natl Acad Sci USA 97:7790–5.

Hoff KG, Ta DT, Tapley TL, et al. 2002. Hsc66 substrate specificity is directed toward a discrete region of the iron-sulfur cluster template protein IscU. J Biol Chem 277:27353–9.

Hwang DM, Dempsey A, Tan KT, Liew CC. 1996. A modular domain of NifU, a nitrogen fixation cluster protein, is highly conserved in evolution. J Mol Evol 43:536–40.

Jensen LT, Culotta VC. 2000. Role of *Saccharomyces cerevisiae* Isa1 and Isa2 in iron homeostasis. Mol Cell Biol 20:3918–27.

Johnson MK. 1994. Iron-sulfur proteins. In: King RB, editor. Encyclopedia of inorganic chemistry. Chichester, UK: Wiley. p 1896–915.

Johnson MK. 1998. Iron-sulfur proteins: new roles for old clusters. Curr Opin Chem Biol 2:173–81.

Johnson MK, Duin EC, Crouse BR, et al. 1998a. Valence-delocalized $[Fe_2S_2]^+$ clusters. In: Solomon EI, Hodgson KO, editors. Spectroscopic methods in bioinorganic chemistry. Washington, DC: American Chemical Society. p 286–301.

Johnson MK, Staples CR, Duin EC, et al. 1998b. Novel roles for Fe-S cluster in stabilizing or generating radical intermediates. Pure Appl Chem 70:939–46.

Jung Y-S, Gao-Sheridan HS, Christiansen J, et al. 1999. Purification and biophysical characterization of a new [2Fe-2S] ferredoxin from *Azotobacter vinelandii*, a putative [Fe-S] cluster assembly/repair protein. J Biol Chem 274:32402–10.

Kaiser JT, Clausen T, Bourenkow GP, et al. 2000. Crystal structure of a NifS-like protein from *Thermotoga maritima*: Implications for iron-sulfur cluster assembly. J Mol Biol 297:451–64.

Kambampati R, Lauhon CT. 2000. Evidence for the transfer of sulfane sulfur from IscS to ThiI during the *in vitro* biosynthesis of 4-thiouridine in *Escherichia coli* tRNA. J Biol Chem 275:10727–30.

Kaut A, Lange H, Diekert K, et al. 2000. Isa1p is a component of the mitochondrial machinery for maturation of cellular iron-sulfur proteins and requires conserved cysteine residues for function. J Biol Chem 275:15955–61.

Kispal G, Csere P, Prohl C, Lill R. 1999. The mitochondrial proteins Atm1p and Nfs1p are essential for the biogenesis of cytosolic Fe/S proteins. EMBO J 18:3981–9.

Krebs C, Agar JN, Smith AD, et al. 2001. IscA, an alternative scaffold for Fe-S cluster biosynthesis. Biochemistry 40:14069–80.

Lange H, Kaut A, Kispal G, Lill R. 2000. A mitochondrial ferredoxin is essential for biogenesis of cellular iron-sulfur proteins. Proc Natl Acad Sci USA 97:1050–5.

Li J, Kogan M, Knight SAB, et al. 1999. Yeast mitochondrial protein, Nfs1p, coordinately regulates iron-sulfur cluster proteins, cellular iron uptake, and iron distribution. J Biol Chem 274:33025–34.

Lill R, Kispal G. 2000. Maturation of cellular Fe-S proteins: an essential function of mitochondria. TIBS 25:352–6.

Malkin R, Rabinowitz JC. 1966. The reconstitution of clostridial ferredoxins. Biochem Biophys Res Commun 23:822–7.

Nishio K, Nakai M. 2000. Transfer of iron-sulfur cluster from NifU to apoferredoxin. J Biol Chem 275:22615–18.

Nuth M, Yoon T, Cowan JA. 2002. Iron-sulfur cluster biosynthesis: Characterization of iron nucleation sites for the assembly of the $[2Fe-2S]^{2+}$ cluster core in IscU proteins. J Am Chem Soc 124:8774–5.

Ollagnier-de-Choudens S, Mattioli T, Takahashi Y, Fontecave M. 2001. Iron-sulfur cluster assembly. Characterization of IscA and evidence for a specific functional complex with ferredoxin. J Biol Chem 276:22604–7.

Pelzer W, Mühlenhoff U, Diekert K, et al. 2000. Mitochondrial Isa2p plays a crucial role in the maturation of cellualr iron-sulfur proteins. FEBS Lett 476:134–9.

Quail MA, Jordan P, Grogan JM, et al. 1996. Spectroscopic and voltammetric characterization of the bacterioferritin-associated ferredoxin from *Escherichia coli*. Biochem Biophys Res Commun 229:635–42.

Schilke B, Voisine C, Beinert H, Craig E. 1999. Evidence for a conserved system for iron metabolism in the mitochondria of *Saccharomyces cerevisiae*. Proc Natl Acad Sci USA 96:10206–11.

Schubert M. 1932. Complex types involved in the catalytic oxidation of thiol acids. J Am Chem Soc 54:4077–85.

Schwartz CJ, Djaman O, Imlay JA, Kiley PJ. 2000. The cysteine desulfurase, IscS, has a major role in *in vivo* Fe-S cluster formation in *Escherichia coli*. Proc Natl Acad Sci USA 97:9009–14.

Schwartz CJ, Giel JL, Patschkowski T, et al. 2001. IscR, an Fe-S cluster-containing transcription factor, represses expression of *Escherichia coli* genes encoding Fe-S cluster assembly proteins. Proc Natl Acad Sci USA 98:14895–900.

Silberg JJ, Hoff KG, Tapley TL, Vickery LE. 2001. The Fe/S assembly protein IscU behaves as a substrate for the molecular chaperone Hsc66 from *Escherichia coli*. J Biol Chem 276:1696–700.

Smith AD, Agar JN, Johnson KA, et al. 2001. Sulfur transfer from IscS to IscU: The first step in iron-sulfur cluster biosynthesis. J Am Chem Soc 123:11103–4.

Sofia HJ, Chen G, Hetzler BG, et al. 2001. Radical SAM, a novel protein superfamily linking unresolved steps in familiar biosynthetic pathways with radical mechanisms: functional characterization using new analysis and information visualization methods. Nucleic Acids Res 29:1097–106.

Spiro TG, Czernuszewicz RS, Han S. 1988. Iron-sulfur proteins and analog complexes, In: Spiro TG, editor. Resonance Raman spectra of heme and metalloproteins. New York: Wiley. p 523–54.

Staples CR, Gaymard E, Stritt-Etter A-L, et al. 1998. Role of the $[Fe_4S_4]$ cluster in mediating disulfide reduction in spinach ferredoxin:thioredoxin reductase. Biochemistry 37:4612–20.

Strain J, Lorenz CR, Bode J, et al. 1998. Supressors of superoxide dismutase (*SOD1*) deficiency in *Saccharomyces cerevisiae*. Identification of proteins predicted to mediate iron-sulfur cluster assembly. J Biol Chem 273:31138–44.

Ta DT, Vickery LE. 1992. Cloning, sequencing, and overexpression of a [2Fe-2S] ferredoxin gene from *Escherichia coli*. J Biol Chem 267:11120–5.

Takahashi Y, Tokumoto U. 2002. A third bacterial system for the assembly of iron-sulfur cluster with homologs in archaea and plastids. J Biol Chem 277: 28380–3.

Urbina HD, Silberg JJ, Hoff KG, Vickery LE. 2001. Transfer of sulfur from IscS to IscU during Fe/S cluster assembly. J Biol Chem 276:44521–6.

Wu G, Mansy SS, Hemann C, et al. 2002a. Iron-sulfur cluster biosynthesis: Characterization of *Schizosaccharomyces* pombe Isa1. J Biol Inorg Chem 7:526–32.

Wu G, Mansy SS, Wu S-P, et al. 2002b. Characterization of an iron-sulfur cluster assembly protein (ISU1) from *Schizosaccharomyces pombe*. Biochemistry 41:5024–23.

Wu S-P, Wu G, Surerus KK, Cowan JA. 2002c. Iron-sulfur cluster biosynthesis: Kinetic analysis of [2Fe-2S] cluster transfer from holo ISU to apo Fd: Role of redox chemistry and a conserved aspartate. Biochemistry 41:8876–85.

Yuvaniyama P, Agar JN, Cash VL, et al. 2000. NifS-directed assembly of a transient [2Fe-2S] cluster within the NifU protein. Proc Natl Acad Sci USA 97:599–604.

Zheng L, Cash VL, Flint DH, Dean DR. 1998. Assembly of iron-sulfur clusters. Identification of an *iscSUA-hscBA-fdx* gene cluster from *Azotobacter vinelandii*. J Biol Chem 273:13264–72.

Zheng L, White RH, Cash VL, et al. 1993. Cysteine desulfurase activity indicates a role for NIFS in metallocluster biosynthesis. Proc Natl Acad Sci USA 90:2754–8.

Zheng L, White RH, Cash VL, Dean DR. 1994. A mechanism for the desulfurization of L-cysteine catalyzed by the *nifS* gene product. Biochemistry 33:4714–20.

6
Genes and Proteins Involved in Nickel-Dependent Hydrogenase Expression

R.J. Maier, J. Olson, and N. Mehta

A key metabolic feature associated with many physiologically diverse bacteria and archaea is the ability to use hydrogen as a growth substrate. This seemingly simple process of hydrogen activation is actually a complex operation, requiring an enzyme capable of multiple redox transitions and a complex metal-containing active site. Apparently little variability can occur in this active site (for hydrogen-oxidizing enzymes) without compromising catalytic efficiency. The major classes of hydrogenases that facilitate energy input into cell metabolism are the NiFe uptake types (Friedrich and Schwartz 1993). These enzymes are usually associated with the membrane in close association with respiratory chain components. Their synthesis depends on a supply of nickel, an input of hydrolyzable nucleotide triphosphate energy, and a mechanism to sense that the substrate hydrogen is available. Also required are a battery of accessory proteins with putative roles in nickel or iron binding, in maintaining the correct structure for insertion/delivery of these metals, and in incorporating CO and CN as ligands to the active center. The maturation process involves other proteins, some of which presumably complex to hydrogenase during the metallo-center assembly process (Drapal and Böck 1998). In addition to all these factors, the regulators that act on free-living hydrogenase can differ from those that function in the same bacterium when it is residing in the nitrogen-fixing symbiosis (Durmowicz and Maier 1998). In the last few years, additional roles for NiFe hydrogenases as sensors of the redox environment have been uncovered. Even as sensors, playing regulatory roles in signal transduction pathways, these regulatory hydrogenases vary little in the important structural features associated with other NiFe hydrogenases (Pierik et al. 1998).

Hydrogenases are not the only nickel-containing enzyme, and researchers must therefore compare the maturation of different nickel proteins to obtain an integrated picture of nickel metabolism. Indeed, similarities between some of the hydrogenase-related nickel-processing proteins with urease and carbon monoxide dehydrogenase maturation factors have been noted, and this has facilitated interpretations of the results for maturation of NiFe hydrogenases. Common themes have developed and have been reviewed (Hausinger 1997; Maroney 1999). This chapter focuses on the recent findings on maturation of NiFe hydrogenases in some of the best-studied aerobic organisms. Extra emphasis is placed on *Bradyrhizobium japonicum*, the symbiont of soybeans; but in some areas, the research on *Rhizobium leguminosarum, Ralstonia eutropha, Rhodobacter capsulatus*, and *Eschericia coli* has been more productive and informative, so they are included for comparison. In addition, some recent results on nickel-metabolism proteins in the peptic ulcer–causing bacterium *Helicobacter pylori* are presented.

The motivation for studying hydrogen use in the symbiotic nitrogen-fixing bacteria comes from studies indicating that legume yield advantages are associated with the hydrogen uptake (Hup) phenotype (Maier and Triplett 1996). Large amounts of hydrogen are produced by diazotrophs under nitrogen-fixing conditions. Some diazotrophs do not evolve the nitrogenase-produced hydrogen; these bacteria (whether symbiotic or not) have been found to contain an uptake hydrogenase that oxidizes hydrogen to protons and electrons. In most cases, the electrons produced by hydrogen activation via hydrogenase are funneled through an efficient electron-transport chain that conserves energy. Most of the energy lost in nitrogenase-dependent hydrogen production can thereby be recovered. This process requires nickel and efficient nickel metabolism in the symbiosis for effective hydrogen oxidation/reductant recycling (Dalton et al. 1988; Brito et al. 2000). Studies on the efficiency provided to legume root nodules by hydrogen oxidation in rhizobia is not confined to the *B. japonicum*–soybean symbiosis. For example, research on the regulation and genetics of hydrogenase in the symbiont of peas, *R. leguminosarum*, has been conducted since the late 1990s (Brito et al. 1997: Hernando et al. 1998; Colombo, et al. 2000).

From sequencing of genes near the structural genes for NiFe hydrogenases, researchers were able to uncover many predicted proteins with poorly defined roles; mutagenesis of these genes revealed that many of these proteins were important in creating a mature enzyme from the precursor form of hydrogenase (Maier and Böck 1996a; Maier and Triplett 1996). Hydrogen-oxidizing organisms devote a considerable amount of enzymatic processing machinery to making a mature hydrogenase, and much of this machinery is related to proper synthesis or insertion of the metallocenter. Thus nickel-binding and -metabolizing proteins with roles in hydrogenase synthesis were discovered. Mutagenesis of the genes involved in metal center processing typically resulted in a loss of hydrogenase activity, although often this phenotype could be, in part, suppressed by the addition of high levels of nickel to the growth medium (Waugh and Boxer 1986; Jacobi et al. 1992; Du and Tibelius 1994). This result suggests that the accessory proteins play roles in nickel transport, and a few such roles have been assigned. However, most of the accessory components seem to play roles in maturation of the nickel-containing enzyme.

Nickel Transport for Hydrogenase Synthesis

Nickel uptake is often mediated by nonspecific divalent cation transport systems with variable nickel specificity and relatively low Ni^{2+} affinity. For organisms in which hydrogen only supplements their array of growth substrates, perhaps this is sufficient for obtaining sufficient nickel. A few bacteria (*R. eutropha, E. coli,* and *B. japonicum*) have been found to contain a high-affinity Ni^{2+} transport system. Considering the importance of hydrogen oxidation to their growth and the fact that nickel can be limiting in many natural environments, a high-affinity nickel transport system, under the control of hydrogenase derepession, would serve the organism well. The best-studied systems are the Nik system for Ni^{2+} transport in *E. coli*, (Navarro et al. 1993) and the HoxN permease of *R. eutropha* (Eitinger and Friedrich 1994; Eitinger et al. 1997). *H. pylori* also transports nickel by a system similar to HoxN (termed NixA), but it is presumed to serve primarily urease synthesis (Mobley et al. 1995). The Nik system (*E. coli*) belongs to the ATP-binding cassette family of transporters, and it consists of five subunits, whereas the HoxN (HupN in *B. japonicum*) is an integral membrane protein that does not require other proteins to act as a nickel permease. From studies of fusions to alkaline phosphotase and β-galactosidase, an eight-transmembrane helix model is proposed for HoxN (Fig. 6.1). The *B. japonicum* homolog is 56% identical to *R. eutropha* HoxN, and a similar model could be presented.

When mutant versions of HoxN are expressed in *E. coli* (Eitinger et al. 1997), it is found that a conserved area within both HoxN type and the integral membrane NikC types of nickel permeases, His-X_4-Asp-His, is crucial for nickel uptake activity and for urease activity. The same mutations in the nickel-binding motif of HoxN in *R. eutropha* adversely affected hydrogenase activities, even when nickel was supplemented to the medium, indicating the histidine-containing motif at residues 62–68 were important for nickel transport. These correspond to residues 80–86 in HupN (Fu et al. 1994). For *B. japonicum*, *hupN* is organized in an operon with two other genes, *hupO* and *hupP*. Based on mutagenesis and complementation studies, Fu et al. (1994) concluded that all three genes were important for nickel metabolism into hydrogenase by *B. japonicum*, but the most deleterious effect on nickel binding and hydrogenase expression was by a mutation in *hupN* alone. The effects of the mutations were evident only in low-nickel conditions. The predicted HupO is also likely a membrane protein, with several potential nickel-binding sites.

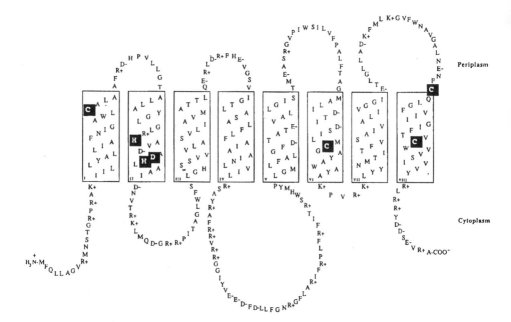

FIGURE 6.1. Topological model for the HoxN nickel permease of *R. eutropha* that is like the HupN of *B. japonicum* (Eitinger et al. 1997). The residues (two histidines and one aspartic acid) in the second transmembrane segment were shown to be key residues for nickel transport, and this motif (His-Xaa$_4$-Asp-His) is conserved in nickel-transport proteins. *Bold*, residues studied by site-directed mutagenesis.

Nickel-Dependent Transcription of Hydrogenase and Hydrogen-Sensing Mechanisms

The synthesis of hydrogenase *in B. japonicum* was shown to depend on the addition of nickel to the culture medium (Kim and Maier 1990). In addition to nickel, hydrogenase derepression requires incubation of cells in a closed gas atmosphere containing hydrogen and oxygen (Kim et al. 1991). The system is oxygen repressed, as the oxygen level cannot exceed 5% partial pressure. Subsequently, it was shown that the three environmental regulators (nickel, oxygen, and hydrogen) act transcriptionally at the same genetic area upstream of the hydrogenase structural genes (Kim et al. 1991). This regulation also required integration host factor (IHF) and σ-54, but the binding sites were farther downstream, nearer the structural genes (Black and Maier, 1995). The three environmental signals for (*B. japonicum*) hydrogenase expression are relayed via molecular signals (Fig. 6.2), which that have been identified in *B. japonicum*, but, in some cases, have been more extensively studied in other systems (see below). Once the original environmental signal is sensed, molecular signals come into play, including a regulatory hydrogenase (HupUV), a histidine kinase sensor (HupT), and a DNA-binding response regulator (HoxA). All three of the environmental signals shown in Figure 6.2 do not necessarily act in the other (hydrogen-oxidizing bacterial) systems, but the components thus far described can account for the sensing and subsequent signal transduction regulatory cascade that originates with all three signals.

The membrane-bound NiFe hydrogenase of *B. japonicum* was ruled out as the sensor for these environmental components, as *hup-lacZ* promoter fusions in a mutant strain lacking hydrogenase still displayed normal regulation (dependency on nickel and hydrogen, and requirement for low levels of oxygen), (Kim et al. 1991). However, an excellent candidiate regulatory sensor for all three components is HupUV. The HupUV genes of *B. japonicum* are located immediately 5′ of the *hupSL* structural genes and, by sequence similarity, encode proteins structurally similar to HupSL. In a mutant strain lacking HupV, neither hydrogenase activity nor hydrogenase peptides accumulated (Black et al. 1994). In addition, regulation (both positive and negative) of hydrogenase transcription (measured with a *hup-lacZ* fusion) by nickel, oxygen, and hydrogen is eliminated in the *hupV* mutant. There is no evidence from either sequence or physiological studies that HupUV itself is a DNA-binding protein. The DNA-binding protein for activation of *B. japonicum hupSL* in free-living conditions is HoxA (Durmowicz and Maier 1997; van Soom et al. 1997), a transcription factor in the NtrC family.

The regulatory hydrogenases HupUV of *R. capsulatus* (Elsen et al. 1996; Vignais et al. 1997) and the *R. eutropha* homolog of HupUV (termed HoxBC) (Lenz and Friedrich 1998; Kleihues et al. 2000) have been partially characterized. They appear to be sensor hydrogenases, functioning as the initial hydrogen-sensor in a two-component signaling pathway. The

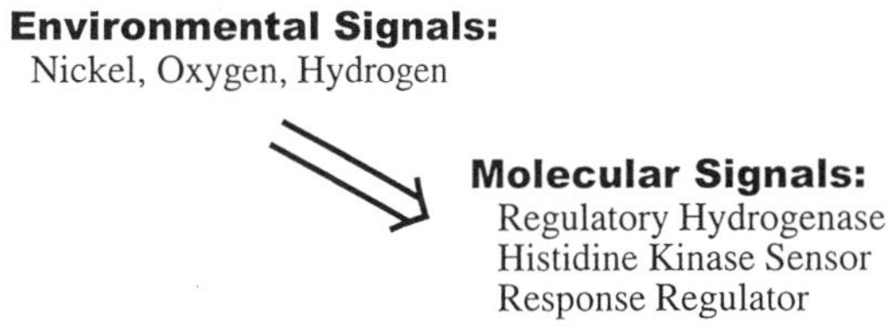

Environmental Signals:
Nickel, Oxygen, Hydrogen

Molecular Signals:
Regulatory Hydrogenase
Histidine Kinase Sensor
Response Regulator

FIGURE 6.2. Regulators of NiFe hydrogenase. The environmental signals that act on hydrogenase (hydrogen uptake) expression are sensed through molecular signals, including regulatory hydrogenase, a histidine kinase, and a response regulator; the latter two represent two-component regulatory factors widespread among bacterial molecular sensory systems.

HupUV-type sensors appear to mediate hydrogenase regulation through HupT (Elsen et al. 1993; van Soom et al. 1999), a histidine protein kinase with autophosphorylation capacity; all three genes (for *R. capsulatus*) were shown to be co-transcribed and solely involved in negative regulation of hydrogenase transcription under approporiate culture conditions (Elsen et al. 1996). A key finding was that the *R. capsulatus* HupUV proteins were shown to interact with hydrogen (Vignais et al. 1997), and thus it was proposed that the transcriptional regulation via HupUV is triggered by its sensing of hydrogen. Because HupUV is similar to NiFe hydrogenases and includes the conserved cysteine-containing motifs involved in liganding of nickel by hydrogenase, it was predicted to be capable of sensing nickel and oxygen as well as.

In the *R. eutropha* system, HoxBC (the regulatory hydrogenase) was shown to be a NiFe hydrogenase (Lenz et al. 1997), which interacts (directly or indirectly) with HoxJ (a histidine protein kinase). Results from a single amino acid substitution in HoxJ are consistent with the idea that autokinase activity of HoxJ is needed for the response to hydrogen and that HoxJ-mediated phosphorylation of HoxA prevents transcriptional activation. Mutations in the phosporyl acceptor residue of the response regulator HoxA still permitted full transcription of hydrogenase structural genes (Lenz and Friedrich 1998), so the model is that the nonphosphorylated form of HoxA is the one that stimulates transcription. This negative type of regulation is unlike other NtrC-like transcription regulators, even though the kinase and the phosphate acceptor DNA-binding molecules are similar to others of the two competing regulatory systems. In addition to this unconventional mode of action of HoxA for phopshorylated response regulators, the regulation (amplification) of and by hoxA is unusually complex (Schwartz et al. 1999).

The demonstration that the regulatory function of HoxBC depends on nickel (Kleihues et al. 2000), along with the clear similarities in HoxBC and HupUV (putative) structure and function, provides an explanation for the nickel- and the hydrogen-dependent regulation described years earlier for the *B. japonicum* hydrogenase, which was based on mutant analysis (in *hupV*) and promoter fusions to study its regulatory properties (Black, et al. 1994). The properties of the new class of NiFe hydrogenases, including low hydrogen-oxidizing activity and a cytoplasmic location, would be expected by their predicted roles as sensory/regulatory hydrogenases. From sequence databases, this type of NiFe hydrogenase–mediated sensing of the redox environment will likely be expanded to include a number of diverse bacteria (Kleihues et al. 2000).

Based on sequence analysis, the HupT proteins of *R. capsulatus* (Elsen et al. 1993) and *B. japonicum* (van Soom et al. 1999) resemble the sensor kinase proteins associated with two component regulatory systems. HupT was, in fact, shown to autophosphorylate at a His residue (Elsen et al. 1996). With

HupUV acting as a sensor, HupT presumably is the signaling link via phosphate transfer to the transcriptional activator, essentially playing the role of HoxJ*, described above. HoxA, a response regulator of the two-component type, is involved with hydrogenase-transcription regulation in free-living *B. japonicum* (Durmowicz and Maier 1997), so it is expected that HupT (and perhaps other proteins) is involved in the HoxA-dependent free-living *hup* expression, whereas HupUV presumably plays the sensor role under both free-living and symbiotic expression of hydrogenase. The oxygen-sensing role mediating oxygen-dependent repression of HoxA-mediated transcriptional activation could be served via a HupUV-dependent redox-sensing mechanism, too. No oxygen binding by HoxBC could be detected; but, in contrast to *B. japonicum*, the *R. eutropha* hydrogenase is not oxygen regulated. Also, many NiFe hydrogenases are known to bind oxygen, changing the environment of the nickel center signal as detected by spectral methods (Albracht 1994). The regulation of hydrogenase in the soybean symbiosis differs from that for free-living *hupSL* expression. For example, the $FixK_2$ protein is involved as a transcription activator uniquely under symbiotic conditions (Durmowicz and Maier 1998). The symbiont of peas, *R. leguminosarum* lacks the sensing or signaling system altogether (Brito et al. 1997).

Accessory Proteins for Nickel Enzyme Synthesis

Nickel is a requirement for the growth of many microorganisms (Hausinger 1994). The biological basis for this is that nickel is a key component of several enzymes: urease, hydrogenase, carbon monoxide dehydrogenase, methyl coenzyme M reductase, and one class of superoxide dismutase. After its initial sequestration, nickel, like most other metals that serve as components for enzymes, must be mobilized by proteins to be incorporated into its final biological sink. As indicated above, these reactions are not well understood and can be complex, especially if other metals or metal clusters are involved. Regarding storage of the metal, our knowledge is likewise scant. The common themes for metallocenter assembly in nickel enzymes include use of a protein that has a nucleotide-binding site and one or more nickel-binding proteins with histidine-rich motifs (Hausinger 1997).

For urease assembly, significant progress has been made in identifying the interacting accessory components by the purification of complexes of accessory proteins, some of which contain apourease within the complex. In vitro, some apourease can be activated by adding nickel ions and carbon dioxide, the latter being incorporated into the protein to form a lysine carbamate metal ligand (Hausinger 1997). However, in vivo three accessory proteins are required (UreD, UreF, UreG), which have been shown to be capable of forming a complex with the apoenzyme (Moncrief and Hausinger 1996). The metallocenter assembly process is facilitated by a fourth protein

A

```
Bj            16 HaHdHHHdHgHdHdHgHdgHHHHHHgHdqdHHHHHdHaH 55
Rl      15 HtHevgddgHgHHHHHdgHHdHdHdHdHHrgdHeHddHHHaedgsvH 60
Ss            15 HsHHHHgdgnfaHsHddHdqqeHHHHH 41
Ac            18 HHHHgydHgHHHdHafvrrpapaeaaplvveglnlH 54
Av            18 HHHHgHdHHHHeHpfvrrpapaeaappaaggpnlH 53
Ms            26 HHHeHdHdHdHpHtHdH 42

Ka UreE      144 HgHHHaHHdHHaHsH 158
Rr CooJ       82 HspfHsHaHsHdHdHaHgHsHdHaHdHcHcHdH 114
```

B

```
                  G1                    G2          G3          G4
Bj      108 AFNLYSSPGAGKT.X₂₇.DAERIRAT.X₄₅.PAAFGLG.X₂₈.LMLINKID 243
Rl      107 ALNFVSSPQSGKT.X₂₇.DAARIRET.X₄₅.PAAFDLG.X₂₈.LMILNKAD 242
Ec      105 VLNLVSSPGSGKT.X₂₇.DAARIRAT.X₄₅.PASFGLG.X₂₈.LMLLNKVD 201
Ac      101 VLNLVSSPGSGKT.X₂₇.DAARIRAT.X₄₅.PAAFDLG.X₂₈.LMLLNKTD 236
Av       99 VLNLVSSPGSGKT.X₂₇.DAARIRAT.X₄₅.PAAFDLG.X₂₈.LMLLNKTD 234
Ms       76 ALNITSSPGAGKT.X₂₇.DADRIRAA.X₄₅.PALFDLG.X₂₈.LVILNKID 211
Ss       90 VMNFLSSPGAGKT.X₂₇.DAQRLRSA.X₄₅.PTTYDLG.X₂₈.VILVTKQD 225

Ka UreG 9 AVGVGGPVGSGKT 21
Rr CooC 2 KIAVTGKGGVGKS 14
```

FIGURE 6.3. Alignments of regions identified to be important in nickel accessory proteins for maturation of the final nickel sink, the nickel-containing enzyme. **A**, The His-rich regions of known or putative HypB proteins and the UreE and CooJ proteins from various organisms. Reprinted with permission from Olson and Maier (2000). **B**, Alignment of the G motifs from known or putative HypB proteins and the related nucleotide-binding P-loop residues (in the G1 region) of CooC and UreG. (Adapled from Olson and Maier (2000).

(UreE), which contains a His-rich area that sequesters nickel and is proposed to deliver nickel to urease in the activation process (Colpas et al. 1999). Furthermore, GTP hydrolysis associated with UreG is important for urease activation (Soriano and Hausinger 1999). Similarly, CODH maturation requires the His-rich protein CooJ and the nucleotide-binding protein CooC (Kerby et al. 1997). Alignment of the His-rich regions and G motifs for a number of nickel metabolism accessory proteins is shown in Figure 6.3. The need for a motif rich in histidine residues is apparently usually required for proper nickel donation (Fig. 6.3A), and a GTP-binding domain associated with the same protein (the His-rich nickel donor) or as a separate GTPase protein is a common theme (Fig. 6.3B). Of course, the metal-locenter assembly (which may involve incorporation of additional metals such as iron) process must maintain the correct folding of the final nickel-containing product (Maier and Böck 1996a; Hausinger 1997). With additional concerns such as the maintenance of a certain redox state, nucleotide binding and release, the involvement of scaffolding proteins, introduction of other (nonmetal) ligands, and proteolytic processing, a seemingly simple process (e.g., metal donation into the apo-enzyme) can be complex.

Many genes proposed to encode factors involving nickel or iron metal-locenter assembly or incorporation for hydrogenases have been sequenced, but their roles in any specific process are poorly defined. For example, even though it is known that each of the proteins HypC, HypD, HypB, HypE, and HypF are required for nickel incorporation into (*E. coli*) hydrogenase, their functions are largely unknown (Maier and Böck 1996a). Still, the properties of HypB, along with the phenotypes of HypB mutants, have resulted in the conclusion that this protein plays a key role in nickel donation to NiFe-hydrogenase (Waugh and Boxer 1986; Maier et al. 1993, 1995; Olson et al. 1997). However, evidence for a direct role in nickel donation to apo-hydrogenase is lacking. Processing of larger precursor forms of NiFe hydrogenase are late events in hydrogenase maturation and, at least for *E. coli* hydrogenase-3, nickel processing into the hydrogenase large subunit occurs before specific proteolytic processing (Maier and Böck 1996b). Nickel-dependent processing of the hydrogenase large subunit was also observed for *A. vinelandii* (Menon and Robson 1994). Extracts from an *E. coli* mutant lacking nickel-transport abilities were used to show that the *hyp* gene products and the protease HycI (the homolog in *B. japonicum* and *R. leguminosarum* is HupD) were required for the nickel-dependent incorporation of nickel into the large subunit (Maier and Böck 1996b). *B. japonicum* and some other organisms that contain membrane-bound hydrogen-uptake NiFe hydrogenases contain an operon apparently lacking in *E. coli* but involved in hydrogenase maturation/processing (Maier and Triplett 1996). This operon is *hupGHIJK* which is usually organized contiguously with *hypAB*. Like hydrogenase genes (*hupSL*) a σ-54 binding site is located just upstream of *hupG*. The organization of the *B. japonicum* hydrogenase genes, including accessory genes, is shown in Figure 6.4. Not shown is *hupNOP* located about 5 kb from *hupUV*. Some of the proteins thought to play metal-processing or enzyme-maturation functions contain sequences that are intriguing to metallo-biochemists (Olson and Maier 1997). These include proteins that contain two zinc-finger motifs near the N terminus and a histidine-rich sequence of HHH(A/D)H conserved in all HypF proteins, a conserved C-X-X-C motif in all HypD, proteins, and two C-X-X-C motifs in Hup I arranged as $C-X_2-C-X_{29}-C-P-X-C$ (Maier et al.

FIGURE 6.4. Organization of the hydrogenase genes of *B. japonicum*. The nickel-transport related genes *hupNOP* located 5′ of *hupUV* are not shown.

1997). Interestingly, a specific chaperone-type role was proposed for HypC in the Ni-dependent maturation of *E. coli* hydrogenase (Magalon and Bock, 2000). The HypC was shown to bind to the precursor form of the large subunit of hydrogenase prior to its maturation (nickel incorporation and proteolytic processing); it was proposed that HypC associates which the metal binding cysteinyl residues within the immature large subunit at the time of metallocenter incorporation.

One of the most intriguing of accessory proteins needed for *B. japonicum* hydrogenase synthesis is encoded by HypB (Fu and Maier 1994). The HypB protein of *B. japonicum* contains an unusually histidine-rich area at its N terminus and GTP-binding areas at the C terminus (Fu et al. 1995). This protein binds a number of divalent metal ions. Evidence for a role of HypB in storage of nickel came from analysis of mutants in which the N-terminal histidine-containing area was deleted in frame (Olson et al. 1997). When supplemented with nanomolar levels of nickel, hydrogenase activity of this mutant strain (JHΔ23H) was lower than the wild type. This phenotype could be cured to nearly wild-type activities by adding micromolar levels of nickel. HypB also plays a role in transcription of hydrogenase, presumably because it is involved in donation of nickel for HupUV, the regulatory hydrogenase. Although the JHΔ23H strain could express *hupSL* promoter activity (Olson et al. 1997), the complete *hypB* deletion strain could not. The role of HypA is not known, but it may work in association with HypB to mobilize nickel into hydrogenase. From analysis of an *R. leguminosarum hypA* mutant, Hernando et al. (1998) concluded that the protein is essential for proper processing of the hydrogenase large subunit.

Bradyrhizobium japonicum and Nickel Storage

Nickel deficiency has been shown to adversely affect the soybean root nodule symbiosis, in particular the enzymes urease and hydrogenase (Dalton et al. 1988), although a plant yield deficiency could not be detected in the low-nickel environment. Limitation of the nickel available to the *R. leguminosarum*–pea symbiosis resulted in lower hydrogenase activities of the bacteroids (Brito et al. 1994) and even of naturally occuring Hup rhizobia, but organisms carrying the *R. leguminosarum* hydrogen-uptake (*hup*) genes responded to nickel supplementation with a clear increase in bacteroid hydrogenase activity (Brito et al. 2000). It was concluded that loss of nickel-dependent processing of hydrogenase in the symbiosis is in part responsible for the lower hydrogenase activities in the low-nickel conditions. The *B. japonicum* genes (*hupNOP*) encoding components responsible for nickel transport (Fu et al. 1994) have been studied by analysis of mutants, but it is not known if these genes function symbiotically. The HypB protein of some hydrogen-oxidizing bacteria contains a histidine-rich domain that has been shown (for *B. japonicum*) to function in the storage of nickel. Other HypB proteins with His-rich portions include those from *A. vinelandii* (Tibelius et al. 1993) and *R. leguminosarum* (Rey et al. 1994).

Another domain of HypB (present in all HypB proteins, including *E. coli*) is responsible for GTP hydrolysis. In *E. coli* and *B. japonicum*, this activity has been strongly correlated with the ability to incorporate nickel into nickel-hydrogenase (Maier et al. 1993; Olson and Maier 2000). Loss of the nickel-binding area of HypB (the His-rich area) results in decreased ability of the cells to store nickel, and when starved of nickel, these nickelin-deficient strains cannot synthesize much hydrogenase. It is important that the JHΔ23H version of nickelin retains the ability to bind a single nickel ion (per HypB monomer); this nickel could be important in the nickel-donation process. The *B. japoncium* HypB appears to serve both a nickel storage role and a nickel mobilization or incorporation role for hydrogenase activity (Olson et al. 1997; Olson and Maier 2000). The two roles of *B. japonicum* HypB (nickelin) could be functionally and structurally separated in vivo and in vitro by studying mutants with lesions introduced separately in the two functional domains, as well as by comparing the properties of the purified mutant versions of the protein (Olson and Maier 2000). The key portion of the sequence of nickelin-mutant strains used in our work are shown in Table 6.1, and the properties of these mutants, or of the purified proteins after their expression in *E. coli*, are shown in Table 6.2. The nickel storage capability and hydrogenase activities of the mutant strains and the GTPase and nickel binding by the pure components are of particular interest. Strain JHKT represents the residue 119–altered version in the G domain for GTP binding and results in a loss of GTPase activity with con-committant loss of nickel-hydrogenase formation. Taken together, the results are consistent with the conclusion that the histidine-rich portion of nickelin plays a role in storage of nickel, and a separate domain (for GTPase activity) is involved in nickel donation.

TABLE 6.1. Partial list of nickelin-mutant strains.

Strain	Amino acid sequence at the N terminus						Total amino acids
	1	10	20	30	40	50	
Wild type	(M) CTVCGCSDGKASIEHAHDHHHDHGHDHD HGHDGHHHHHHGHDQDHHHHHHDHAHGX$_{247}$						302
Δ23H	(M) CTVCGCSDGKASIE ------------------------------------ --HGX$_{247}$						264
ΔEg	First 67 amino acids like wild type; amino acids 68 to 216 deleted in frame, last 6 amino acids (at C terminus) like wild type						73
K119T	Lysine at position 119 changed to threonine						302

TABLE 6.2. Characteristics of full-length and mutated hypB proteins.

		In vivo			In vitro	
Strain	HypB expressed	Hydrogenase activity (%)	Promoter activity (%)	Nickel storage	GTPase (per mm)	Nickel binding
JH	HypB	100	100	Yes	0.35	8.7
JHΔEg	None	0	0	No	—	—
JHΔ23H	HypBΔ23H	40 (1 μM nickel)	30 (1 μM nickel)	No	0.35	1
		80 (100 μM nickel)	60 (100 μM nickel)			
JHKT	HypBK119T	0	0	—	0.03	8.5

Requirement for Nickel-Metabolism Proteins in *Helicobacter pylori*

The peptic ulcer–causing bacterum *H. pylori* contains two distinct nickel-containing enzymes. These are a membrane-bound hydrogen-uptake-type hydrogenase (Maier et al. 1996) and urease, which is an important virulence factor for colonization of the stomach and combating acidity (McGee et al. 1999). The membrane-bound enzyme functions in the hydrogen uptake direction and is associated with a cytochrome-containing respiratory chain, but whether the bacterium gleans additional energy for growth by respiring hydrogen is not known. The accessory proteins for urease synthesis have been well studied (McGee and Mobley 1999; McGee et al. 1999) and include components for nickel-dependent processing of the apoenzyme. From the genome sequences of *H. pylori* (Tomb et al. 1997; Alm et al. 1999), it is clear that *H. pylori* contains urease and hydrogenase accessory genes that are homologous to the known *ure* and *hyp* genes, respectively. In contrast to most other organisms, the *hyp* genes of *H. pylori* are not clustered but are widely dispersed throughout the chromosome. To examine the roles of *hyp* genes in hydrogenase synthesis of *H. pylori*, gene-directed mutations were created. Their phenotype regarding Ni-enzyme activity is shown in Table 6.3. Mutation of *hyp*A or *hyp*B resulted in a hydrogenase-negative

TABLE 6.3. Effect of nickel supplementation on hydrogenase and urease activities.

	Hydrogenase		Ureas	
Strain	0	5 μM	0	5 μM
Wild type	0.6 ± 0.2	0.7 ± 0.1	64 ± 23	24 ± 7
hypA	<0.01	0.14 ± 0.06	0.3 ± 0.1	25 ± 11
hypB	<0.01	0.07 ± 0.01	1.6 ± 0.5	29 ± 9

Immunoblot of Urease Large Subunit: Cells

Grown Without Nickel Supplementation

Wt   *hypA*

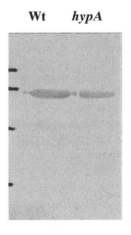

Activity: 58   0   (μmoles/min/mg)

FIGURE 6.5. Immunoblot of the urease large subunit. Extracts of *H. pylori* wild type (WT) and a *hypA* mutant from cells grown without nickel supplementation were subjected to sodium dodecyl sulfate–polyacrylamide gel electrophoresis (SDS-PAGE) and identified by blotting with an anti-urease large-subunit antiserum. Urease activity was $58\,\mu mol\,min^{-1}mg^{-1}$ for the wild type and $0\,\mu mol\,min^{-1}mg^{-1}$ for the *hypA* mutant.

phenotype, which could be partially restored by adding nickel to the growth medium. It was surprising that the mutants were also urease deficient and this phenotype could be cured by adding nickel to the medium (Table 6.3 and Olson, Mehta, Maier, 2001). Mutation of the *H. pylori* hydrogenase structural genes did not yield a urease-deficient phenotype, so the effect of *hyp* gene loss affecting urease is not the result of an indirect affect of lack of functional hydrogenase.

The urease-deficient phenotype caused by *hyp* gene mutation could be due to a lack of synthesis of hydrogenase protein, or to a loss of a maturation step. Figure 6.5 shows an immunoblot analysis to identify the large subunit of urease from extracts of the wild type and a *hyp*A mutant of *H. pylori*. Equal amounts of extract protein were loaded onto the gel, so it is clear that the *hyp*A mutant is fully capable of making apourease. Still, the urease activities clearly demonstrate that the mutant is devoid of enzyme activity. Because other results implicate a nickel-metabolism deficiency in the phenotype of *hyp*A, the most reasonable explanation for this result is that the mutant makes a nickel-deficient urease enzyme. It seems that some of the hydrogenase accessory proteins, previously connected to nickel donation for hydrogenase synthesis in other bacteria, are also playing a role in nickel urease maturation. This surprising connection between two dissimilar nickel enzymes needs to be rigorously investigated by genetic and biochemical approaches.

Conclusion

The seemingly simple process of hydrogen activation by Ni-containing hydrogenases requires an enzyme with complex redox transition attributes and precision with respect to synthesis of its metal center. The maturation of these enzymes requires a source of nickel oftentimes acquired by concentration of the metal against a gradient, and a battery of maturation accessory proteins to ensure proper metal mobilization and incorporation. Some of the maturation proteins associate with immature hydrogenase during the process, and accessory proteins have roles ranging from sequestering/storing of nickel to serving as chaperones to maintain proper folding presumably during metallocenter incorporation into the large subunit of hydrogenase. The properties and roles of some Ni-metabolizing hydrogenase accessory proteins have similarities to the maturation proteins needed for other nickel-containing enzymes. Indeed, in the gastric pathogen *H. phlori*, some of the hydrogenase maturation proteins serve the Ni-dependent maturation of urease as well as that for hydrogenase. Sensor-regulator NiFe-hydrogenases are sometimes used to ensure that the primary hydrogenase involved in hydrogen-oxidizing energy generation is synthesized only when the proper redox environment (such as low oxygen and presence of hydrogen) is encountered. Due to the similarities in the mechanisms of Ni-enzyme maturation among different bacteria and among different Ni-enzyme sinks (urease, hydrogenase, carbon monoxide dehydrogenase), an integrated picture of the requirements for Ni-enzyme maturation should be anticipated within the next few years.

References

Albracht SP. 1994. Nickel hydrogenases: in search of the active site. Biochim Biophys Acta 1188:167–204.

Alm RA, Ling LS, Moir DT, et al. 1999. Genomic-sequence comparison of two unrelated isolates of the human gastric pathogen *Helicobacter pylori*. Nature 397: 176–80.

Black LK, Maier RJ. 1995. IHF- and RpoN-dependent regulation of hydrogenase expression in *Bradyrhizobium japonicum*. Mol Microbiol 16:405–13.

Black LK, Fu C, Maier RJ. 1994. Sequence and characterization of *hupU* and *hupV* genes of *Bradyrhizobium japonicum* encoding a possible nickel-sensing complex involved in hydrogenase expression. J Bacteriol 176:7102–6.

Brito B, Martinez M, Fernandez D, et al. 1997. Hydrogenase genes from *Rhizobium leguminosarum* bv. vicae are controlled by the nitrogen fixation regulatory protein NifA. Proc Natl Acad Sci USA 94:6019–24.

Brito B, Monza J, Imperial J, et al. 2000. Nickel availability and *HupSL* activation by heterologous regulators limit symbiotic expression of the *Rhizobium leguminosarum* bv. viciae hydrogenase system in hup rhizobia. Appl Environ Microbiol 66:937–42.

Brito B, Palacios JM, Hidalgo E, et al. 1994. Nickel availability to pea (*Pisum sativum* L.) plants limits hydrogenase activity of *Rhizobium leguminosarum* bv. bacteroids by affecting the processing of the hydrogenase structural subunits. J Bacteriol 176:5297–303.

Colombo MV, Gutierrez D, Palacios JM, et al. 2000. A novel autoregulation mechanism of *fnrN* expression in *Rhizobium leguminosarum* bv viciae. Mol Microbiol 36:477–86.

Colpas GJ, Brayman TG, Ming L-J, Hausinger RP. 1999. Identification of metal-binding residues in the *Klebsiella aerogenes* urease nickel metallochaperone, UreE. Biochemistry 38:4078–88.

Dalton DA, Russell SA, Evans HJ. 1988. Nickel as a micronutrient element for plants. Biofactors 1:11–16.

Drapal N, Böck A. 1998. Interaction of the hydrogenase accessory protein HypC with HycE, the large subunit of *Escherichia coli* hydrogenase 3 during enzyme maturation. Biochemistry 37:2941–8.

Du L, Tibelius KH. 1994. The *hupB* gene of the *Azotobacter chroococcum* hydrogenase gene cluster is involved in nickel metabolism. Curr Microbiol 28:21–4.

Durmowicz MC, Maier RJ. 1997. Roles of HoxX and HoxA in biosynthesis of hydrogenase in *Bradyrhizobium japonicum*. J Bacteriol 179:3676–82.

Durmowicz MC, Maier RJ. 1998. The FixK$_2$ protein is involved in regulation of symbiotic hydrogenase expression in *Bradyrhizobium japonicum*. J Bacteriol 180:3253–6.

Eitinger T, Friedrich B. 1994. A topological model for the high-affinity nickel transport of *Alcaligenes eutrophus*. Mol Microbiol 12:1025–32.

Eitinger T, Wolfam L, Degen O, Anthon C. 1997. A Ni^{2+} binding motif is the basis of high affinity transport of the *Alcaligenes eutrophus* nickel permease. J Biol Chem 272:17139–44.

Elsen S, Colbeau A, Chabert J, Vignais PM. 1996. The *hupTUV* operon is involved in negative control of hydrogenase synthesis in *Rhodobacter capsulatus*. J Bacteriol 178:5174–81.

Elsen S, Richaud R, Colbeau A, Vignais PM. 1993. Sequence analysis and interposon mutagenesis of the *hupT* gene, which encodes a sensor protein involved in repression of hydrogenase synthesis in *Rhodobacter capsulatus*. J Bacteriol 175:7404–12.

Friedrich B, Schwartz E. 1993. Molecular biology of hydrogen utilization in aerobic chemolithotrophs. Annu Rev Microbiol 47:351–83.

Fu C, Maier RJ. Nucleotide sequences of two hydrogenase-related genes (*hypA* and *hypB*) from *Bradyrhizobium japonicum*, one of which (*hypB*) encodes an extremely histidine-rich region and guanine nucleotide-binding domains. Biochem Biophys Acta 1184:135–8.

Fu C, Javedan S, Moshiri F, Maier RJ. 1994. Bacterial genes involved in incorporation of nickel into a hydrogenase enzyme. Proc Natl Acad Sci USA. 91: 5099–103.

Fu C, Olson JW, Maier RJ. 1995. HypB protein of *Bradyrhizobium japonicum* is a metal-binding GTPase capable of binding 18 divalent nickel ions per dimer. Proc Natl Acad Sci USA 92:2333–7.

Hausinger R. 1994. Nickel enzymes in microbes. Sci Total Environ 148:157–66.

Hausinger RP. 1997. Metallocenter assembly in nickel-containing enzymes. J Biol Inorg Chem 2:279–86.

Hernando Y, Palacios J, Imperial J, Ruiz-Argüeso T. 1998. *Rhizobium leguminosarum* bv. viciae *hypA* gene is specifically expressed in pea (*Pisum sativum*) bacteroids and required for hydrogenase activity and processing. FEMS Microbial Lett 169:295–302.

Jocobi A, Rossmann R, Böck A. 1992. The *hyp* operon gene products are required for the maturation of catalytically active hydrogenase isoenzymes in *Escherichia coli*. Arch Microbiol 158:444–51.

Kerby RL, Ludden PW, Roberts GD. 1997. In vivo nickel insertion into the carbon monoxide dehydrogenase of *Rhodospirillum rubrum:* molecular and physiological characterization of *cooCTJ*. J Bacteriol 179:2259–66.

Kim H, Maier RJ. 1990. Transcriptional regulation of hydrogenase synthesis by nickel in *Bradyrhizobium japonicum*. J Biol Chem 265:18729–32.

Kim H, Yu C, Maier RJ. 1991. Common *cis*-acting region responsible for transcriptional regulation of *Bradyrhizobium japonicum* hydrogenase by nickel, oxygen, and hydrogen. J Bacteriol 173:3993–9.

Kleihues L, Lenz O. Bernhard M, et al. 2000. The H_2 sensor of *Ralstonia eutropha* is a member of the subclass of regulatory (NiFe) hydrogenases. J Bacteriol 182: 2716–24.

Lenz O, Friedrich B. 1998. A novel multicomponent regulatory system mediates H_2 sensing in *Alcaligenes eutrophus*. Proc Natl Acad Sci USA 95:12474–9.

Lenz O, Strack A, Tran-Betcke A, Friedrich B. 1997. A hydrogen-sensing system in transcriptional regulation of hydrogenase gene expression in *Alcaligenes* species. J Bacteriol 179:1655–63.

Maier T, Böck A. 1996b. Generation of active (NiFe) hydrogenase in vitro from a nickel-free precursor form. Biochemistry 35:10089–93.

Maier T, Böck A. 1996a. Nickel incorporation into hydrogenases. In: Hausinger RR, Eichhorn GL, Marzilli LG, editors. Advances in inorganic biochemistry: mechanisms of metallocenter assembly. New York: VHC Publishers. 173–92.

Maier RJ, Triplett EW. 1996. Toward more productive, efficient, and competitive nitrogen-fixing symbiotic bacteria. Crit Revs Plant Sci 15:191–234.

Maier RJ, Fu C, Gilbert J, et al. 1996. Hydrogen uptake hydrogenase in *Helicobacter pylori*. FEMS Microbiol Lett 141:71–6.

Maier RJ, Olson JW, Fox J. 1997. Nickel-dependent expression and maturation of hydrogenase. In: Ludden PW. Burris JE, editor. Biosynthesis and function of metal clusters for enzymes: the 25th Steenbock Symposium Proceedings. Madison: University of Wisconsin. p 133–42.

Maier T, Jacobi A, Sauter M, Böck A. 1993. The product of the *hypB* gene, which is required for nickel incorporation into hydrogenases, is a novel guanine nucleotide-binding protein. J Bacteriol 175:630–5.

Maier T, Lottspeich F, Böck A. 1995. GTP hydrolysis by HypB is essential for nickel insertion into hydrogenases of *Escherichia coli*. Eur J Biochem 230:133–8.

Magalon A, Bock A. 2000. Analysis of the HypC-hycE complex, a key intermediate in the assembly of the metal center of the Escherichia coli hydrogenase 3. J Biol Chem 275:21114–20.

Maroney MJ. 1999. Structure/function relationships in nickel metallobiochemistry. Curr Opin Chem Biol 3:188–99.

McGee DJ, Mobley HL. 1999. Mechanisms of *Helicobacter pylori* infection: bacterial factors. Curr Topics Microbiol Immunol 241:155–180.

McGee DJ, May CA, Garner RM, et al. 1999. Isolation of *Helicobacter pylori* genes that modulate urease activity. J Bacteriol 181:2477–84.

Menon AL, Robson RL. 1994. In vivo and in vitro nickel-dependent processing of the (NiFe) hydrogenase in J Bacteriol 176:291–5.

Mobley HLT, Garner RM, Bauerfeind P. 1995. *Helicobacter pylori* nickel-transport gene *nixA*: synthesis of catalytically active urease in *Escherichia coli* independent of growth conditions. Mol Microbiol 16:97–109.

Moncrief MBC, Hausinger RP. 1996. Purification and activation properties of UreD-UreF-Urease apoprotein complexes. J Bacteriol 178:5417–21.

Navarro C, Wu LF, Mandrand-Berthelot MA. 1993. The *nik* operon of *Escherichia coli* encodes a periplasmic binding-protein-dependent transport system for nickel. Mol Microbiol 9:1181–91.

Olson JW, Maier RJ. 1997. The sequences of *hypF*, *hypC* and *hypD* complete the *hyp* gene cluster required for hydrogenase activity in *Bradyrhizobium japonicum*. Gene 199:93–99.

Olson JW, Maier. RJ. 2000. Dual roles of *B. japonicum* nickelin protein in nickel storage and GTP-dependent Ni mobilization. J Bacteriol 182:1702–5.

Olson JW, Fu C, Maier RJ. 1997. The HypB protein from *Bradyrhizobium japonicum* can store nickel and is required for the nickel-dependent transcriptional regulation of hydrogenase. Mol Microbiol 24:119–28.

Olson JW, Mchta NS, Maier RJ. 2001. Requirement of nickel metabolism proteins HypA and HypB for full activity of both hydrogenase and urease in *Helicobacter pylori*. Mol Microbiol 39:176–182.

Pierik AJ, Smelz M, Lenz O, et al. 1998. Characterization of the active site of a hydrogen sensor from *Alcaligenes eutrophus*. FEBS Lett 438:231–5.

Rey L, Imperial J, Palacios JM, Ruiz-Argüeso T. 1994. Purification of *Rhizobium leguminosarum* HypB, a nickel-binding protein required for hydrogenase synthesis. J Bacteriol 176:6066–73.

Schwartz E, Buhrke T, Gerischer U, Friedrich B. 1999. Positive transcriptional feedback controls hydrogenase expression in *Alcaligenes eutrophus* H16. J Bacteriol 181:5684–92.

Soriano A, Hausinger RP. 1999. GTP-dependent activation of urease apoprotein in complex with the UreD, UreF, and UreG accessory proteins. Proc Natl Acad Sci USA 96:11140–4.

Tibelius KH, Du L, Tito D, Stejskal F. 1993. The *Azotobacter chroococcum* hydrogenase gene cluster: sequences and genetic analysis of four accessory genes, *hupA*, *hupB*, *hupY*, and *hup C*. Gene 127:53–61.

Tomb JF, White O, Kerlavage AR, et al. 1997. The complete genome sequence of the gastric pathogen *Helicobacter pylori*. Nature 388:539–47.

van Soom C, de Wilde D, Vanderleyden J. 1997. HoxA is a transcriptional relator for expression of the *hup* structural genes in free-living *Bradyrhizobium japonicum*. Mol Microbiol 23:967–77.

van Soom C, Lerouge I, Vanderleyden J, et al. 1999. Identification and characterization of *hupT*, a gene involved in negative regulation of hydrogen oxidation in *Bradyrhizobium japonicum*. J Bacteriol 181:5085–9.

Vignais PM, Dimon B, Zorin NA, et al. 1997. HupUV proteins of *Rhodobacter capsulatus* can bind H_2: evidence from the H-D reaction. J Bacteriol 179:290–2.

Waugh R, Boxer DH. 1986. Pleiotropic hydrogenase mutants of *Escherichia coli* K-12: growth in the presence of nickel can restore hydrogenase activity. Biochimie 68:157–66.

Suggested Reading

Rey L, Murillo J, Hernando Y, et al. 1993. Molecular analysis of a microaerobically induced operon required for hydrogenase synthesis in *Rhizobium leguminosarum* biovar vicae. Mol Microbiol 8:471–81.

Vignais PM, Billoud B, Meyer J. 2001. Classification and phyloseny of hydrogenases. FEMS Microbiol Rev 25:455–501.

7
Genes and Genetic Manipulations of *Desulfovibrio*

JUDY D. WALL, CHRISTOPHER L. HEMME, BARBARA RAPP-GILES,
JOSEPH A. RINGBAUER, JR., LAURENCE CASALOT, and TARA GIBLIN

An increased understanding of the environmental roles of the sulfate-reducing bacteria (SRB) has stimulated interest in the metabolism of these anaerobic bacteria. This interest has, in turn, driven the development of molecular tools to examine the genetic regulation of these various processes. Much of the metabolic work has been reviewed in two books (Odom and Singleton 1993; Barton 1995), a volume of *Methods in Enzymology* (Peck and LeGall 1994), and additional reviews (Hansen 1994; Voordouw 1995; Widdel 1988), but none replaces Postgate's (1984) earlier monograph for providing a real feel for these bacteria. Summaries of the advances in genetic and molecular approaches, which are essentially limited to work with *Desulfovibrio* strains, can be found in Voordouw (1993, 1995), Wall (1993), van Dongen et al. (1994) and van Dongen (1995). Here, we will focus on recent molecular tools that have been successfully used with SRB and present some exciting observations abstracted from the preliminary genome sequencing of *Desulfovibrio vulgaris* Hildenborough.

Vectors

In 1989, two groups—Powell and co-workers and van den Berg and colleagues—published the first gene transfer into *Desulfovibrio* strains employing the broad host range IncQ plasmids. These plasmids were mobilized via conjugation from *Escherichia coli* and were replicated stably in the SRB recipients. The isolation of a small cryptic plasmid, pBG1, from *Desulfovibrio desulfuricans* G100A (Wall et al. 1993) allowed the construction of a series of vectors compatible with the IncQ plasmids (Rousset et al. 1998) (Table 7.1). These vectors have been used both in *D. desulfuricans* G20 (a plasmid-cured, nalidixic acid resistant derivative of G100A) (Wall et al. 1993) and *Desulfovibrio fructosivorans* (formerly *Desulfovibrio fructosovorans*) (Rousset et al. 1998). The facility of conjugation between SRB strains and *E. coli* has allowed the introduction into the SRB of cloned genes for production of redox proteins in an anaerobic, highly reducing

TABLE 7.1. Vectors for use in *Desulfovibrio*.

Plasmid	Replicon	Markers[a]	Use	Reference
pSC27	pMB1, pBG1	Mob[+], Km[R]	Cloning	Rapp-Giles et al. (2000)
pSC270	pMB1	Mob[+], Km[R]	Unstable, marker delivery	Wall
pBMC6	pMB1, pBG1	Cm[R]	Cloning, blue-white in *E. coli*	Rousset et al. (1998)
pBMK6	pMB1, pBG1	Km[R]	Cloning, blue-white in *E. coli*	Rousset et al. (1998)
pBMS6	pMB1, pBG1	Sm[R]	Cloning, blue-white in *E. coli*	Rousset et al. (1998)
pBMC7	pMB1, pBG1	Mob[+], Cm[R]	Cloning, blue-white in *E. coli*	Rousset et al. (1998)
pBMK7	pMB1, pBG1	Mob[+], Km[R]	Cloning, blue-white in *E. coli*	Rousset et al. (1998)
pBMS7	pMB1, pBG1	Mob[+], Sm[R]	Cloning, blue-white in *E. coli*	Rousset et al. (1998)
pDDCAT	pMB1, pBG1	Mob[+], Km[R]	Promoter probe, Cm[R] reporter	English and Wall
pJAR23	pMB1, pBG1	Mob[+], Km[R]	Promoter probe, lacZ reporter	Ringbauer and Wall
pDDC1	IncP, pBG1	Mob[+], Km[R], Tc[R]	Cosmid cloning	Wickman and Wall
pSTW1	pMB1	Mob[+], Cm[R], *sacB*	Marker exchange	Wickman and Wall
pJRC2	IncQ	Mob[+], Cm[R], Km[S]	Cloning	Rousset et al. (1998)

[a] *Mob[+]*, can be mobilized during conjugation with broad-host-range plasmids; *Km[R]*, kanamycin resistance; *Km[3]*, kanamycin sensitivity; *Cm*, chloramphenicol; *Sm*, streptomycin; *Tc*, tetracycline.

environment. In particular, the *D. desulfuricans* strain G20 has proven valuable for the production of cytochromes from *D. vulgaris* (Voordouw et al. 1990; Pollock et al. 1991; Blanchard et al. 1993; Pollock and Voordouw 1994). The success of this heterologous expression is likely due to the reasonably close phylogenetic relationship between these strains (Fig. 7.1). Phylogenetic trees based on 16S rDNA gene sequences of the family Desulfovibrionaceae show a distinguishable grouping of *Desulfovibrio* strains (Castro et al. 2000).

It is interesting that several *Desulfovibrio* strains have not yet been demonstrated to participate in conjugation with either IncQ- or pBG1-based plasmids for unknown reasons (Argyle et al. 1992; van Dongen 1995). Therefore alternative techniques for gene transfer are necessary. While progress has been made in electroporation of *Desulfovibrio* (Rousset et al. 1991, 1998), the development of reliable and efficient procedures deserves further attention. Evidence has been obtained for the presence of a restriction barrier that limits transformation in *D. vulgaris* (Fu and Voordouw 1997). The efficiency of electroporation of plasmids into *D. vulgaris* was increased from an undetectable level to 10^3 transformants

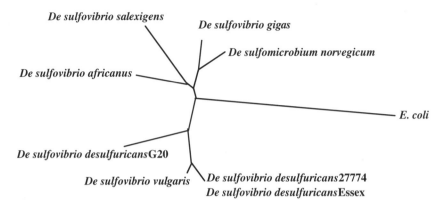

De sulfovibrio salexigens

De sulfovibrio gigas

De sulfomicrobium norvegicum

De sulfovibrio africanus

E. coli

De sulfovibrio desulfuricans G20

De sulfovibrio vulgaris   *De sulfovibrio desulfuricans* 27774
De sulfovibrio desulfuricans Essex

FIGURE 7.1. Phylogenetic tree of 16S rDNA sequences for select strains of sulfate-reducing bacteria. The tree was constructed with *Phylip* by alignment of a 190-bp segment of each 16S rDNA sequence by *Clustalw*. Bootstrap analysis was repeated 100 times with *SEQBOOT*, and the distance matrix was prepared with *DNADIST*. Sequences were loaded randomly into *NEIGHBOR* for neighbor-joining analysis. The GenBank accession numbers for the sequences used were *Desulfomicrobium norvegicum* M37315, *Desulfovibrio africanus* M37315, *D. desulfuricans* 27774 M34113, *D. desulfuricans* Essex AF192153, *D. desulfuricans* G20 (to be submitted), *Desulfovibrio gigas* M34400, *Desulfovibrio salexigens* M34401, *D. vulgaris* M34399, and *E. coli* Z83204.

per microgram DNA when plasmids were prepared from *D. vulgaris* versus *E. coli* (Fu and Voordouw 1997). In similar experiments with *D. desulfuricans* G20, electroporation with plasmids prepared from G20 yielded 3×10^4 transformants per microgram DNA compared to about 20 for plasmids prepared from *E. coli* (Ringbauer and Wall). Mutant derivatives lacking this restriction barrier for use in gene manipulation experiments have yet to be identified in either strain.

Gene Fusions

A powerful genetic tool for analyzing the expression of genes and operons is gene fusion. Unfortunately, many of the reporter genes commonly used require aerobic conditions for appropriate protein folding or have oxygen as a substrate (Table 7.2). Thus with strict anaerobes, a number of transcription or translation fusion vectors are not readily applicable. Exceptions are the *cat* reporter gene encoding chloramphenicol acetyl transferase, which provides a relative level of antibiotic resistance, and *gus* when used with the substrate 4-methylumbelliferyl glucuronide, which produces a somewhat diffusible fluorescent product. Unfortunately, the versatile *lacZ*

TABLE 7.2. Reporter genes commonly used in gram-negative bacteria.

Gene	Enzyme	Activity conditions
lacZ	β-galactosidase	X-gal (oxygen is needed for color development)
phoA	Alkaline phosphatase	Oxidizing conditions for enzyme folding
xylE	Catechol 2,3-dioxygenase	Oxygen is substrate
lux	Luciferase	Oxygen is substrate
gfp	Green fluorescent protein	Oxygen is needed for protein folding
Gus	β-glucuronidase	4-Methylumbelliferyl glucuronide
cat	Chloramphenicol acetyltransferase	Growth in presence of chloramphenicol

gene system with its chromogenic substrate (X-gal, 5-bromo-4-chloro-3-indolyl-β-D-galactopyranoside) needs oxygenic conditions for blue color formation. Gene expression in the SRB can be visualized with *lacZ* fusions, but color development requires transfer of SRB colonies onto aerobic plates where the cells are no longer growing (Ringbauer and Wall). The transfer eliminates the possibility of comparing the intensities of color development among colonies and causes the number of colonies screened to be limited by practical considerations. The fluorescent substrate (4-methylumbelliferyl β-D-galactoside) is effective under anaerobic conditions; however, diffusion of the fluorescent product limits the utility of this substrate (Casalot and Wall).

Geesey and co-workers (personal communication, 2000) obtained exciting results on the introduction of *gfp* on an IncQ plasmid into *D. desulfuricans* G20. Contrary to expectation, a low level of fluorescence was seen in G20 cells carrying this gene under conditions of biofilm formation on hematite (α-Fe_2O_3) surfaces. This result suggests that either conditions favorable to folding of the green fluorescent protein were achieved under anaerobiosis in these SRB or small amounts of oxygen were accessible to the protein inside of the bacteria. The resolution of these possibilities may provide an additional approach to studies of gene expression in the SRB. As yet, the field of SRB genetics awaits the introduction of a facile visible screen for gene expression.

Mutant Construction

For genetic studies in bacteria that are not easily manipulated, a transposon capable of random mutagenesis and tagging of mutated genes is extremely valuable. Random mutagenesis of *D. desulfuricans* G20 was documented when derivatives of Tn5 that have increased expression of transposase (Simon 1984; Wolk et al. 1991) were used (Wall et al. 1996).

Unfortunately, the efficiency was 10^{-6} at best. Tn7 was then discovered to transpose in G20 into a unique site (Wall et al. 1996). Such site-specific transposition was described for *E. coli* (Craig 1991), and the recognition sequences for insertion from the SRB and *E. coli* are strongly conserved (Wall et al. 1996). The transposition efficiency of Tn7 was 100 times better than that determined for Tn5 in G20. Therefore, a mutant of Tn7 was generated to eliminate the insertion site specificity with the hope of obtaining a facile mutagenesis tool. Such a mutant was obtained that exhibited random transposition. Unfortunately, the efficiency of transposition, $1–2 \times 10^{-6}$, was again unacceptably low for creating a library of mutations for screening (Wall et al. 1996).

An alternative to using existing transposons for mutagenesis was successfully demonstrated for coryneform bacteria (Vertes et al. 1994). A naturally occurring insertion element that encoded its own transposase was modified to contain a selectable marker. Upon introduction into a naive background, transposition mutagenesis could be demonstrated. A candidate insertion element from *D. vulgaris*, IS*D1*, was identified and sequenced by Fu and Voordouw (1998), who also recognized the possibility of this special application. Because IS*D1* was found to have interrupted a gene being used for counterselection during genetic constructions, it was clearly capable of active transposition. Its target sequence, AANN (Fu and Voordouw 1998), would appear to offer little restriction on insertion sites. Although the frequency of transposition of this insertion element is not known, it provides the genetic material for the development of an artificial transposon for *Desulfovibrio*.

To determine the function of a gene once identified, it is important to be able to construct well-defined mutations in the gene. Preliminary characterization can be gained from plasmid insertion mutagenesis (Rapp-Giles et al. 2000). An internal fragment of the gene to be mutagenized is cloned in a plasmid that is unable to replicate in the target bacterium. After introduction into the cell, selection for maintenance of the plasmid antibiotic marker identifies cells in which the plasmid has integrated into the chromosome by recombination in the region of the gene fragment. The chromosomal copy of the gene is partially duplicated, and the two incomplete copies are separated by plasmid sequences. This technique offers a relatively quick method for testing the feasibility of creating mutations in genes that may prove to be important in metabolism. However, revertants and supressors can be generated in these plasmid-interrupted mutants, because none of the sequences of the gene is deleted from the cell. Counter to expectations, a mutation in *cycA*, the gene encoding cytochrome c_3, was successfully constructed with this technique in *D. desulfuricans* G20. Although the mutant was not significantly impaired in growth rate with lactate as carbon and reductant source, the extent of growth was reduced by about 15%. It was inferred from these results that this cytochrome plays an important role in electron flow from lactate but that alternative pathways also exist (Rapp-Giles et al. 2000).

Rousset et al. (1991) reported the first directed gene deletion created by marker exchange in *D. fructosivorans*. A plasmid containing a chromosomal fragment spanning the genes for the [NiFe] hydrogenase was manipulated to remove the structural genes and replace them with a kanamycin-resistance cassette. Electroporation of *D. fructosivorans* with this genetic construct resulted in a double homologous recombination, eliminating the chromosomal copy of the hydrogenase genes. Single recombinants in which the plasmid was integrated into the chromosome were rare (Casalot, personal communication, 1999). In *D. vulgaris* and *D. desulfuricans* G20, double recombination is the rare event. To circumvent this limitation to stable mutant construction, Fu and Voordouw (1997) had excellent success with a marker exchange mutagenesis procedure adapted from *Pseudomonas* (Schweizer 1992). In this procedure, a construct like that made by Rousset et al. (1991) was used—an antibiotic resistance cassette replaced all or part of the gene to be mutated in a cloned fragment of the chromosome. The vector carrying this construct was unable to replicate in the SRB and, importantly, carried a counterselectable marker, *sacB*. The *Bacillus subtilis sacB* encodes levansucrase, which generates a product toxic for Gram-negative bacteria when the enzyme is expressed in the presence of sucrose (Reid and Collmer 1987). After introduction of the plasmid, selection of the antibiotic resistance identified cells that had integrated the plasmid, and subsequent enrichment for resistance to sucrose revealed cells in which a second recombination event had eliminated plasmid sequences. Sucrose resistance can result from any mechanism that inactivates the *sacB* gene. Therefore, the desired constructs were identified by Southern blot analysis (Fu and Voordouw 1997). This procedure has been quite successful for *D. vulgaris* (Fu and Voordouw 1997; Keon et al. 1997; Voordouw and Voordouw 1998), but other *D.* should be amenable to mutant construction by this procedure as well.

Genome Sequence Searches

A preliminary sequence for the *D. vulgaris* Hildenborough genome was released by the Institute for Genomic Research (supported by the U.S. Department of Energy; www.tigr.org). This is a rich resource for gaining a more comprehensive understanding of the metabolism of this bacterium and, by extrapolation, that of physiologically similar microbes (Cordwell 1999). It must be kept in mind that the identification of a putative gene by sequence gazing does not prove that the gene is expressed or that the enzyme is active. However, finding possible genes for a pathway offers justification for further exploration.

An apparently incongruous metabolic observation was the inability of most *Desulfovibrio* strains to grow on hexoses, even though some strains were shown to store polyglucose (Stams et al. 1983). This storage polymer can be mobilized and supports NTP synthesis both aerobically and anaerobically (Santos et al. 1993) but apparently not growth (van Niel and Gottschal 1998). Under aerobic conditions, this polymer may provide reducing power for oxygen consumption, thereby protecting the cell in this toxic environment. Despite this metabolic capacity, the cells are still unable to use glucose as a substrate. From these results it is inferred that uptake and/or activation of hexoses must be defective. In fact, assays of all the glycolytic enzymes of *D. gigas* have revealed a complete complement of activities necessary for fermentation of glucose, with the exception of a system for hexose transport (Fareleira et al. 1997). Searches of the *D. vulgaris* genome revealed homologs of all the glycolytic enzymes (Fig. 7.2), a result completely compatible with the biochemistry of *D. gigas*. Fu and Voordouw (1998) suggested that *D. vulgaris* was defective in glycolysis because the endogenous insertion element IS*D1* likely prevented expression of two genes essential for glycolysis. IS*D1* was found near *gap*, the gene encoding glyceraldehyde-3-phosphate dehydrogenase, and, by extrapolation from the gene arrangements in other bacteria, the position of IS*D1* would cause a polar mutation–blocking expression of *pgk* (phosphoglycerate kinase) and *tpi* (triosephosphate isomerase). However, an examination of the two *gap* genes found in the *D. vulgaris* genome revealed that *pgk* and *tpi* were not present in an operon with either *gap* gene. The position of IS*D1* was confirmed to be just downstream of one of the *gap* genes Fu and Voordouw (1998), but a conserved gene arrangement was not found. Thus again transport or activation of hexoses may be the limitation for their use as growth substrates by *D. vulgaris*.

It is interesting that an examination of TCA cycle genes (Fig. 7.3) suggests an incomplete cycle. Possible genes for pyruvate dehydrogenase, α-ketoglutarate dehydrogenase, and citrate synthase were not identified. These may not be present in the chromosome, or the missing genes may not be recognized because of low sequence similarity or because the genes providing these functions are actually novel. If both α-ketoglutarate dehydrogenase and citrate synthase are not present, how then does the cell make the essential intermediate α-ketoglutarate for glutamate production? Further exploration of the genome uncovered sequences with strong similarity to the ferredoxin-dependent enzyme α-ketoglutarate synthase (E value 1.7E-74), an enzyme first described in the phototroph *Chlorobium limicola* forma *thiosulfatophilum* (Evans et al. 1966; Buchanan and Arnon 1990). If functional enzymes are made from the putative genes identified (Fig. 7.3), the partial cycle could function to make all the necessary intermediates for biosynthesis, including α-ketoglutarate. The sequences found provide justification for the conjecture that most of the cycle can operate in reverse.

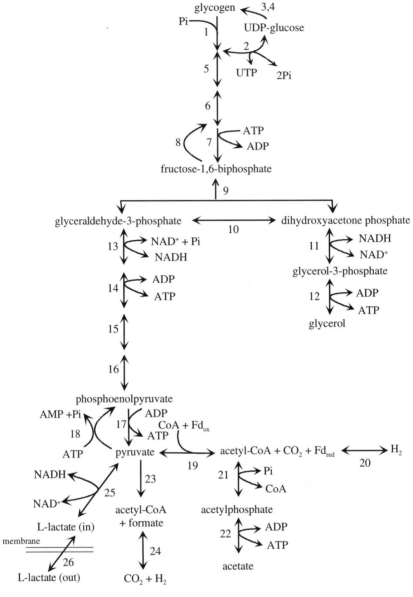

FIGURE 7.2. Putative pathway for central carbon metabolism in *D. vulgaris* Hildenborough. The numbers designate the enzyme catalyzing the step (*arrow*). The E values, indicating the likelihood that the similarity to the query sequence occurred by chance, are given in parentheses below. All sequences were compared to *E. coli* genes, except where indicated. *1*, glycogen phosphorylase (1.7E-15); *2*, UTP-glucose-1-phosphate uridylyltransferase (2.3E-54, *Sinorhizobium meliloti*); *3*, glycogen synthase (4.1E-75); *4*, glycogen-branching enzyme (1.4E-175); *5*, phosphoglucomutase (2.0E-175); *6*, phosphoglucose isomerase (3.7E-39, *Bacillus stearothermophilus*); *7*, phosphofructokinase (2.9E-22, *Clostridium acetobutylicum*);

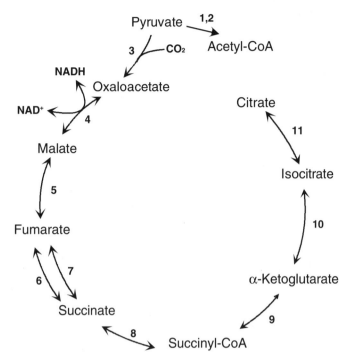

FIGURE 7.3. Putative tricarboxylic acid cycle in *D. vulgaris* Hildenborough. The numbers designate the enzyme catalyzing the step (arrow). The E values, indicating the likelihood that the similarity to the query sequence occurred by chance, are given in parentheses below. All sequences were compared to *E. coli* genes except for *A. fulgidus* genes. *1*, pyruvic-ferredoxin oxidoreductase (2.8E-46, *A. fulgidus*); *2*, pyruvate formate-lyase (7.2E-91); *3*, pyruvate carboxylase (5.0E-93, *A. fulgidus*); *4*, malate dehydrogenase (9.5E-12); *5*, fumarase class II, anaerobic (1.7E-138); *6*, fumarate reductase, flavoprotein (9.6E-86); *7*, succinate dehydrogenase (2.5E-78); *8*, succinyl-coA synthetase (1.7E-64); *9*, α-ketoglutarate synthase (1.7E-74, *A. fulgidus*); *10*, isocitrate dehydrogenase (1.3E-106); *11*, aconitase (2.0E-40).

Figure 7.2. *Continued.* *8*, fructose-1,6-bisphosphatase (2.0E-70); *9*, fructose-bisphosphate aldolase (4.7E-08); *10*, triosephosphate isomerase (1.0E-36); *11*, glycerol-3-phosphate dehydrogenase (3.1E-96); *12*, glycerol kinase (3.4E-156); *13*, glyceraldehyde-3-phosphate dehydrogenase (7.5E-78); *14*, phosphoglycerate kinase (1.0E-128); *15*, phosphoglyceromutase (3.3E-69); *16*, enolase (5.9E-133); *17*, pyruvate kinase (6.4E-65); *18*, pyruvate (H_2O) dikinase (2.0E-66, *D. vulgaris*); *19*, pyruvic-ferredoxin oxidoreductase (2.8E-46, *Archaeoglobus fulgidus*); *20*, hydrogenase (0.E-00, *D. vulgaris*); *21*, phosphotransacetylase (3.1E-112); *22*, acetate kinase (8.3E-82); *23*, pyruvate formate-lyase (7.2E-91); *24*, formate dehydrogenase (1.0E-118, *D. vulgaris*); *25*, L-lactate dehydrogenase (1.2E-30); *26*, lactate permease (1.0E-121, *D. vulgaris*).

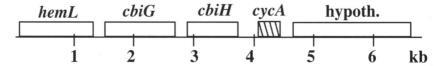

FIGURE 7.4. Arrangement of genes in the region of the cytochrome c_3 structural gene, *cycA*. The boxes represent open reading frames that could potentially code for the genes indicated. *hemL*, glutamate-1-semialdehyde 2,1-aminomutase; *cbiG*, protein of unknown function; *cbiH*, precorrin-3-methylase; *hypoth.*, hypothetical protein from *Caenorhabditis elegans*. All genes are transcribed from left to right. No information about operon structure is available, except that *cycA* is monocistronic in the related strain *D. desulfuricans* G20 (Rapp-Giles et al. 2000).

An examination of the genetic neighborhood of the cytochromes provided another intriguing inference (Fig. 7.4). Immediately upstream of the tetraheme cytochrome c_3 gene are two putative genes involved in heme modifications. These genes have strong homology with *Salmonella* genes *cbiH* and *cbiG* (E values 3E-57 and 1E-30, respectively) (Roth et al. 1993). In *Salmonella*, *cbiH* encodes precorrin-3-methylase, and the function of CbiG is unknown. It is intriguing to speculate that these genes might provide methylations or other modifications for the hemes in the various cytochromes. Generally these genes are part of large operons devoted to cobalamin biosynthesis; however, in *D. vulgaris*, homologs of other genes in this pathway are apparently not nearby. In fact, immediately upstream of these two genes is a homolog of *hemL*, encoding glutamate-1-semialdehyde 2,1-aminomutase (E value 1E-133), necessary for δ-aminolevulinate biosynthesis in the glutamate pathway. This observation, along with an apparent absence of a classical δ-aminolevulinate synthase, suggests that the C-5 pathway may be that used for heme biosynthesis in *D. vulgaris*.

Dilling and Cypionka (1990) reported oxygenic respiration in *D. vulgaris* that was coupled to phosphorylation but could not support growth. Although pathways involving rubredoxin have been proposed for oxygen reduction (Chen et al. 1993a, 1993b; LeGall and Xavier 1996), gene sequencing of *D. vulgaris* Miyazaki unexpectedly turned up a gene for a putative cytochrome *c* oxidase-like protein (Kitamura et al. 1995). Biochemical and biophysical examination of membrane-bound cytochromes did not support expression of such a gene (Ozawa et al. 1997). In searching the sequences of the *D. vulgaris* Hildenborough genome, not only was a gene for subunit I of an aa_3-type cytochrome *c* oxidase (*col*) (Kitamura et al. 1995) observed but, apparently downstream, genes for subunits III, IV, and II of that oxidase (E values 2E-49, 9E-9 and 1E-59, respectively) were seen. These genes could hypothetically produce a proton-pumping heme-copper respiratory oxidase homologous to the cytochrome bo_3 of *E.*

coli, which operates under high aeration. *E. coli* has a second respiratory oxidase, cytochrome *bd*, which is maximally produced under limited aeration (Gennis and Stewart 1996). Further exploration of the *D. vulgaris* genome revealed putative genes for a cytochrome *d* ubiquinol oxidase subunit I and a cytochrome *d* oxidase subunit, which have homology with the corresponding *E. coli* genes (E values 1E-107 and 3E-71, respectively). Studies by van Niel and Gottschal (1998) of the oxygen consumption of several *D. desulfuricans* strains led these researchers to propose that multiple systems were likely to exist for oxygen reduction that functioned differentially with respect to oxygen concentration. The presence of these cytochrome oxidase genes stimulates the question of whether conditions can be found under which they might be functional in this strict anaerobe and whether these systems could contribute to oxygen consumption.

Conclusion

These examples of observations of sequences leave more questions than answers but provide intriguing hints that the metabolism of this important group of bacteria is versatile and complex.

Acknowledgments. We thank John Heidelberg for making preliminary sequence information of the *D. vulgaris* genome available, and Gill Geesey for unpublished information concerning the expression of the green fluorescent protein in *Desulfovibrio*. This work was supported by the Natural and Accelerated Bioremediation Research Program and the Basic Research Program of the U.S. Department of Energy (grant nos. DE-FG02–97ER62495 and DE-FG02–87-ER13713, respectively) and the Missouri Agricultural Experiment Station.

References

Argyle JL, Rapp-Giles BJ, Wall JD. 1992. Plasmid transfer by conjugation in *Desulfovibrio desulfuricans*. FEMS Microbiol Lett 94:255–62.

Barton LL, editor. 1995. Sulfate-reducing bacteria. New York: Plenum Press.

Blanchard L, Marion D, Pollock B, et al. 1993. Overexpression of *Desulfovibrio vulgaris* Hildenborough cytochrome c_{553} in *Desulfovibrio desulfuricans* G200: evidence of conformational heterogeneity in the oxidized protein by NMR. Eur J Biochem 218:293–301.

Buchanan BB, Arnon DI. 1990. A reverse KREBS cycle in photosynthesis: consensus at last. Photosyn Res 24:47–53.

Castro HF, Williams NH, Ogram A. 2000. Phylogeny of sulfate-reducing bacteria. FEMS Microbiol Ecol 31:1–9.

Chen L, Liu M-Y, LeGall J, et al. 1993b. Purification and characterization of an NADH-rubredoxin oxidoreductase involved in the utilization of oxygen by *Desulfovibrio gigas*. J Biochem 216:443–8.

Chen L, Liu M-Y, LeGall J, et al. 1993a. Rubredoxin oxidase, a new flavo-hemo-protein, is the site of oxygen reduction to water by the "strict anaerobe" *Desulfovibrio gigas*. Biochem Biophys Res Comm 193:100–5.

Cordwell SJ. 1999. Microbial genomes and "missing" enzymes: redefining biochemical pathways. Arch Microbiol 172:269–79.

Craig NL. 1991. Tn7: a target site-specific transposon. Mol Microbiol 5:2569–73.

Dilling W, Cypionka H. 1990. Aerobic respiration in sulfate-reducing bacteria. FEMS Microbiol Lett 71:123–8.

Evans MCW, Buchanan BB, Arnon DI. 1966. A new ferredoxin-dependent carbon reduction cycle in a photosynthetic bacterium. Proc Natl Acad Sci USA 55:928–34.

Fareleira P, LeGall J, Xavier AV, Santos H. 1997. Pathways for utilization of carbon reserves in *Desulfovibrio gigas* under fermentative and respiratory conditions. J Bacteriol 179:3972–80.

Fu R, Voordouw G. 1997. Target gene-replacement mutagenesis of *dcrA*, encoding an oxygen sensor of the sulfate-reducing bacterium *Desulfovibrio vulgaris* Hildenborough. Microbiology 143:1815–26.

Fu R, Voordouw G. 1998. IS*D1*, an insertion element from the sulfate-reducing bacterium *Desulfovibrio vulgaris* Hildenborough: structure, transposition, and distribution. Appl Environ Microbiol 63:53–61.

Gennis RB, Stewart V. 1996. Respiration. In: Neidhardt FC, Curtiss R III, Ingraham JL, et al. editors. Volume 1, *Escherichia coli* and *Salmonella*. 2nd ed. Washington, DC: ASM Press. p 217–61.

Hansen TA. 1994. Metabolism of sulfate-reducing prokaryotes. Antonie Leeuwenhoek 66:165–85.

Keon RG, Fu R, Voordouw G. 1997. Deletion of two downstream genes alters expression of the *Desulfovibrio vulgaris* subsp. *vulgaris* Hildenborough. Arch Microbiol 167:376–83.

Kitamura M, Mizugai K, Taniguchi M, et al. 1995. A gene encoding a cytochrome *c* oxidase-like protein is located closely to the cytochrome *c*-553 gene in the anaerobic bacterium, *Desulfovibrio vulgaris* (Miyazaki F). Microbiol Immunol 39:75–80.

LeGall J, Xavier AV. 1996. Anaerobes response to oxygen: the sulfate-reducing bacteria. Anaerobe 2:1–9.

Odom JM, Singleton R Jr, editors. 1993. The sulfate-reducing bacteria: contemporary perspectives. New York: Springer-Verlag.

Ozawa K, Mogi T, Suzuki M, et al. 1997. Membrane-bound cytochromes in a sulfate-reducing strict anaerobe *Desulfovibrio vulgaris* Miyazaki F. Anaerobe 3:339–46.

Peck HD Jr, LeGall J, editors. 1994. Volume 243, Inorganic microbial sulfur metabolism. Methods in enzymology. San Diego, CA: Academic Press.

Pollock WBR, Voordouw G. 1994. Molecular biology of *c*-type cytochromes from *Desulfovibrio vulgaris* Hildenborough. Biochimie 76:554–60.

Pollock WBR, Loutfi M, Bruschi M, et al. 1991. Cloning, sequencing and expression of the gene encoding the high-molecular-weight cytochrome *c* from *Desulfovibrio vulgaris* Hildenborough. J Bacteriol 173:220–8.

Postgate JR. 1984. The sulphate-reducing bacteria. 2nd ed. Cambridge, UK: Cambridge University Press.

Powell B, Mergeay M, Christofi N. 1989. Transfer of broad host-range plasmids to sulphate-reducing bacteria. FEMS Microbiol Lett 59:269–74.

Rapp-Giles BJ, Casalot L, English RS, et al. 2000. Cytochrome c_3 mutants of *Desulfovibrio desulfuricans*. Appl Environ Microbiol 66:671–7.

Ried JL, Collmer A. 1987. An *nptI-sacB-sacR* cartridge for constructing directed, unmarked mutations in Gram-negative bacteria by marker exchange-eviction mutagenesis. Gene 57:239–46.

Roth JR, Lawrence JG, Rubenfield M, et al. 1993. Characterization of the cobalamin (vitamin B_{12}) biosynthetic genes of *Salmonella typhimurium*. J Bacteriol 175:3303–16.

Rousset M, Casalot L, Rapp-Giles BJ, et al. 1998. New shuttle vectors for the introduction of cloned DNA in *Desulfovibrio*. Plasmid 39:114–22.

Rousset M, Dermoun Z, Chippaux M, Belaich J-P. 1991. Marker exchange mutagenesis of the *hydN* genes in *Desulfovibrio fructosovorans*. Mol Microbiol 5:1735–40.

Santos H, Fareleira P, Xavier AV, et al. 1993. Aerobic metabolism of carbon reserves by the "obligate anaerobe" *Desulfovibrio gigas*. Biochem Biophys Res Commun 195:551–7.

Schweizer HP. 1992. Allelic exchange in *Pseudomonas aeruginosa* using novel ColE1-type vectors and a family of cassettes containing a portable *oriT* and the counter-selectable *Bacillus subtilis sacB* marker. Mol Microbiol 6:1195–204.

Simon R. 1984. High frequency mobilization of Gram-negative bacterial replicons by the *in vitro* constructed Tn5-Mob transposon. Mol Gen Genet 196:413–20.

Stams FJM, Veenhuis M, Weenk GH, Hansen TA. 1983. Occurrence of polyglucose as a storage polymer in *Desulfovibrio* species and *Desulfobulbus propionicus*. Arch Microbiol 136:54–9.

van den Berg WAM, Stokkermans JPWG, van Dongen WMAM. 1989. Development of a plasmid transfer system for the anaerobic sulphate reducer, *Desulfovibrio vulgaris*. J Biotechnol 12:173–84.

van Dongen WMAM. 1995. Molecular biology of redox-active metal proteins from *Desulfovibrio*. In: Barton LL, editor. Sulfate-reducing bacteria. New York: Plenum Press. p 185–215.

van Dongen WMAM, Stokkermans JPWG, van den Berg WAM. 1994. Genetic manipulation of *Desulfovibrio*. Methods Enzymol 243:319–30.

van Niel EWJ, Gottschal JC. 1998. Oxygen consumption by *Desulfovibrio* strains with and without polyglucose. Appl Environ Microbiol 64:1034–9.

Vertes AA, Asai Y, Kobayashi M, et al. 1994. Transposon mutagenesis of coryneform bacteria. Mol Gen Genet 245:397–405.

Voordouw G. 1993. Molecular biology of the sulfate-reducing bacteria. In: Odom JM, Singleton R Jr, editors. The sulfate-reducing bacteria: contemporary perspectives. New York: Springer-Verlag. p 88–130.

Voordouw G. 1995. The genus *Desulfovibrio*: the centennial. Appl Environ Microbiol 61:2813–9.

Voordouw JK, Voordouw G. 1998. Deletion of the *rbo* gene increases the oxygen sensitivity of the sulfate-reducing bacterium *Desulfovibrio vulgaris* Hildenborough. Appl Environ Microbiol 64:2882–7.

Voordouw G, Pollock WBR, Bruschi M, et al. 1990. Functional expression of *Desulfovibrio vulgaris* Hildenborough cytochrome c_3 in *Desulfovibrio desulfuricans* following conjugational gene transfer from *Escherichia coli*. J Bacteriol 172: 6122–6.

Wall JD. 1993. Genetics of the sulfate-reducing bacteria. In: Odom JM, Singleton R Jr, editors. The sulfate-reducing bacteria: contemporary perspectives. New York: Springer-Verlag. p 77–87.

Wall JD, Murnan T, Argyle J, et al. 1996. Transposon mutagenesis in *Desulfovibrio desulfuricans*: development of a random mutagenesis tool from Tn7. Appl Environ Microbiol 62:3762–7.

Wall JD, Rapp-Giles BJ, Rousset M. 1993. Characterization of a small plasmid from *Desulfovibrio desulfuricans* and its use for shuttle vector construction. J Bacteriol 175:4121–8.

Widdel F. 1988. Microbiology and ecology of sulfate- and sulfur-reducing bacteria. In: Zehnder AJB, editor. Biology of anaerobic microorganisms. New York: Wiley-Interscience. p 469–585.

Wolk CP, Cai Y, Panoff J-M. 1991. Use of a transposon with luciferase as a reporter to identify environmentally responsive genes in a cyanobacterium. Proc Natl Acad Sci USA 88:5355–9.

8
Function and Assembly of Electron-Transport Complexes in *Desulfovibrio vulgaris* Hildenborough

GERRIT VOORDOUW

The bioenergetics of *Desulfovibrio* spp. have been studied extensively but are as yet poorly understood. Peck contributed enormously to this topic by showing that *Desulfovibrio* spp. couple proton-driven ATP synthesis to sulfate respiration; by advancing models for the bioenergetics of *Desulfovibrio* spp., such as the hydrogen cycling model (Odom and Peck 1981); and by studying the molecular biology of *Desulfovibrio* spp., leading to the cloning and sequencing of genes for nickel-containing hydrogenases. The sequence of the *Desulfovibrio vulgaris* Hildenborough genome (www.tigr.org) will soon complete our knowledge of all redox proteins involved in energy transformation associated with sulfate respiration. This chapter reviews the structure and assembly of enzymes and protein complexes that are active in the electron-transport pathway from hydrogen to sulfate in *Desulfovibrio*.

The Components: Hydrogenases, Cytochrome c_3, and the Hmc Complex

Desulfovibrio spp. have periplasmic, nickel-containing and iron-only hydrogenases. A cytoplasmic NADP-reducing hydrogenase was described for *Desulfovibrio fructosovorans* (Malki et al. 1995) but is not widely distributed. Two different classes of nickel-containing hydrogenases, the NiFe- and NiFeSe-enzymes, are found in *Desulfovibrio*. The structures of NiFe-hydrogenases from *Desulfovibrio gigas* and *D. vulgaris*, which contain a large (L; 60 kDa) and a small (S; 30 Da) subunit, have been determined (Volbeda et al. 1995; Higuchi et al. 1997). It appears that the periplasmic NiFe-hydrogenase is widespread in *Desulfovibrio* spp., whereas the periplasmic iron-only and NiFeSe hydrogenases have a more limited distribution (Voordouw et al. 1990). The function of all periplasmic hydrogenases is likely the uptake of hydrogen to serve as electron donor for sulfate reduction.

Malki et al. (1997) provided genetic evidence for this. Deletion of both the *hyn* genes, encoding periplasmic NiFe-hydrogenase, and the *hnd* genes, encoding the cytoplasmic NADP-reducing hydrogenase, did not eliminate the ability of the organism to use hydrogen as electron donor for sulfate reduction. This was credited to the presence of the *hyd* genes, encoding periplasmic iron-only hydrogenase. These experiments indicated a role in hydrogen uptake for both the periplasmic NiFe and the periplasmic iron-only hydrogenase, although this function for the latter must still be confirmed by making the appropriate *hyd* deletion mutations. In contrast, van den Berg et al. (1991) presented evidence that iron-only hydrogenase in *D. vulgaris* Hildenborough has an essential role in hydrogen production in lactate metabolism. This was not shown by deletion mutagenesis, because the authors were unable to construct a *hyd* replacement mutant, but by expression of *hyd* antisense RNA. This decreased iron-only expression about three-fold and led to a decreased growth rate and growth yield on hydrogen-sulfate medium. Expression of antisense RNA also decreased the size of the hydrogen burst (the concentration of hydrogen in the culture head space) during the initial stages of growth on lactate-sulfate media (van den Berg et al. 1991).

There are two problems with the proposal that iron-only hydrogenase is essential for lactate metabolism in *Desulfovibrio* spp. First, the enzyme is not present in all species and, second, it is not clear why the enzyme would have to be located in the periplasm if its main function is to produce hydrogen. In their hydrogen cycling hypothesis, Odom and Peck (1981) proposed that hydrogen production from lactate is catalyzed by an as yet unidentified cytoplasmic hydrogenase and that subsequent oxidation of this cytoplasmically produced hydrogen to protons by a periplasmic hydrogenase provides the proton gradient that can drive ATP synthesis. Thus, if iron-only hydrogenase in *Desulfovibrio* were essential for hydrogen production, one would expect the enzyme to be a cytoplasmic, not a periplasmic, enzyme, similar to CpI, the 60 kDa iron-only hydrogenase of *Clostridium pasteurianum* (Peters et al. 1998). Instead, iron-only hydrogenase in *Desulfovibrio* consists of two subunits (α, 46 kDa; β, 10 kDa), which are homologous to the N- and C-terminal parts of the CpI polypeptide, respectively (Nicolet et al. 1999). This splitting is essential for export of the enzyme to the periplasm. The β-subunit is synthesized with a 34 amino acid twin-arginine signal sequence that allows export of redox proteins to the periplasm through a special pore, the characteristics of which were identified (Settles et al. 1997; Sargent et al. 1998; Weiner et al. 1998). This structural alteration of iron-only hydrogenase in *Desulfovibrio*, compared to the enzyme in *Clostridium*, suggests an altered function. Indeed, studies on the phenotype of an iron-only hydrogenase-deletion mutant of *D. vulgaris* Hildenborough indicated that growth with lactate and hydrogen as electron donor for sulfate reduction was not affected, whereas lower growth yields per mole of sulfate were obtained when hydrogen was the sole electron donor of sulfate reduction (Pohorolic et al. 2002). Thus the primary function of

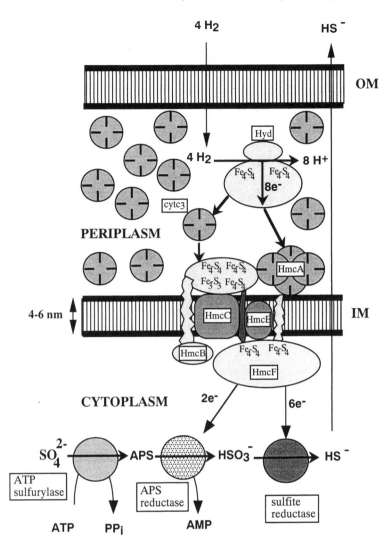

FIGURE 8.1. Redox proteins catalyzing electron transfer from hydrogen to sulfate in *Desulfovibrio*. The cytoplasmic uptake of protons associated with the reduction of sulfate is not shown. *Hyd,* hydrogenase; *cyctc₃,* cytochrome c_3; *HmcA, HmcB, HmcC, HmcE, HmcF,* components of the Hmc complex; *IM,* inner membrane; *OM,* outer membrane.

periplasmic NiFe- and iron-only hydrogenases in *Desulfovibrio* appears to be hydrogen uptake for sulfate reduction (Fig. 8.1).

In vitro studies established that electrons are transferred efficiently from both NiFe- and iron-only hydrogenases to cytochrome c_3. This small 13 kDa *c*-type cytochrome has four hemes (reduction potential −200 to −300 mV) and is usually abundantly present in the *Desulfovibrio* periplasm. Its structure–function properties have been studied extensively. A *cyc* mutant was constructed for *D. desulfuricans* which showed impaired growth with pyruvate as electron donor. Growth with lactate and, surprisingly, with hydrogen as electron donor for sulfate reduction were not affected (Rapp-Giles et al. 2000). This appears to be in conflict with the notion, derived from in vitro studies, that electrons are transferred from hydrogenases to cytochrome c_3 for subsequent transfer to the sulfate reduction pathway. A possible explanation for this dilemma is to propose that electrons are transported from hydrogenases directly to the cytochrome c_3 domains of the high molecular weight cytochrome from *D. vulgaris* (HmcA in Fig. 8.1). Indeed, in vitro studies showed that such electron transport is possible, although at a much reduced rate compared to electron transport to cytochrome c_3. It is interesting that addition of small catalytic amounts of cytochrome c_3 accelerate greatly electron transfer from hydrogenases to the hemes of HmcA (Pereira et al. 1998). Thus the presence of large amounts of cytochrome c_3 may not be essential, and it is conceivable that transport from hydrogen to sulfate may be able to bypass cytochrome c_3 in vivo (Fig. 8.1).

The Hmc complex of *D. vulgaris* Hildenborough consists of six redox proteins (HmcA–HmcF) (Rossi et al. 1993; Voordouw 1995; Keon and Voordouw 1996). HmcA is a periplasmic high molecular weight cytochrome containing 16 *c*-type hemes. Different forms of this cytochrome may be present in other *Desulfovibrio* spp. For instance, *D. desulfuricans* ATCC 27774 has a periplasmic multiheme cytochrome with 9 hemes, arranged in two tetra-heme clusters, with a protein fold similar to that of cytochrome c_3 and with the 9th heme located between the two c_3 domains (Matias et al. 1999). This smaller cytochrome is homologous to the C-terminal region of HmcA. HmcA also shares homology with MtrA, a *c*-type cytochrome with 10 hemes, thought to be part of the Fe(III) and Mn(IV) reduction system of *Shewanella putrefaciens* (Beliaev and Saffarini 1998). HmcA is encoded by the *hmcA* gene, the first of the *hmc* operon, which also includes the *hmcB*, *hmcC*, *hmcD*, *hmcE*, and *hmcF* genes (Fig. 8.2). Analysis of protein hydrophobicity reveals that HmcB–HmcE are integral membrane proteins (Rossi et al. 1993). These contain iron–sulfur clusters, or *b*-type heme, for transmembrane electron transport. HmcF is a cytoplasmic protein containing iron–sulfur clusters. It is probably widely distributed, because *Archaeoglobus fulgidus* has several homologs to this protein, one of which is encoded as part of an operon that is similar to the *hmc* operon (Fig. 8.2). The *hmc*-like operon of *A. fulgidus* is located at base pair 1,600,820 of the genomic sequence (Klenk et al. 1997). Its gene order is different from *D. vulgaris*, e.g., the *hmcA* homolog is the last gene of the operon. The *A. fulgidus* HmcA-like protein is greatly reduced in size and contains only 6 *c*-type heme-binding sites. Nevertheless, the evidence that *A. fulgidus* has an Hmc complex with a structure similar to that of *D. vulgaris* (Fig. 8.1) is strong, because all other proteins encoded by the *A. fulgidus* operon share

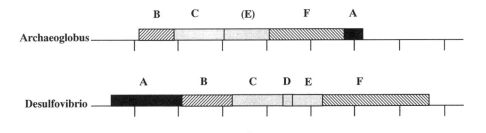

```
                 HEAKANDCRTCHHVRIDTCTACHTVNGTADSKFVQLEKAMHQPDSMRSCVGCHNT-RVQQPTCAG 364
Desulfovibrio    + A   N C +CH  + + C  CH   G             +H      C  CH T V++P  +G
                 YHASTNTCWSCHDSKEEFCDQCHDYVG------------IHP-----ECWDCHYTPSVEKPHYSG 1630807


                 FHIEKGTLCQGCHHNSPASLTPPKCASCHGKPFDADRGDRPGLKAAYHQQCMGCHDRMKIEKP 531
Archaeoglobus    +H     T C  CH +         C  CH       D  G+     H +C  CH      +EKP
                 YHASTNT-CWSCHDSKEEF-----CDQCH---------DYVGI----HPECWDCHYTPSVEKP 1630819
```

FIGURE 8.2. The gene arrangement of the *hmc* operon in *D. vulgaris* Hildenborough and a putative *hmc* operon in *A. fulgidus*. The sequence homology of the two HmcA proteins, which are multiheme *c*-type cytochromes, is also shown.

homology with their *D. vulgaris* counterparts. The presence of a membrane-bound Hmc complex with a proposed function in electron transport from hydrogen to sulfate may thus be a universal feature of sulfate-reducing bacteria.

Biochemical and Genetic Evidence for Hmc Complex Function in Hydrogen Oxidation

Pereira et al. (1998) provided biochemical evidence that electron transfer from either iron-only or NiFe-hydrogenases to HmcA is possible, although at a slow rate. The increase in electron-transfer rate by addition of cytochrome c_3 indicates that a more probable electron-transport path is from hydrogen through hydrogenase and cytochrome c_3 to HmcA.

Keon et al. (1997) showed that the Hmc complex is present in *D. vulgaris* at its highest concentration when hydrogen is the sole electron donor for sulfate reduction. The presence of lactate or pyruvate led to a decreased content of the Hmc complex, as shown by immunoblotting studies. Deletion of two genes downstream from the *hmc* operon, *rrf*1 and *rrf*2, led to a twofold increased content of the Hmc complex and a more rapid growth with hydrogen as the electron donor.

More direct evidence that the Hmc complex is important for hydrogen metabolism was provided by Dolla et al. (2000), who replaced a 5 kb DNA

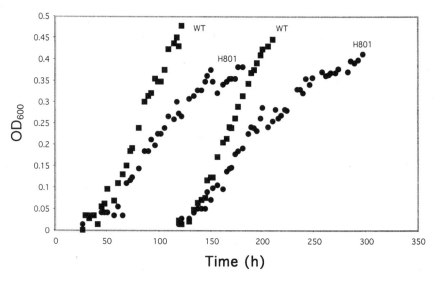

FIGURE 8.3. Growth of *D. vulgaris* wild type (*WT*) and H801 in liquid culture with hydrogen as sole electron donor for sulfate reduction. Growth was at 32°C in unstirred 300 mL nephelometer flasks with 50 mL defined medium under conditions of gas exchange. Cell density in OD_{600} units is plotted as a function of time. Cells of the first culture were used to inoculate a second culture at $t = 100$ h. Wildtype: $\mu = 0.050 \, h^{-1}$; H801: $\mu = 0.038 \, h^{-1}$.

fragment encoding most of the Hmc complex with a chloramphenicol resistance (*cat*) marker. The resulting mutant, *D. vulgaris* H801, did not express either HmcA or HmcF, as judged by immunoblotting. Measurement of hydrogen uptake activity by whole cells indicated a threefold decreased uptake activity in the mutant strain. Yet the level of iron-only hydrogenase, responsible for 90% of hydrogenase activity in *D. vulgaris* Hildenborough (Lissolo et al. 1986), was unchanged, as shown by immunoblotting, activity staining of native gels, and direct measurement of hydrogenase activity in broken cells using methylviologen as an artificial electron acceptor. These results indicate that the H801 strain is deficient in electron transport from hydrogenases to the sulfate-reduction pathway, not in the hydrogenase activity.

Specific growth rates and growth yields of the H801 strain were identical to the wild-type strain when lactate or pyruvate served as electron donor for sulfate reduction. Only with hydrogen as electron donor was a clearly reduced growth rate observed: $\mu = 0.038 \, h^{-1}$ for H801 and $\mu = 0.050 \, h^{-1}$ for wild type. (Dolla et al. 2000) (Fig. 8.3). These results indicate that the Hmc complex is important for growth on hydrogen, as suggested in Figure 8.1 and that there must be other transmembrane complexes that allow electron transport from periplasmic hydrogenases to the cytoplasmic sulfate-reduction pathway. The nature of these other complexes remains to be characterized. Now that the genome sequence of *D. vulgaris* is being completed, these complexes can in principle be identified by genomic approaches, i.e., by determining whether certain gene products are upregulated in the H801 strain compared to the wild-type strain to compensate for the loss of the *hmc* operon genes.

During growth of *D. vulgaris*, the increase in cell density (A_{600}) as a function of time is linearly related to a decrease in the sulfate concentration. The H801 and wild-type strains appear to be equally efficient in making cell mass during sulfate respiration. From the slopes of the lines in Figure 8.4, a similar growth yield per mole of sulfate ($Y = 9.7 \, g \, mol^{-1}$) can be derived for the H801 and wild-type strains. This result indicates that the Hmc complex does not contribute to greater bioenergetic efficieny in *D. vulgaris*, e.g., by coupling the transport of electrons to the pumping of additional protons to the periplasm. This could indicate that the proton gradient derives solely from the fact that hydrogen oxidation (producing protons) is periplasmic and sulfate reduction (consuming protons) is cytoplasmic, as proposed originally by Odom and Peck (1981) and Peck (1993). Other transmembrane electron-transport complexes that remain once the Hmc complex is deleted may operate similarly, giving the same bioenergetic efficiency at a lower growth rate.

A final interesting aspect of the growth phenotype of the H801 strain is that it is much slower in forming colonies on agar plates, especially when hydrogen serves as the electron donor for sulfate reduction. Assuming that colony formation involves reduction of the redox potential of the environ-

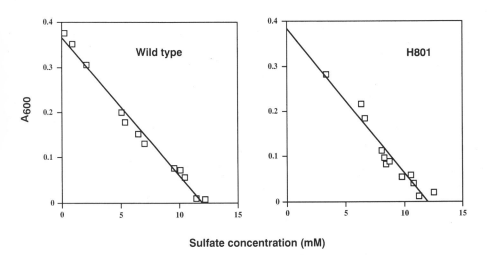

FIGURE 8.4. Molar growth yields of *D. vulgaris* Hildenborough wild type and H801 strains.

ment surrounding a cell on the agar surface, one concludes that the H801 strain is deficient in its ability to establish such a reduced redox potential.

Assembly of Redox Protein Complexes in *Desulfovibrio vulgaris*

The existence of a specialized system for the export of hydrogenases to the periplasm, now referred to as the twin-arginine translocation (Tat) system was first suspected through research on the iron-only hydrogenase of *D. vulgaris*. After cloning and sequencing of the *hydA* and *hydB* genes, encoding the α and β subunits of this hydrogenase, researchers were at first not at all clear how this hydrogenase was exported to the periplasm, because a typical Sec-dependent signal sequence was lacking (Voordouw and Brenner 1985). The Sec system for export of proteins to the periplasm consists of the SecA, SecE, SecG, and SecY proteins. These translocate their protein substrates in unfolded form through a threading mechanism, driven by SecA-catalyzed ATP hydrolysis. The translocation is initiated by an N-terminal signal sequence, usually 20–30 amino acids in length, which generally has a positively charged N terminus but is otherwise hydrophobic throughout. The signal peptidase cleavage site is often at an Ala-Ala sequence. Examples of Sec-system-dependent signal sequences are provided by those for the *c*-type cytochromes of *D. vulgaris* (Fig. 8.5). After export to the periplasm, *c*-type cytochromes fold and combine covalently with heme, which is also exported, to give the functional holo-proteins. The details of the mechanism of *c*-type cytochrome assembly are not yet clear. However, this mehanism must be similar in all prokaryotes and has Sec-system-mediated export as an important feature, because *c*-type cytochromes are periplasmic, or membrane-bound with the *c*-type heme on the external side of the cytoplasmic membrane in all prokaryotes.

In the case of iron-only hydrogenase neither the HydA not the HydB subunit has a typical Sec-system-dependent signal sequence. However, studies by Prickril et al. (1986) established that HydB had a complex signal sequence of 34 amino acid residues (Fig. 8.5). Subsequent sequence analysis of genes for NiFe hydrogenase of *Desulfovibrio* and many other species and comparison with the N termini of the mature proteins established that all small subunits of periplasmic hydrogenases have a complex signal sequence of 30–50 amino acid residues (Fig. 8.5). Different from Sec-system signal peptides these sequences all shared a conserved motif RRXFXK, (Fig. 8.5, bold face type). These observations suggest a unique mechanism for hydrogenase transport that involves a specialized pore, containing at least one protein component that recognizes the conserved motif. It soon became apparent that many other redox proteins use the Tat system rather than the Sec system for translocation to the periplasm. The reasons for this are thought to be that insertion of cofactors such as nickel or iron in iron-sulfur clusters is a cytoplasmic process; the insertion of *c*-type

```
Dv-Cyc                                  MRKLFFCGVLALAVAFALPVVA
Dv-Cyf                                  MKRVLLLSSLCAALSFGLAVSGVA
Dv-HmcA                             MRNGRTLLRWAGVLAATAIIGVGGFWSQGTT

DvHHydB                         MQIVNLTRRGFLKAACVVTGGALISIRMTGKAVA
DvMHydB                         MQIASITRRGFLKVACVTTGAALIGIRMTGKAVA

Dg-HynB             MKCYIGRGKNQVEERLERRGVSRRDFMKFCTAVAVAMGMGPAFAPKVAEA
Dv-HynB             MRFSVGLGKEGAEERLARRGVSRRDFLKFCTAIAVTMGMGPAFAPEVARA
Df-HynB             MNFSVGLGRMNAEKRLVQNGVSRRDFMKFCATVAAAMGMGPAFAPKVAEA

Db-HysB                          MSLSRREFVKLCSAGVAGLGISQIYHPGIVHA

Ec-FdnG                           MDVSRRQFFKICAGGMAGTTVAALGFAPKQALA
Ws-FdhA            MSEALSGRGNDRRKFLKMSALAGVAGVSQAVGSDQSKVLRPA
Dv-HmcB                            MDRRRFLTLLGSAGLTATVATAGTAKAA
```

FIGURE 8.5. Signal sequences of redox proteins translocated through either the Sec or the Tat system. *Dv-Cyc*, *Dv-Cyf*, *Dv-HmcA*, Sec-dependent signal peptides of cytochrome c_3, cytochrome c_{553}, and the 16-heme, high molecular weight cytochrome (HmcA) from *D. vulgaris*; DvHHydB, DvMHydB, Tat-dependent signal sequences of iron-only hydrogenases of *D. vulgaris* Hildenborough and *D. vulgaris oxamicus* Monticello; *Dg-HynB*, *Dv-HynB*, *Df-HynB*, NiFe hydrogenases of *D. gigas*, *D. vulgaris*, and *D. fructosovorans*; *Db-HysB*, NiFeSe hydrogenase of *Desulfovibrio baculatus*; *Ec-FdnG*, catalytic subunit of nitrate-inducible formate dehydrogenase from *E. coli*; *Ws-FdhA*, catalytic subunit of formate dehydrogenase from *W. succinogenes*; *Dv-HmcB*, electron-transferring subunit of the Hmc complex of *D. vulgaris*. See text for details.

heme is the only known exception. Because cofactor insertion requires partial or complete folding of the polypeptide chain, redox proteins cannot be translocated through the threading mechanism of the Sec system. Instead, the Tat system is thought to allow export of proteins in partially or completely folded form. Early studies on export of *D. vulgaris* iron-only hydrogenase expressed from a plasmid containing the *hydA* and *hydB* or separate genes in *Escherichia coli* indicated that export was cooperative, i.e., required the presence of both genes (van Dongen et al. 1988). This implies that the signal sequence of the small subunit was able to translocate the small-large subunit complex. This has been referred to as the hitchhiker mechanism (Rodrigue et al. 1999). In the case of translocation of NiFe hydrogenases, the mechanism also involves C-terminal processing of the large subunit, required for cytoplasmic nickel insertion (see Chapter 6). Hatchikian et al. (1999) showed that the large subunit of iron-only hydrogenase is also processed at the C terminus, presumably also for correct assembly of the active-site iron-sulfur cluster or for export to the periplasm.

The general nature of the Tat system became apparent when it was found that FdhA, the catalytic subunit of formate dehydrogenase from *Wolinella succinogenes*, also has a twin-arginine signal peptide (Bokrantz et al. 1991)

(Fig. 8.5). FdH requires a molybdopterin guanine dinucleotide (MGD) cofactor for activity, which is apparently inserted in the cytoplasm. The fully functional enzyme is encoded by an operon that includes the *fdhA*, *fdhB*, and *fdhC* genes. FdhB has sequence homology to HmcB. Both are elecron-transferring subunits, containing four Fe_4S_4 clusters. This assignment is based on the homology of these subunits with DmsB, the electron-transferring subunit of *E. coli* dimethysulfoxide (DMSO) reductase, which is encoded by a *dmsA,B,C.* operon similar to *fdhA,B,C.* The assignment of four Fe_4S_4 clusters in DmsB is derived from the positioning of 16 conserved Cys residues in the sequence (Bilous et al. 1988; Keon and Voordouw 1996) and from extensive biophysical studies. FdhC is a *b*-type heme-containing membrane-bound subunit similar to HmcC, HmcD, and HmcE in the Hmc complex. Berg et al. (1991) showed similarly that FdnG, the catalytic subunit of nitrate-inducible formate dehydrogenase from *E. coli*, contains a twin-arginine signal sequence (Fig. 8.5). The FdnG, FdnH, and FdnI subunits of this enzyme are homologous to FdhA, FdhB, and FdhC. FdnG binds and MGD cofactor for activity, FdnH has four Fe_4S_4 clusters for its electron-transfer function, and FdnI is a membrane-bound *b*-type heme-containing subunit. A closer inspection of the HmcB sequence also indicates the presence of a twin-arginine signal peptide (Fig. 8.5) (Keon and Voordouw 1996), indicating the electron-transferring part of HmcB to be periplasmic, as indicated in Figure 8.1. This subunit is firmly anchored by a large membrane-spanning, hydrophobic sequence at its C terminus. Regarding HmcA as the catalytic subunit of the Hmc complex (i.e., the subunit that may accept electrons from periplasmic hydrogenases), I note that in the assembly of the Hmc complex the catalytic subunit is exported by a Sec-system-dependent mechanism. Because the Sec-translocation system allows no hitchhiking, export of the HmcB subunit, which requires cytoplasmic iron-sulfur insertion, needs its own twin-arginine signal sequence. In contrast, in the case of the FdhABC and FdnGHI transmembrane redox protein complexes, the catalytic subunit requires cytoplasmic assembly. As demonstrated for periplasmic hydrogenase assembly, it is possible that the electron-transferring FdhB and FdnH subunits, which lack their own twin-arginine signal peptides, are exported together with the FdhA and FdnG subunits through a hitchhiker mechanism. Thus the general topology of the FdhABC and FdnGHI complexes may resemble that shown for the Hmc complex in Figure 8.1, the latter being structurally more elaborate and involving three membrane-bound subunits and one cytoplasmic subunit (HmcF).

By searching protein databases Berks (1996) uncovered 90 twin-arginine signal peptide sequences, translocating a wide variety of redox proteins. Sequence comparison of these signal sequences indicated that they group functionally, not according to species (Voordouw 2000). For instance all NiFe hydrogenase twin-arginine signal peptides group together, indicating co-evolution of enzyme and signal peptide. Genetic studies have since uncovered the nature of the protein components of the pore-catalyzing twin-arginine translocation. TatB and TatC appear to be essential for this translocation system. Mutations in the genes for these proteins are highly pleiotropic and prevent the export of a wide variety of redox proteins in *E. coli* (Sargent et al. 1998; Weiner et al. 1998; Sargent et al. 1999). Examination of the emerging *D. vulgaris* genome sequence indicates the presence of *tatB* and *tatC* homologs in close proximity, whereas *tatA* and *tatE* homologs are located elsewhere on the chromosome. The finding of a *tatB,C* operon in *Desulfovibrio* confirms the essental cooperative role of TatB and TatC proteins (Sargent et al. 1999). The important conclusion from these studies for the physiology of hydrogen metabolism in *Desulfovibrio* is that all hydrogenases with twin-arginine sequences on the small subunit must be regarded as peripasmic. This includes the iron-only, the NiFe, and the NiFeSe hydrogenases, which have sometimes been proposed to be cytoplasmic. Thus, with the exception of the NADP-reducing hydrogenase of *D. fructosovorans*, *Desulfovibrio* spp. may not contain a cytoplasmic hydrogenase, and the main function of all known Tat-system-translocated peripasmic hydrogenases appears to be in hydrogen uptake.

Conclusion

The *hmc* operon of *D. vulgaris* Hildenborough encodes a transmembrane redox protein complex (the Hmc complex) that consists of redox proteins HmcA to HmcF. HmcA is the periplasmic high molecular weight cytochrome containing 16 *c*-type hemes. HmcB to HmcE are integral membrane proteins, containing iron–sulfur clusters, or *b*-type heme. HmcF is a cytoplasmic protein containing iron-sulfur clusters. All protein components of the Hmc complex are either membrane bound or membrane associated. Its proposed physiologic function is to catalyze electron transport from periplasmic hydrogen oxidation to cytoplasmic sulfate reduction reactions. A mutant strain, *D. vulgaris* H801, was constructed in which a 5 kb DNA fragment containing most of the *hmc* operon is replaced by the *cat* gene. Growth of *D. vulgaris* H801 and of the wild-type strain with lactate or pyruvate as electron donors for sulfate reduction was similar. However, growth of strain H801 with hydrogen as an electron donor for sulfate reduction (acetate and CO_2 as the carbon source) was significantly slower than of the wild-type strain. The growth yields per mole of sulfate were similar for strains H801 and wild-type. These results prove the importance of the Hmc complex in electron transport from hydrogen to sulfate. Electron transport through the Hmc complex is not linked to energy conservation, leading to additional ATP synthesis and *Desulfovibrio* must have additional transmembrane complexes catalyzing this same function.

Mutant H801 was also found to be deficient in establishing a low-redox-potential niche. On minimal medium plates in which hydrogen serves as the sole electron donor, colonies of the wild-type and H801 strains formed after 14 and 30 days, respectively. These results suggest that, in addition to transmembrane electron transport from hydrogen to sulfate, the redox reactions catalyzed by the Hmc complex are important for establishment of the low-redox-potential environment that allows single cells to grow into colonies.

Assembly of the Hmc complex requires transport of polypeptides across or insertion of polypeptides into the cytoplasmic membrane. For periplasmic hydrogenases from *Desulfovibrio*, a mechanism was proposed more than 10 years ago in which a single signal peptide at the N terminus of the small subunit a allowed transport of a folded and assembled complex of the large and small subunits. This has been referred to as the hitchhiker mechanism. Strong evidence for a unique mechanism was provided by the fact that all small-subunit hydrogenase signal sequences share a consensus sequence (RRxFxK). Database searching and genetic studies with *E. coli* showed that membrane transport of all proteins binding redox cofactors (except *c*-type cytochromes) is catalyzed by the twin arginine translocation (Tat) system. Mutations in the *E. coli tat* operon are pleiotropic and show defects in the export and assembly of many enzymes involved in anaerobic respiration. Assemble of the Hmc complex requires protein export of HmcA through the standard Sec and of HmcB through the Tat system.

Acknowledgments. This chapter is based on my contribution to the Symposium on the Power of Anaerobes and is dedicated to the memory of Harry Peck. This work was supported by a research grant from the Natural Science and Engineering Research Council of Canada and by a fellowship from the Hanse Wissenschaftskolleg in Delmenhorst, Germany. A database of the *D. vulgaris* Hildenborough genome was searched at the Web site of the Institute for Genomic Research (www.tigr.org.) Sequencing of this genome is financially supported by the U.S. Department of Energy.

References

Beliaev AS, Saffarini DA. 1998. *Shewanella putrfaciens mtrB* encodes and outer membrane protein required for Fe(III) and Mn(IV) reduction. J Bacteriol. 180: 6292–7.

Berg BL, Li J, Heider J, Stewart V. 1991. Nitrate-inducible formate dehydrogenase in *Escherichia coli* K12. J. Biol Chem 266:22380–5.

Berks BC. 1996. A common export pathway for proteins binding complex redox cofactors? Mol Microbiol 22:393–404.

Bilous PT, Cole ST, Anderson WF, Weiner JH. 1988. Nucleotide sequence of the *dmsABC* operon encoding the anaerobic dimethylsulfoxide reductase of *Escherichia coli*. Mol Microbiol 2:785–95.

Bokrantz M, Gutmann M, Körtner C, et al. 1991. Cloning and nucleotide sequence of the structural genes encoding the formate dehydrogenase of *Wolinella succinogenes*. Arch Microbiol 156:119–28.

Dolla A, Pohorelic BKJ, Voordouw JK, Voordouw G. 2000. Deletion of the *hmc*-operon of *Desulfovibrio vulgaris* subsp. *vulgaris* Hildenborough hampers hydrogen metabolism and low-redox potential niche establishment. Arch Microbiol 174:143–51.

Hatchikian EC, Magro, Forget N, et al. 1999. Carboxy-terminal processing of the large subunit of [Fe] hydrogenase from *Desulfovibrio desulfuricans* ATTC 7757. J Bacteriol 181:2947–52.

Higuchi Y, Yagi T, Yasuoka N. 1997. Unusual ligand structure in Ni-Fe active center and an additional Mg site in hydrogenase revealed by high resolution X-ray structure analysis. Structure 5:1671–80.

Keon RG, Voordouw G. 1996. Identification of the HmcF and topology of the HmcB subunit of the Hmc complex of *Desulfovibrio vulgaris*. Anaerobe 2:231–8.

Keon RG, Fu R, Voordouw G. 1997. Deletion of two downstream genes alters expression of the *hmc* operon of *Desulfovibrio vulgaris* subsp. *vulgaris* Hildenborough. Arch Microbiol 167:376–83.

Klenk HP, Clayton RA, Tomb J, et al. 1997. The complete genome sequence of the hyperthermophilic, sulphate-reducing archaeon *Archaeoglobus fulgidus*. Nature 390:364–70.

Lissolo T, Choi, ES, LeGall J, Peck HD Jr. 1986. The presence of multiple intrinsic nickel-containing hydrogenases in *Desulfovibrio vulgaris* Hildenborough. Biochem Biophys Res Commun 139:701–8.

Malki S, De Luca G, Fardeau ML, et al. 1997. Physiological characteristics and growth behavior of single and double hydrogenase mutants of *Desulfovibrio fructosovorans*. Arch Microbiol 167:38–45.

Malki S, Saimmaime I, De Luca G, et al. 1995. Characterization of an operon encoding an NADP-reducing hydrogenase in *Desulfovibrio fructosovorans*. J Bacteriol 177:2628–36.

Matias PM, Coelho R, Pereira IA, et al. 1999. The primary and three-dimensional structures of a nine haem cytochrome *c* from *Desulfovibrio desulfuricans* ATCC 27774 reveal a new member of the Hmc family. Structure 7:119–30.

Nicolet Y, Piras C, Legrand D, et al. 1999. *Desulfovibrio desulfuricans* iron hydrogenase: the structure shows unusual coordination to an active site Fe binuclear center. Structure 7:13–23.

Odom JM, Peck HD Jr. 1981. Hydrogen cycling as a general mechanism for energy coupling in the sulfate-reducing bacteria, *Desulfovibrio* sp. FEMS Microbiol Lett 12:47–50.

Peck HD Jr. 1993. Bioenergetic strategies of the sulfate-reducing bacteria. In: Odom JM, Singleton R Jr, editors. The sulfate-reducing bacteria: contemporary perspectives. New York: Springer-Verla. p 41–76.

Pereira IAC, Romao CV, Xavier AV, et al. 1998. Electron transfer between hydrogenases and mono- and multiheme cytochromes in *Desulfovibrio* spp. J Biol Inorg Chem 3:494–8.

Peters JW, Lanzilotta WN, Lemon BJ, Seefeldt LC. 1998. X-ray crystal structure of the Fe-only hydrogenase (CpI) from *Clostridium pasteurianum* to 1.8 Ångstrom resolution. Science 282:1853–8.

Pohorelic BKJ, Voordouw JK, Lojou E, et al. 2002. Effects of deletion of genes encoding Fe-only hydrogenase of *Desulfovibrio vulgaris* Hildenhorough on hydrogen and lactate metabolism. J Bacteriol 184:679–686.

Prickril BC, Czechowski MH, Przybyla AE, et al. 1986. Putative signal peptide on the small subunit of the periplasmic hydrogenase from *Desulfovibrio vulgaris*. J Bacteriol 167:722–5.

Rapp-Giles BJ, Caselot L, English RS, et al. 2000. Cytochrome c_3 mutants of *Desulfvibrio desulfuricans*. Appl Environ Microbiol 66:671–7.

Rodrigue A, Chanal A, Beck K, et al. 1999. Co-translocation of a periplasmic enzyme complex by a hitchhiker mechanism through the bacterial *tat* pathway. J Biol Chem 274:13223–8.

Rossi M, Pollock WBR, Reij MW, et al. 1993. The *hmc* operon of *Desulfovibrio vulgaris* subsp. *vulgaris* Hildenborough encodes a potential transmembrane redox protein complex. J Bacteriol 175:4699–711.

Sargent F, Bogsch EG, Stanley NR, et al. 1998. Overlapping functions of components of a bacterial Sec-independent protein export pathway. EMBO J 17: 3640–50.

Sargent F, Stanley NR, Berks BC, Palmer T. 1999. Sec-independent protein translocation in *Escherichia coli*. A distinct and pivotal role for the TatB protein. J Biol Chem 274:36073–82.

Settles AM, Yonetani A, Baron A, et al. 1997. Sec-independent protein translocation by the maize Hcf106 protein, Science 278:1467–1470.

Van den Berg WAM, van Dongen, WMAM, Veeger C. 1991. Reduction of the amount of periplasmic hydrogenase in *Desulfovibrio vulgaris* (Hildenborough) with antisense RNA: direct evidence for an important role of this hydrogenase in lactate metabolism. J Bacteriol 173:3688–94.

Van Dongen W, Hagen W, van den Berg W, Veeger C. 1988. Evidence for an unusual mechanism of membrane translocation of the periplasmic hydrogenase of *Desulfovibrio vulgaris* (Hildenborough), as derived from expression in *Escherichia coli*. FEMS Microbiol Lett 50:5–9.

Volbeda A, Charon M-H, Piras C, et al. 1995. Crystal structure of the nickel-iron hydrogenase from *Desulfovibrio gigas*. Nature 373:580–7.

Voordouw G. 1995. The genus *Desulfovibrio*: the centennial. Appl Environ Microbiol 61:2813–19.

Voordouw G. 2000. A universal system for the transport of redox proteins: early roots and latest developments. Biophys Chem 86:131–40.

Voordouw G, Brenner S. 1985. Nucleotide sequence of the gene encoding the hydrogenase from *Desulfovibrio vulgaris* (Hildenborough). Eur J Biochem 148:515–20.

Voordouw G, Niviere V, Ferris FG, et al. 1990. Distribution of hydrogenase genes in *Desulfovibrio* spp. and their use in identification of species from the oil field environment. Appl Environ Microbiol 56:3748–54.

Weiner JH, Bilous PT, Shaw GM, et al. 1998. A novel and ubiquitous system for membrane targeting and secretion of cofactor-containing proteins. Cell 93: 93–101.

9
Iron-Sulfur Proteins in Anaerobic Eukaryotes

RICHARD CAMMACK, DAVID S. HORNER, MARK VAN DER GIEZEN, JAROSLAV KULDA, and DAVID LLOYD

Certain species of eukaryotes live in anaerobic environments. Some, such as the protozoa *Trichomonas vaginalis*, *Tritrichomonas foetus*, and *Giardia intestinalis* are pathogenic in humans or animals. Other ciliate protozoa and anaerobic fungi, such as the rumen fungus *Neocallimastix frontalis*, inhabit the digestive tracts of animals and insects. Some of these anaerobic eukaryotes contain intracellular organelles known as hydrogenosomes instead of mitochondria. Like the mitochondria, hydrogenosomes oxidize substrates such as pyruvate, but release hydrogen instead of consuming oxygen. Gene sequencing indicates that the hydrogenosomes share a common origin with the mitochondria and obtained their enzymes by lateral gene transfer. Hydrogen-producing metabolism involves the iron-sulfur proteins pyruvate:ferredoxin oxidoreductase (PFOR), ferredoxin, and hydrogenase. PFOR and [Fe] hydrogenase are similar to those seen in anaerobic bacteria, but the [2Fe-2S] ferredoxins of the trichomonads are similar to those associated with monooxygenases. The iron-sulfur proteins in representative species have been surveyed by electron-paramagnetic resonance (EPR) spectroscopy. Hydrogenosomes from the trichomonads and *N. frontalis* appear to contain similar membrane-bound proteins of the [2Fe-2S] type. Signals have been associated with hydrogenase, PFOR, and possibly fumarate reductase. The low redox potentials generated by PFOR are the basis for the selective toxicity of the broad-spectrum nitroimidazole drug metronidazole toward the anaerobic pathogens. Reduction of metronidazole leads to the formation of free radicals. Pathogenic strains of *T. vaginalis* that are resistant to metronidazole are now emerging, and more resistant strains have been developed in the laboratory by growth in the presence of increasing concentrations of the drug. Examination of the highly resistant strains shows that ferredoxin and other iron-sulfur proteins are at lower levels or, in extreme cases, absent.

Anaerobic environments, such as swamps, marine and freshwater sediments, and the body cavities of animals, are teeming with large numbers of single-celled eukaryotes (protists) that, like all cells, must produce ATP to survive. Some of these protists, such as ciliates, trichomonads, and chytrid fungi, contain hydrogenosomes. These organelles, with double membranes, lack cytochromes and the apparatus of the respiratory chain of mitochondria. They produce energy through the metabolism of pyruvate or malate and have the unusual property, for a eukaryotic cell, of excreting molecular hydrogen as a by-product (Embley et al. 1993; Müller 1997). In some protists, such as *Giardia* and *Entamoeba*, the evidence for hydrogenosomes is less clear; they may survive by anaerobic fermentations in the cytosol, or they may contain hydrogenosomes that are more difficult to detect.

Iron-sulfur proteins are a group of enzymes and other electron carriers that contain clusters of iron and sulfide linked directly to amino-acyl side chains, usually cysteines. They are widely distributed in nature. Soon after their discovery, Hall et al. (1971) proposed that they could be used to follow the course of evolution. Studies of genome sequences have revealed that iron-sulfur cluster binding motifs are among the most commonly recognized sequences.

Iron-sulfur proteins are constituents of hydrogenosomes. It has been shown that iron is a virulence factor *T. foetus* (Kulda et al. 1999) and that the expression of metabolic enzymes of the hydrogenosome is controlled by iron availability (Vanacova et al. 2001). Genes for the pyridoxal-5′-phosphate-dependent cysteine desulfurase (IscS), similar to the enzyme responsible for sulfur insertion into mitochondrial iron-sulfur clusters, were found in *T. vaginalis* and *G. intestinalis* (Tachezy et al. 2001). Targeting sequences in the genes have been taken as evidence that the mature proteins are situated in the hydrogenosomes.

We examined the iron-sulfur proteins involved in production of hydrogen in hydrogenosomes. The anaerobic eukaryotes investigated include the trichomonads *T. vaginalis* and *T. foetus*, which infect humans and cattle, respectively; the diplomonad *Giardia lamblia*, an intestinal parasite (Fig. 9.1); and the chytrid fungus *N. frontalis*, a commensal organism isolated from llama feces. The iron-sulfur proteins we detected by EPR are PFOR, ferredoxin, and hydrogenase. PFOR is a central enzyme in the metabolism of many anaerobes. The enzymes from different species use different forms of ferredoxin or flavodoxin as electron carriers. Two types of metal-containing hydrogenases are found in bacteria: the nickel-iron ([NiFe]-type)

FIGURE 9.1. Whole-cell electron scanning and transmission micrographs of *G. intestinalis* **a**, **b**, *T. vaginalis* **c**, **d**, and *T. foetus* **e**, **f**. *A*, axostyle; *Aa*, anterior axoneme; *Af*, anterior flagella; *C*, costa; *Cf*, caudal flagella; *F*, funis; *G*, golgi; *H*, hydrogenosomes; *K*, kinetosomes; *Lc*, lateral crest; *N*, nucleus; *P*, pelta; *Pa*, posterolateral axoneme; *PAH*, paraxostylar hydrogenosomes; *PAJ*, pelta-axostylar junction; *PC*, periflagellar canal; *PCH*, paracostal hydrogenosomes; *Pf*, posterolateral flagella; *Pff*, posterior free flagellum; *Pv*, peripheral vesicles; *RF*, recurrent flagellum; *UM*, undulating membrane; *Vd*, ventral disk; *Vf*, ventral flagella; *Vg*, ventral groove; *VLF*, ventrolateral flange.

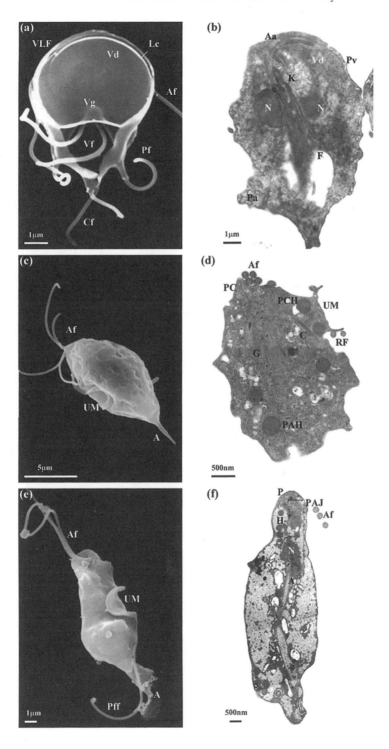

and the iron only ([Fe]-type) (Cammack et al. 2001; Vignais et al. 2001). So far, only the [Fe]-type has been found in protozoa and fungi. The active center of [Fe] hydrogenases, the H cluster, is a complex structure with six iron atoms and numerous diatomic ligands. (Peters et al. 1998; Nicolet et al. 1999). In addition, hydrogenases from different sources contain various arrangements of secondary electron-transfer components, the [4Fe-4S] and [2Fe-2S] clusters and flavins.

Iron-sulfur proteins can be observed by EPR spectroscopy, either in their oxidized or in their reduced state. As a method of observing iron-sulfur clusters, EPR is discriminating but not particularly sensitive; lack of a detectable EPR signal cannot be taken as evidence of absence. However, a positive EPR signal is good evidence for the intactness of an iron-sulfur cluster in a protein. Moreover, EPR can be used to follow reduction of the clusters and, by use of mediated electrochemical titrations, to estimate redox potentials.

Ferredoxins are electron-transfer proteins that can mediate between pyruvate:ferredoxin oxidoreductase and hydrogenase. It appears that during the course of the evolution, different types of ferredoxin were recruited for this purpose. In clostridia, ferredoxins of the 2[4Fe-4S] type are used (Uyeda and Rabinowitz 1971). In *T. vaginalis* (Chapman et al. 1986) and *T. foetus* (Marczak et al. 1983), [2Fe-2S] ferredoxins are used. Their axial EPR spectra at $g = 1.94, 2.02$ (Fig. 9.2) resemble those of the ferredoxins that are involved in P450 monooxygenase systems. Similar ferredoxins, with various functions, have been isolated from

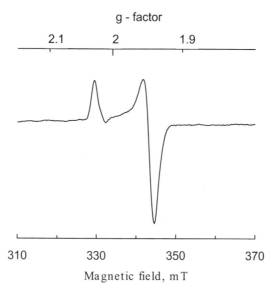

FIGURE 9.2. EPR spectrum of *T. vaginalis* ferredoxin, recorded at 30 K and microwave frequency 9.34 GHz.

organisms as diverse as humans and *Escherichia coli*. The structure of *T. vaginalis* ferredoxin showed that it contains a unique surface cavity, which exposes one of the inorganic sulfur atoms of the cluster to solvent. This cavity was proposed to be the site of interaction with metronidazole (see below). The ferredoxins from *Giardia* spp. appear to be of the [3Fe-4S] (Townson et al. 1994) or, more probably, the [4Fe-4S] type (Ellis et al. 1993); note that [4Fe-4S] clusters are easily modified to [3Fe-4S] clusters during extraction.

Further examination by EPR of *T. foetus* and *T. vaginalis* hydrogenosomes showed the presence of multiple iron-sulfur proteins, which could be resolved by the redox potential and temperature dependence of their signals (Ohnishi et al. 1980; Chapman et al. 1986). Altogether, eight EPR signals of different clusters were identified in *T. vaginalis*, designated centers A–H. The $g = 1.94$ signal was shown to include two components; center A was membrane bound, and center B corresponded to the soluble ferredoxin. Some indication of the type of cluster can be deduced from the temperature dependence of the signals. [4Fe-4S] and [3Fe-4S] clusters tend to be detectable only at temperatures <30 K; [2Fe-2S] clusters are seen at 77 K or higher. [3Fe-4S] clusters are detected in their oxidized state; [2Fe-2S] and [4Fe-4S] clusters are seen in their reduced state. Some tentative assignments of the signals are given in Table 9.1.

In the hydrogenosomal membranes, EPR spectra showed no trace of the highly characteristic features of the iron-sulfur clusters of complex I (NADH:ubiquinone reductase) and the Rieske protein of complex III of the mitochondrial respiratory chain. This is consistent with the absence of

TABLE 9.1. EPR-detectable iron-sulfur clusters in *T. vaginalis* hydrogenosomes.

Cluster	g factors	Maximum temperature[a] (K)	$E_m{}^b$ (mV)	Type	Assignment[c]
A	2.018, 1.938	140	−125	[2Fe-2S]	Fumarate reductase (?)
B	2.025, 1.942	150	−300	[2Fe-2S]	Ferredoxin
C	2.048, 1.94, 1.86	30	−285	[4Fe-4S]	
D	2.085, 1.88, 1.83	25	−270	[4Fe-4S]	PFOR (1)
E	2.05, 1.95. 1.89	40	−390	[4Fe-4S]	PFOR (1)
F	2.05, 1.95, 1.874	25	−340	[4Fe-4S]	
G	2.07, 1.95, 1.904	35		[4Fe-4S]	Hydrogenase (?)
H	2.03, 1.995	25		[3Fe-4S]	Fumarate reductase (?)
[Fe]	2.10, 2.04, 2.0	100		H cluster	Hydrogenase (2)

[a] Highest temperature at which the EPR signal was detected.
[b] Redox potentials are expressed relative to the standard hydrogen electrode.
[c] Assignment of clusters is based on Chapman et al. (1986) (1), Payne et al. (1993), and data from Williams et al. (2) (?) tentative assignments.

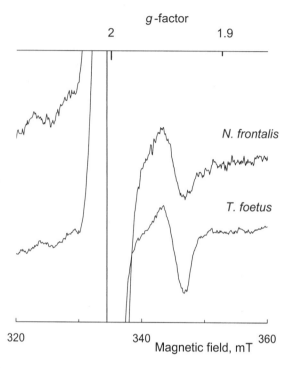

FIGURE 9.3. EPR spectra of membrane preparations of *T. foetus* and *N. frontalis*, reduced with dithionite and recorded at 30 K and microwave frequency 9.38 GHz.

a mitochondrial-type respiratory chain. There is some suggestion of an enzyme resembling complex II (succinate:ubiquinone reductase) (Ohnishi et al. 2000; Beinert 2002). In view of the anaerobic nature of the metabolism of these species, a more likely assignment would be as a fumarate reductase; however, this activity has not been reported. The EPR signals from these enzymes—the [2Fe-2S] center at $g = 1.94$—have the characteristic property that their signal increases in two phases as the redox potential becomes more negative. The second phase is due to the reduction of a [4Fe-4S] cluster in the same protein subunit, which enhances the electron-spin relaxation (Johnson et al. 1985; Cammac et al. 1986). EPR signals with the appropriate characteristics were detected in membranes of *T. foetus* (Ohnishi et al. 1980) and *Trichomonas vaginalis* (centers A and H) (Chapman et al. 1986). It is interesting that a similar signal has been observed in membranes of *N. frontalis* (Fig. 9.3).

Giardia intestinalis contains PFOR and has been shown to produce hydrogen (Lloyd et al. 2002) but has not been found to contain hydrogenosomes. Its iron-sulfur composition appears to be quite distinct from those of the trichomonads. EPR showed both soluble and membrane-bound iron-sulfur proteins of the [4Fe-4S] type (Ellis et al. 1993).

Further identification of these clusters required separation of the proteins. Hydrogenase was extracted from *T. vaginalis* by Payne et al. (1993). The EPR signal of this protein showed the unique EPR signature of the actives-site cluster of [Fe-] hydrogenases, together with center G. The assignment as an [Fe] hydrogenase was confirmed by the sensitivity of the enzyme activity to inhibition by carbon monoxide.

Trichomonads are sensitive to the nitroimidazole drug metronidazole (Flagyl), one of the few drugs that is effective against anaerobic pathogens. It forms a stable free radical on reduction, which can be observed by EPR at room temperature and is associated with antimicrobial action (Moreno et al. 1984; Chapman et al. 1985). Only organisms that use PFOR can produce sufficiently low redox potentials to produce this radical, hence the drug is selectively toxic to fermentative anaerobes (Land and Johnson, 1997). Some clinical isolates of *T. vaginalis* are now resistant to metronidazole. Kulda et al. (1993) showed that *T. vaginalis* cells resistant to metronidazole can be selected by culture in the presence of increasing concentrations of the drug. The first stage is aerobic resistance, in which iron-sulfur proteins are still present (Fig. 9.4). In the final stage (anaerobic

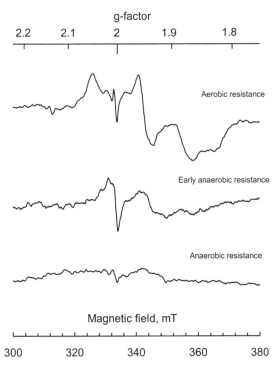

FIGURE 9.4. EPR spectrum of hydrogenosomal membranes from *T. vaginalis* cells with induced resistance to metronidazole, recorded at 15 K and microwave frequency 9.35 GHz.

resistance), the cells show decreasing levels of EPR-detectable iron-sulfur proteins, consistent with the loss of their PFOR/hydrogenase pathway. The resistant cells show decreasing levels of radical formation from metronidazole (Rasoloson et al. 2002). The cells can use other fermentative pathways for growth but are increasingly sensitive to oxygen damage (Rasoloson et al. 2001).

Evolution of Anaerobic Protists

Shortly after the discovery of hydrogenosomes in *Trichomonas* (Lindmark and Müller 1973), it was suggested that they were the remains of an endosymbiotic anaerobic Gram-positive *Clostridium*-like bacterium (Whatley et al. 1979). The hypothesis was founded on the biochemical similarities in pathways of hydrogen production between hydrogenosomes and clostridia. This clostridium hypothesis received some indirect support from trees based on ribosomal RNA sequences. These trees suggested that *Trichomonas*, and other parasitic protozoa, including *Giardia* and *Microsporidia* (which had not been shown to contain any organelle at all), branched from other eukaryotes before the symbiosis that gave rise to mitochondria and aerobic eukaryotic metabolism (Embley and Hirt 1998). However, more recent data suggest that *Trichomonas*, *Giardia*, and Microsporidia all contain mitochondria-derived genes (Horner et al. 1996; Hirt et al. 1997; Embley and Hirt 1998) and that mitochondrial chaperonins are located within *Trichomonas* hydrogenosomes (Bui et al. 1996). These new data have been taken to support an alternative hypothesis, first proposed by Cavalier-Smith (1987), that *Trichomonas* hydrogenosomes are biochemically modified mitochondria. This would also explain why proteins are imported into *Trichomonas* hydrogenosomes using mitochondrial-like N-terminal targeting sequences (Bradley et al. 1997).

Most available data support a mitochondrial origin for ciliate hydrogenosomes. In ciliates, hydrogenosomes occur in multiple lineages (Embley et al. 1995), which are sometimes closely related to ciliates with mitochondria. In these cases, mitochondria and hydrogenosomes closely resemble each other in their ultrastructure and intracellular locations (Finlay and Fenchel 1989). Furthermore, the ciliate *Nyctotherus ovalis* was discovered to contain an associated genome that is unique among hydrogenosomes studied (Akhmanova et al. 1998). Data are preliminary at present, but in *Nyctotherus* there is evidence of mitochondria-derived protein translation machinery inside the hydrogenosome.

The situation for chytrid fungi such as *Neocallimastix* is more controversial. Originally, fungal hydrogenosomes were reported to be surrounded by a single membrane (Marvin-Sikkema et al. 1993), which prompted the theory that fungal hydrogenosomes were derived from peroxisomes rather than mitochondria (which have a double membrane). Further high-resolution electron microscopic studies revealed a double membrane for fungal hydrogenosomes as well (Benchimol et al. 1997; van der Giezen et al. 1997a). Physiologically and morphologically, fungal hydrogenosomes resemble mitochondria rather than peroxisomes. Properties shared with mitochondria, but not peroxisomes, include a transmembrane pH gradient (Biagini et al. 1997; Marvin-Sikkema et al. 1994), an alkaline rather than an acid lumen, and a transmembrane electrochemical potential. In addition, free Ca^{2+} and calcium phosphate precipitates have been detected in fungal hydrogenosomes, suggesting that fungal hydrogenosomes, like mitochondria, accumulate this important intracellular messenger (Biagini et al. 1997). They share a similar type of ATP/ADP carrier (van der Giezen et al. 2002). Also, proteins depend on a mitochondrial-like N-terminal extension for proper import into hydrogenosomes (Brondijk et al. 1997; van Der Giezen et al. 1997a, 1997b, 1998; Durand et al. personal communication 1996).

Anaerobic eukaryotes that contain hydrogenosomes are not all phylogenetically closely related to each other. In phylogenetic trees, they are surrounded by eukaryotes that contain mitochondria and, therefore, carry out aerobic energy generation (Embley et al. 1995; Embley and Hirt 1998). Thus hydrogenosomes have occurred repeatedly throughout eukaryote evolution, and, if we are to better understand anaerobic eukaryote energy metabolism, it is important to know if all hydrogenosomes have originated in the same way. Moreover, anaerobic energy metabolism of the kind seen in contemporary hydrogenosomes has figured strongly in recent hypotheses concerning the origin of eukaryotic cells. For example, both the hydrogen (Martin and Müller 1998) and the syntrophic hypotheses (Moreira and Lóez-García 1998) posit that host dependence on a hydrogen-producing symbiont is the selective principle that forged the common ancestor of eukaryotic cells.

Unrooted phylogenetic trees of PFOR sequences show that they fall into two classes: the four-subunit types (from archaea and some eubacteria) and the homodimeric types (found in eubacteria) (Fig. 9.5A). In the case of PFOR of eukaryotes, the most extensive phylogenetic study suggests monophyly of the genes of Diplomonads, Trichomonads, and Entamoebae. The apicomplexan parasite *Cryptosporidium parvum* has a eubacterial/ eukaryote type PFOR fused at the C terminus to a cytochrome P450 reductase/sulfite reductase-type flavoprotein component. The PFOR-like portion appears to be monophyletic with other eukaryotic PFOR. Biochemical data suggest that this fusion facilitates the use of $NADP^+$ as an acceptor, rather than ferredoxin. It is interesting that several domains (notably a well-conserved C-terminal region) homologous to PFOR are present in the yeasts *Saccharomyces cerevisiae* and *Schizosaccharomyces pombe*, where they have been incorporated into an atypical sulfite-reductase complex.

The history of eukaryote hydrogenases and hydrogenase-derived genes is perhaps even more fascinating (Fig. 9.5B). Hydrogen production has been described in only a few eukaryote lineages: parabasalids, chytrid fungi,

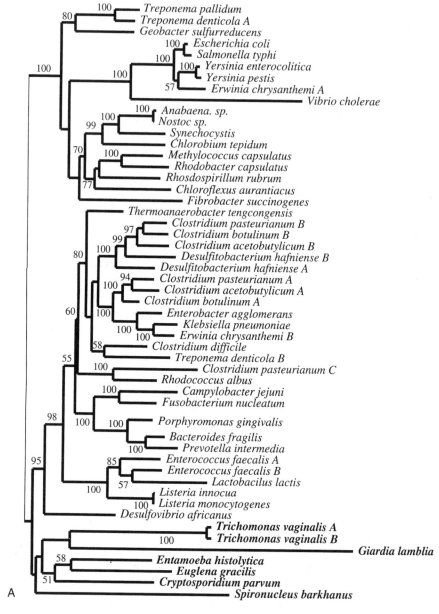

FIGURE 9.5. Phylogenetic trees of PFOR **A** and [Fe] hydrogenases **B**.

some hydrogenosomal ciliates, the Percolozoan *Psalteromonas lanterna* (in which the activity is hydrogenosomal) and plastids of some green algae such as *Chlamydomonas* (in which the activity is photosynthetically linked). Of these organisms, putative hydrogenase genes have been characterised from

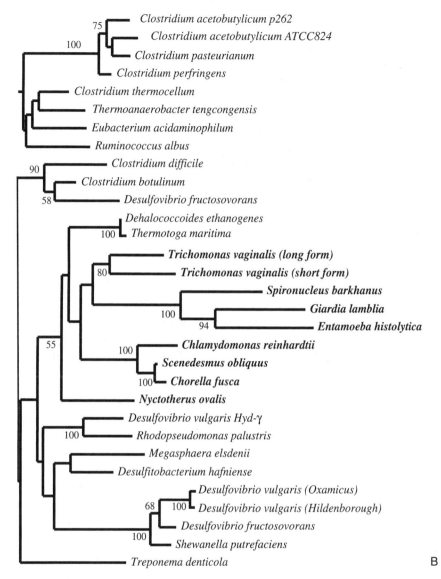

FIGURE 9.5. *Continued.*

Trichomonas spp., the ciliate *Nyctotherus ovalis*, and several algae. A gene corresponding to an [Fe] hydrogenase was observed in the anaerobic chytridiomycete fungus *Neocallimastix* sp. L2 (Voncken et al. 2002). The genome of *Spironucleus barkhanus*, a flagellate closely related to *Giardia*, also contains a hydrogenase-like gene.

The gene sequences of the hydrogenases appear to have developed into other proteins. Some of the hydrogenase proteins contain domains that are homologous to subunits of mitochondrial complex I, NADH:ubiquinone reductase, which suggests that they had a common ancestral electron-transfer protein. In addition, the increase in genome sequencing of eukaryotes revealed the presence of genes related to hydrogenases, currently known as the NARF family. These genes have conserved all the features thought to be required for the coordination of active-site clusters and are present on all aerobic eukaryote genomes for which significant amounts of genome data have been generated (*Saccharomyces*, *Schizosaccharomyces*, human, *Drosophila*, fugu, *Arabidopsis*, *Glycine*, etc.). In the case of *Saccharomyces*, gene-disruption experiments have shown the gene knockout to be lethal. Clearly these organisms do not produce hydrogen metabolically, so the function of these gene products is intriguing.

Conclusion

The phylogeny and unexpected distribution of hydrogenase-related genes among extant organisms suggest an ancient and crucial role for members of this gene family in eukaryotes and are consistent with hydrogen syntrophy based models of eukaryogenesis. For the protists studied, the available data are consistent with a mitochondrial origin for the hydrogenosomes. But one needs to realize that the origin of the enzymes need not necessarily be the same as the origin of the organellar structure. The hydrogenosomes may have originated several times in the course of evolution, with lateral transfer of genes from other eukaryotes or anaerobic bacteria. At present, the lack of a well-supported sister group to the eukaryotes prevents discrimination between the ideas presented in the hydrogen and syntrophic hypotheses. Furthermore, the most parsimonious explanation of these observations is that many eukaryotes, such as the green algae, retained ancestral energy metabolic genes that were available for recruitment during adaptation to anaerobic and, in at least one case, photosynthetic lifestyles, a remarkable demonstration of the plasticity of eukaryote molecular physiology.

Acknowledgments. We thank Ruth Williams, Alan Chapman, and Eva Tomkova for the EPR measurements. This work was supported by grants to R.C. by the BBSRC (SBD07568); to J.K. by the Grant Agency of the Czech Republic (GACR 204/97/0263) and the Ministry of Education of the Czech Republic (VS 96142).

References

Akhmanova A, Voncken F, van Alen T, et al. 1998. A hydrogenosome with a genome. Nature 396:527–8.

Beinert H. 2002. Spectroscopy of succinate dehydrogenases, a historical perspective. Biochim Biophys Acta Bioeng 1553:7–22.

Benchimol M, Durand R, Almeida JCA. 1997. A double membrane surrounds the hydrogenosomes of the anaerobic fungus *Neocallimastix frontalis*. FEMS Microbiol Lett 154:277–82.

Biagini GA, Finlay BJ, Lloyd D. 1997. Evolution of the hydrogenosome. FEMS Microbiol Lett 155:133–40.

Bradley PJ, Lahti CJ, Plumper E, Johnsons PJ. 1997. Targeting and translocation of proteins into the hydrogenosome of the protist *Trichomonas*: similarities with mitochondrial protein import. EMBO J 16:3484–93.

Brondijk THC, Durand R, van der Giezen M, et al. 1997. scsb, a cDNA encoding the hydrogenosomal beta subunit of succinyl-CoA synthetase from the anaerobic fungus *Neocallimastix frontalis*. Mol Gen Genet 253:315–23.

Cammack R, Frey M, Robson R. 2001. Hydrogen as a fuel: learning from nature. London: Taylor & Francis.

Cammack R, Patil DS, Weiner JH. 1986. Evidence that centre 2 in *Escherichia coli* fumarate reductase is a [4Fe-4S] cluster. Biochim Biophys Acta 870:545–51.

Cavalier-Smith T. 1987. The simultaneous symbiotic origin of mitochondria, chloroplasts, and microbodies. Ann NY Acad Sci 503:55–71.

Chapman A, Cammack R, Linstead D, Lloyd D. 1985. The generation of metronidazole radicals in hydrogenosomes isolated from *Trichomonas vaginalis*. J Gen Microbiol 131:2141–4.

Chapman A, Cammack R, Linstead DJ, Lloyd D. 1986. Respiration of *Trichomonas vaginalis*: components detected by electron paramagnetic resonance spectroscopy. Eur J Biochem 156:193–8.

Ellis JE, William R, Cole D, et al. 1993. Electron-transport components of the parasitic protozoan *Giardia lamblia*. Febs Lett 325:196–200.

Embley TM, Hirt RP. 1998. Early branching eukaryotes? Curr Opin Genet Develop 8:624–9.

Embley TM, Finlay BJ, Dyal PL, et al. 1995. Multiple origins of anaerobic clliates with hydrogenosomes within the radiation of aerobic clliates. Proc R Soc Lond Ser B Biol Sci 262:87–93.

Embley TM, Horner DA, Hirt RP. 1997. Anaerobic eukaryote evolution: hydrogenosomes as biochemically modified mitochondria? Trends Ecol Evol 12:437–41.

Finlay BJ, Fenchel T. 1989. Hydrogenosomes in some anaerobic protozoa resemble mitochondria. FEMS Microbiol Lett 65:311–14.

Hall DO, Cammack R, Rao KK. 1971. A role for ferredoxins in the origin of life and biological evolution. Nature 233:136–8.

Hirt RP, Healy B, Vossbrinck CR, et al. 1997. A mitochondrial Hsp70 orthologue in *Vairimorpha necatrix*: molecular evidence that microsporidia once contained mitochondria. Curr Biol 7:995–8.

Horner DS, Hirt RP, Kilvington S, et al. 1996. Molecular data suggest an early acquisition of the mitochondrion endosymbiont. Proc R Soc Lond Ser B Biol Sci 263:1053–9.

Johnson MK, Morningstar JE, Cecchini G, Ackrell BAC. 1985. Detection of a tetranuclear iron sulphur centre in fumarate reductase from *Escherichia coli* by EPR. Arch Microbiol 131:756-62.

Kulda J, Poislova M, Suchan P, Tachezy J. 1999. Iron enhancement of experimental infection of mice by *Tritrichomonas foetus*. Parasitol Res 85:692–9.

Kulda J, Tachezy J, Cerkasovova A. 1993. In vitro induced anaerobic resistance to meteronidazole in *Trichomonas vaginalis*. J Eukary Microbiol 40:262–9.

Land KM, Johnson PJ. 1997. Molecular mechanisms underlying metronidazole resistance in trichomonads. Exper Parasitol 87:305–8.

Lindmark DG, Müller M. 1973. Hydrogenosome, a cytoplasmic organelle of the anaerobic flagellate *Tritrichomonas foetus*, and its role in pyruvate metabolism. J Biol Chem 248:7724–8.

Lloyd D, Ralphs JR, Harris JC. 2002. *Giaridia intestinalis*, a eukaryote without hydrogenosomes, produces hydrogen. Microbiol SGM 148:727–33.

Marczak R, Gorrell TE, Müller M. 1983. Hydrogenosomal ferredoxin of the anaerobic protozoon, *Tritrichomonas foetus*. J Biol Chem 258:12427–33.

Martin W, Müller M. 1998. The hydrogen hypothesis for the first eukaryote. Nature 392:37–41.

Marvin-Sikkema FD, Driessen AJM, Gottschal JC, Prins RA. 1994. Metabolic energy generation in hydrogenosomes of the anaerrobic fungus *Neocallimastix*: evidence for a functional relationship with mitochondria. Mycol Res 98:205–12.

Marvin-Sikkema FD, Gomes TMP, Grivet JP, et al. 1993. Characterization of hydrogenosomes and their role in glucose metabolism of *Neocallimastix* sp 12. Arch Microbiol 160:388–96.

Moreira D, López-García P. 1998. Symbiosis between methanogenic archaea and delta-proteobacteria as the origin of eukaryotes: The syntrophic hypothesis. J Mol Evol 47:517–30.

Moreno SNJ, Mason RP, Docampo R. 1984. Distinct reduction of nittrofurans and metronidazole to free radical metabolites by *Tritrichomonas foetus* hydrogenosomal and cytosolic enzymes. J Biol Chem 259:8252–9.

Müller M. 1993. The gydrogenosome. J Gen Microbiol 139:2879–89.

Nicolet Y, Piras C, Legrand P, et al. 1999. Desulfovibrio desulfuricans iron hydrogenase: the structure shows unusual coordination to an active-site Fe binuclear center. Structure 7:13–23.

Ohnishi T, Lloyd D, Lindmark DC, Müller M. 1980. Respiration of *Tritrichomonas foetus*: components detected by hydrogenosomes and in intact cells by electron paramagnetic resonance spectrometry. Mol Biochem Parasitol 2:39–50.

Ohnishi T, Moser CC, Page CC, et al. 2000. Simple redox-linked proton-transfer design: new insights from structures of quinol-fumarate reductase. Structure Folding Design 8:R23–R32.

Payne MJ, Chapman A, Cammack R. 1993. Evidence for an [Fe]-type hydrogenase in the parasitic protozoan *Trichomonas vaginalis*. Febs Lett 317:101–4.

Peter JW, Lanzilotta WN, Lemon BJ, Seefeldt LC. 1998. X-ray crystal structure of the Fe-only hydrogenase (Cpl) from *Clostridium pasteurianum* to 1.8 Angstrom resolution. Science 282:1853–8.

Rasoloson D, Tomkova E, Cammack R, et al. 2001. Metronidazole-resistant strains of *Trichomonas vaginalis* display increased susceptibility to oxygen. Parasitology 123:45–56.

Rasoloson D, Vanacova S, Tomkova E, et al. 2002. Mechanisms of in vitro development of resistance to metronidazole in *Trichomonas vaginalis*. Microbiol Mol Biol Rev 148:2467–77.

Tachezy J, Sanchez LB, Müller M. 2001. Mitochondrial type iron-sulfur cluster assembly in the amitochondriate eukaryotes *Trichomonas vaginalis* and *Giardia intestinalis*, as indicated by the phylogeny of IscS. Mol Biol Evo 18:1919–28.

Townson SM, Hanson GR, Upcroft JA, Upcroft P. 1994. A purified ferredoxin from *Giardia duodenalis*. Eur J Biochem 220:439–46.

Uyeda K, Rabinowitz JC. 1971. Pyruvate-ferredoxin oxidoreductase III. Purification and properties of the enzyme. J Biol Chem 246:311–19.

van der Giezen M, Kiel JAKW, Sjollema KA, Prins RA. 1998. The hydrogenosomal malic enzyme from the anaerobic fungus *Neocallimastix frontalis* is targeted to mitochondria of the methylotrophic yeast *Hansenula polymorpha*. Curr Genet 33:131–5.

van der Giezen M, Rechinger KB, Svendsen I, et al. 1997b. a mitochondrial-like targeting signal on the hydrogenosomal malic enzyme from the anaerobic fungus *Neocallimastix frontalis*: Support for the hypothesis that hydrogenosomes are modified mitochondria. Mol Microbiol 23:11–21.

van der Giezen M, Sjollema KA, Art RRE, et al. 1997a. Hydrogenosomes in the anaerobic fungus *Neocallimastix frontalis* have a double membrane but lack an associated organelle genome. Febs Lett 408:147–50.

van der Giezen M, Slotboom DJ, Horner DS, et al. 2002. Conserved properties of hydrogenosomal and mitochondrial ADP/ATP carriers: a common origin for both organelles. EMBO J 21:572–9.

Vanacova S, Rasoloson D, Razga J, et al. 2001. Iron-induced changes in pyruvate metabolism of *Tritrichomonas foetus* and involvement of iron in expression of hydrogenosomal proteins. Microbiology 147:53–62.

Vignais PM, Billoud B, Meyer J. 2001. Classification and phylogeny of hydrogenases. FEMS Microbiol Rev 25:455–501.

Voncken FGJ, Boxma B, van Hoek A, et al. 2002. A hydrogenosomal Fe-hydrogenase from the anaerobic chytrid *Neocallimastix* sp L2. Gene 284:103–12.

Whatley JM, John P, Whatley FR. 1979. From extracellular to intracellular: the establishment of mitochondria and chloroplasts. Proc R Soc Lond Ser B Biol Sci 204:165–87.

Suggested Reading

Bui ETN, Bradley PJ, Johnson PJ. 1996. A common evolutionary origin for mitochondria and hydrogenosomes. Proc Nat Acad Sci USA 93:9651–6.

10
Oxygen and Anaerobes

Donald M. Kurtz, Jr.

Oxidative Stress in Bacteria

Aerobic microorganisms use the high-reduction potential of molecular oxygen to generate energy during respiration. However, this high-reduction potential also leads to adventitious production of the reactive and toxic-reduced dioxygen species, superoxide and hydrogen peroxide (Storz and Imlay 1999; Touati 2000). Aerobic microorganisms have developed robust defense mechanisms to combat this oxidative stress. Two principal enzymes are used by aerobes to deal with the immediate products of dioxygen reduction during periods of oxidative stress. Superoxide dismutases (SODs) catalyze the dismutation of superoxide to hydrogen peroxide and dioxygen (Eq. 10.1).

$$2O_2^- + 2H^+ \rightarrow O_2 + H_2O_2 \qquad \text{(SOD)} \qquad (10.1)$$

Bacterial SODs typically contain either nonheme iron (FeSODs) or manganese (MnSODs) at their active sites, although bacterial copper/zinc and nickel SODs are also known (Imlay and Imlay 1996; Chung et al. 1999). Catalases are usually heme-containing enzymes that catalyze disproportionation of hydrogen peroxide to water and molecular oxygen (Eq. 10.2) (Zámocky and Koller 1999; Loewen et al. 2000).

$$2H_2O_2 \rightarrow O_2 + 2H_2O \qquad \text{(catalase)} \qquad (10.2)$$

Both SODs and catalases are ubiquitous among aerobic microorganisms.

Obligately anaerobic bacteria cannot grow via dioxygenic respiration and usually require complete anaerobicity for growth in the laboratory. Thus oxidative stress defenses were historically thought to be minimal in anaerobes. This view applied to the sulfate-reducing bacteria, which were among the first obligate anaerobes to undergo detailed physiologic and biologic characterization. These bacteria characteristically derive energy for growth using sulfate rather than dioxygen as terminal electron acceptor. Only more recently was it realized that many species of sulfate-reducing bacteria are aerotolerant. Some species have been shown to swim toward and grow near oxic/anoxic interfaces (Marschall et al. 1993; Fu and Voordouw 1997; Johnson et al. 1997). Evidence for dioxygenic respiration of sulfate reducers was obtained, albeit not for growth (Cypionka 2000). In hindsight, this aerotolerance is fully consistent with the natural growth habitats of sulfate reducers. Thus sulfate-reducing bacteria must also deal with oxidative stress resulting from adventitious dioxygen reduction.

Oxidative Stress Protection in Sulfate-Reducing Bacteria

Some sulfate-reducing bacteria have been shown to contain SODs and catalases, but other species do not demonstrate the expected activities for these enzymes (Hatchikian et al. 1977; van Niel and Gottschal 1998; Dos Santos et al. 2000). The apparent absence of these enzymes can be rationalized on the basis that their catalytic dismutation reactions (Eqs. 10.1 and 10.2) generate dioxygen, which may be disadvantageous for strict anaerobes but raises the question of how these organisms protect themselves against transient air exposure.

Evidence for an alternative oxidative stress protection mechanism in sulfate-reducing bacteria has begun to emerge. Table 10.1 provides data on the proteins implicated in this alternative system. All but one of these proteins contain distinctive types of nonheme iron active sites. This chapter describes recent results on three of these novel proteins: DcrH, Rbo, and Rbr, all from *Desulfovibrio vulgaris* Hildenborough.

TABLE 10.1. Proteins implicated in an alternative oxidative stress protection system in sulfate-reducing bacteria.

Name (acronym)	Metal site(s)	Probable function	References
Desulfovibrio chemoreceptors A,H (DcrA,DcrH)	c-type heme (DcrA); $[(Fe_2\{\mu\text{-}O\})(His)_5(Glu)\text{-}(Asp)]$ (DcrH)	dioxygen or redox sensors	Fu and Voordouw (1997)
Rubredoxin oxidoreductase (Rbo)	$[Fe(Cys)_4]$ (I); $[Fe(His)_4(Cys)]$ (II)	SOR[a]	Lumppio et al. (2001)
Neelaredoxin (Nlr)	$[Fe(His)_4(Cys)]$	SOR[a]	Silva et al. (1999)
Rubrerythrin (Rbr)	$[Fe(Cys)_4]$; $[(Fe_2\{\mu\text{-}O\})(His)(Glu)_5]$	Peroxidase	Lumppio et al. (2001)
Rubredoxin:oxygen oxidoreductase (ROO)	$[(Fe_2\{\mu\text{-}O\})(His)_5(Glu)\text{-}(Asp)_2]$	dioxygen reduction	Frazão et al. (2000)

[a] Superoxide reductase.

Desulfovibrio Chemoreceptor H (DcrH)

Voordouw and co-workers identified at least 12 *Desulfovibrio* chemore-ceptor (*dcr*) genes in the genome of *Desulfovibrio vulgaris* Hildenborough (Deckers and Voordouw 1994). The chemoreceptor attribution was based primarily on the presence of excitation and methylation amino acid sequence motifs that are homologous to proteins of known methyl-accepting chemotaxis proteins in enteric bacteria. Thus the *dcr* gene family was proposed to include a set of innermembrane-spanning proteins used by *D. vulgaris* to sense and respond to specific molecules in or physical states of its environment. These chemotaxis proteins typically contain a periplasmic N-terminal sensing domain, a C-terminal cytoplasmic transmitting domain, and a transmembrane domain linking the other two. Transduction from the sensing to transmitting domains triggers a phospho-rylation/methylation cascade, which results ultimately in a change in fre-quency of flagellar motor reversal, causing the bacteria to swim either up or down a concentration gradient of the environmental stimulus (Jurica and Stoddard 1998; Mowbray and Sandgren 1998).

One of the hypothetical *dcr* gene products encodes the 959-residue DcrH. The proposed membrane topology of DcrH is shown in Figure 10.1 (Deckers and Voordouw 1996). The presumed periplasmic-sensing domain shows no amino acid sequence homologies to proteins of known function. However, an amino acid sequence motif at the C terminus of DcrH was rec-ognized that is homologous to the consensus sequence of the dioxygen-

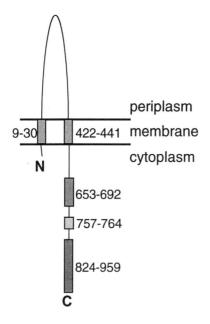

FIGURE 10.1. The membrane topology proposed for *D. vulgaris* Hildenborough DcrH based on hydrophathy analysis and sequence homologies to other bacterial chemoreceptors (Deckers and Voordouw 1996). *Shaded boxes*, putative membrane-spanning (residues 9–30 and 422–431), exci-tation (residues 653–692), and methylation (residues 757–764) regions. The C-terminal box (residues 824–959) indicates the Hr-like region. Reprinted with permission from Xiong et al. (2000), copyright 2000 American Chemical Society.

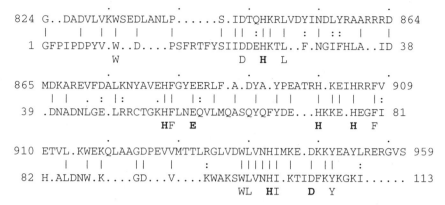

```
824 G..DADVLVKWSEDLANLP......S.IDTQHKRLVDYINDLYRAARRRD 864
     |   | | | |     |     | || :|| |  : | :: | |
  1 GFPIPDPYV.W..D....PSFRTFYSIIDDEHKTL..F.NGIFHLA..ID 38
              W                 D  H   L

865 MDKAREVFDALKNYAVEHFGYEERLF.A.DYA.YPEATRH.KEIHRRFV 909
     | |   . : |:    .|| |: | | | | | | | || | |:
 39 .DNADNLGE.LRRCTGKHFLNEQVLMQASQYQFYDE...HKKE.HEGFI 81
             HF  E                         H    H  F

910 ETVL.KWEKQLAAGDPEVVMTTLRGLVDWLVNHIMKE.DKKYEAYLRERGVS 959
     | | |   || |    :   ||||||| | || :
 82 H.ALDNW.K....GD...V....KWAKSWLVNHI.KTIDFKYKGKI...... 113
                              WL  HI   D  Y
```

FIGURE 10.2. Amino acid sequence alignment of *D. vulgaris* Hildenborough DcrH residues 824–959 (*top*) with *Phascolopsis gouldii* Hr (*bottom*). Letters indicate residues that are common to both sequences, which are also conserved in all known Hr sequences; boldface indicates residues furnishing iron ligands in Hr. Reprinted with permission from Xiong et al. (2000), copyright 2000 American Chemical Society.

carrying protein hemerythrin (Hr) (Fig. 10.2) (Stenkamp 1994). In Hr, this sequence motif is contained in a four-helix bundle tertiary structure that surrounds an oxo-bridged diiron active site (Fig. 10.3). Hr was previously reported only in a few marine and terrestrial invertebrates. Xiong et al. (2000) showed that the C-terminal domain of DcrH, when expressed separately in recombinant form in *Escherichia coli*, folds into a stable protein, DcrH-Hr. The UV-vis absorption and resonance Raman spectra of DcrH-Hr and of its azide adduct provide clear evidence for an oxo-bridged diiron(III) site that is similar to that found in Hr. Based on UV-vis absorption spectra, exposure of the reduced DcrH-Hr to air resulted in formation of an dioxygen adduct that is also similar to that of Hr. Finally, Xiong et al. (2000) demonstrated by immunoblotting, using antibodies raised against the recombinant DcrH-Hr, that a full-length DcrH is expressed in *D. vulgaris*. Given the air-sensitive nature of *D. vulgaris*, an obvious physiologic function to assign DcrH is dioxygen sensing.

Rubredoxin Oxidoreductase (Rbo)

Rbo, which has also been named desulfoferrodoxin (Moura et al. 1990), is the product of the *rbo* gene in several sulfate-reducing bacteria (Brumlik and Voordouw 1989). Rbo from the sulfate-reducing bacterium *Desulfoarculus baarsii* was shown to complement an SOD deficiency in *E. coli* (Pianzzola et al. 1996), even though this Rbo showed almost no detectable SOD activity. In fact, Rbo has no detectable amino acid sequence homologies to known

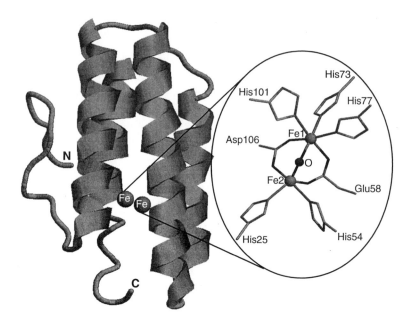

FIGURE 10.3. The Hr subunit and its diiron site. Dioxygen binds reversibly to the iron labeled Fe2 (Stenkamp 1994). Reprinted with permission from Xiong et al. (2000), copyright 2000 American Chemical Society.

SODs. Nevertheless, the *D. baarsii* Rbo was reported to reduce the steady-state level of intracellular superoxide in the SOD-deficient *E. coli* strain (Liochev and Fridovich 1997). This Rbo in vitro showed evidence for superoxide reductase (SOR) activity (Lombard et al. 2000), i.e., reduction of O_2^-, presumably to H_2O_2, without dismutation (Eq. 10.3).

$$e^- + O_2^- + 2H^+ \rightarrow H_2O_2 \quad (SOR) \qquad (10.3)$$

Rbo is a homodimeric protein, each subunit of which contains two distinct mononuclear nonheme iron centers in separate domains (Fig. 10.4) (Coehlo et al. 1997). Center I contains a distorted rubredoxin-type [Fe(SCys)$_4$] coordination sphere. [Fe(SCys)$_4$] sites in proteins are known to catalyze exclusively electron transfer, which is, therefore, the putative function for center I. Center II contains a unique [Fe(NHis)$_4$(SCys)] site that is rapidly oxidized by O_2^-, and is, therefore, the likely site of superoxide reduction (Lombard et al. 2000). A blue nonheme iron protein, neelaredoxin (Nlr) from *Desulfovibrio gigas* (Silva et al. 1999), contains an iron center closely resembling that of Rbo center II (Table 10.1). The blue color is due to the oxidized (i.e., Fe(III)) form [Fe(NHis)$_4$(SCys)] site, which, in both Nlr and Rbo, has a prominent absorption feature at ~650 nm. Reduction of center II to its Fe(II) form fully bleaches its visible absorption. These absorption features have been used to probe the reactivity of Rbo with superoxide.

Rbo is typically isolated with center I oxidized (i.e., Fe(III)) and center II reduced (i.e., Fe(II)). This semireduced form is called pink Rbo (Rbo_{pink}) to distinguish it from the gray color of fully oxidized Rbo, in which both centers I and II are Fe(III) (Rbo_{gray}). Using pulse radiolysis coupled with a rapid response spectrophotometer, Coulter et al. (2000a) confirmed that *D. vulgaris* Rbo has no SOD activity, at least none that can outcompete spontaneous disproportionation of superoxide. Nevertheless, these authors found that center II of Rbo_{pink} reduces superoxide at a nearly diffusion-controlled rate ($1.5 \times 10^9 M^{-1} s^{-1}$) at pH 7.8, forming a transient intermediate with an absorption feature at 600nm. This intermediate decays at ~$40 s^{-1}$ to a form that is indistinguishable from Rbo_{gray}. This latter step must involve processes internal to Rbo, since its rate was found to be independent of superoxide concentration. These authors also determined a one-to-one stoichiometry for superoxide generated by pulse radiolysis to Rbo_{gray} produced, which is consistent with the one-electron SOR reaction (Eq. 10.3). The proposed SOR reaction cycle for center II based on these kinetics is shown in Scheme 10.1. The 600nm intermediate is formulated as a ferric-peroxo species that decays to the resting Rbo_{gray} form with loss of H_2O_2. Center II could then be re-reduced by center I, although the rates and electron donors for this portion of the SOR catalytic cycle have not yet been characterized.

The $40 s^{-1}$ conversion of the 600nm intermediate to Rbo_{gray} (Scheme 10.1) might seem too low to compete with spontaneous disproportionation of superoxide, which occurs with a second-order rate constant of ~$5 \times 10^5 M^{-1} s^{-1}$ at pH 7 (Bielski and Cabelli 1991). However, in aerobic bacterial cells such as *E. coli*, the steady-state concentration of superoxide is esti-

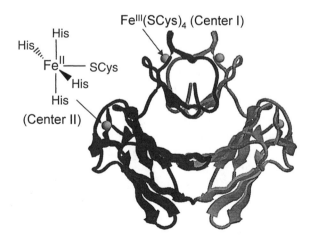

FIGURE 10.4. The *D. vulgaris* Rbo dimer (Coehlo et al. 1997). *Spheres*, iron atoms in centers I and II. The model was generated via *RASMOL* (Sayle and Milner-White 1995) and coordinates from 1DFX in the Protein Databank.

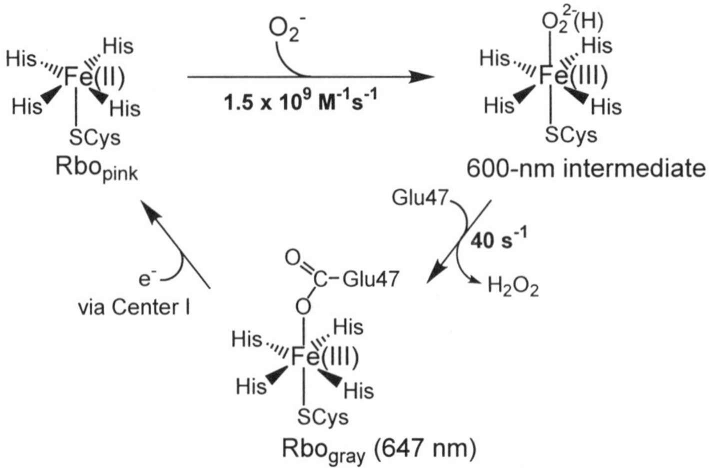

SCHEME 10.1.

mated to be $\sim 10^{-10}$ M (Gort and Imlay 1998). Even if we assume that *D. vulgaris* could tolerate an order of magnitude higher (e.g., $\sim 10^{-9}$ M) steady-state intracellular superoxide, the first-order rate constant for spontaneous superoxide disproportionation would be $\sim 5 \times 10^{-4}$ s^{-1}, or several orders of magnitude lower than 40 s^{-1}. Furthermore, since 10^{-9} M is equivalent to ~ 1 molecule per 10^{-15} L bacterial cell (Gort and Imlay 1998), the intracellular Rbo concentration is expected to be at least an order of magnitude higher than the steady-state superoxide concentration. Under these conditions, the first-order rate constant for reaction of Rbo$_{pink}$ with superoxide, even at the diffusion-controlled rate of 1.5×10^9 M^{-1}s^{-1}, would then be only ~ 2 s^{-1}, or an order of magnitude slower than the subsequent first-order conversion of the 600 nm intermediate to Rbo$_{gray}$ (Scheme 10.1). Thus, based on these estimates, this internal conversion of the 600 nm intermediate is unlikely to limit SOR turnover of Rbo in vivo.

While Rbo had previously been shown to protect against superoxide stress when expressed heterologously in *E. coli*, this role for Rbo had not been established in any native organism. The availability of a Δrbo strain of *D. vulgaris* allowed Lumpio et al. (2001) to investigate whether Rbo could protect against superoxide stress in the native organism. Figure 10.5 shows that, under the study's conditions of air exposure, the Δrbo strain was significantly more sensitive to paraquat-induced superoxide stress than to air alone relative to wild type (Lumppio et al. 2001). The survival of wild-type cells exposed to [air + 10 µM paraquat] for 4 h decreased by a factor of ~ 2 relative to the same cells exposed to air alone. However, the survival of the Δrbo [air + paraquat]-exposed cells after 4 h decreased by a factor of ~ 500 relative to air-only exposed cells. Paraquat had no effect on the anaerobic growths of either strain, nor did the Δ*rbo* strain show any hypersensitivity to hydrogen peroxide exposure. Thus, while the *D. vulgaris* Δ*rbo* strain had previously been found to be more air-sensitive than wild type (Voordouw and Voordouw 1998), this air sensitivity is greatly magnified by intracellular production of superoxide. The SOD activities of cell extracts of the Δ*rbo* strain were indistinguishable from those of wild type. These results were the first to demonstrate a specific role for Rbo in counteracting superoxide stress in a native organism and, together with the results presented above, suggest that Rbo does so via its SOR activity.

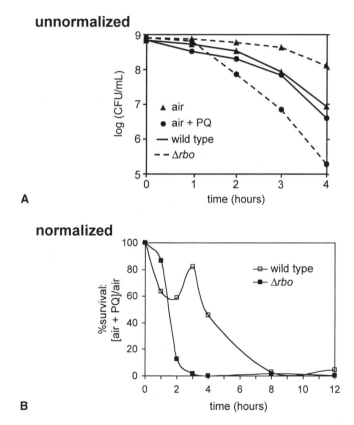

FIGURE 10.5. Viabilities of *D. vulgaris* wild-type and Δ*rbo* strains following exposure to either air or air plus 10 μM paraquat. **A,** the surviving colony-forming units (CFUs) vs the times of either air or air plus paraquat exposure. **B,** The same data as percent survival of the air plus paraquat–exposed cells normalized to the survivals of the air-exposed cells. The greater absolute survivals for the air-only exposed Δ*rbo* strain relative to wild type is an artifact due probably to the greater initial cell density used for the Δ*rbo* than for the wild-type strain in these experiments. Reprinted with permission from Lumppio et al. 2001, copyright 2001 American Society for Microbiology.

Rubrerythrin (Rbr)

Another novel nonheme iron protein, Rbr, has also been implicated in oxidative stress defense in anaerobic microbes. Rbr is a homodimeric protein that contains both a rubredoxin-like [Fe(SCys)$_4$] center and a non-sulfur, oxo-bridged diiron site (deMaré et al. 1996). The best characterized Rbr is that from the anaerobic sulfate-reducing bacterium, *D. vulgaris* Hildenborough. Prominent features of the tertiary structure and metal sites of this representative Rbr are shown in Figure 10.6. *Desulfovibrio vulgaris* contains a second Rbr-like protein, called nigerythrin (Ngr) (Pierik et al. 1993), and the Rbr and Ngr genes are separately transcribed (Lumppio et al;. 1997).

A

Fe(SCys)$_4$ diiron-oxo

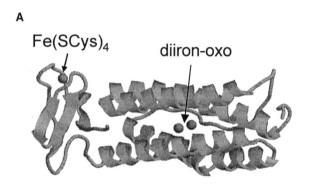

B

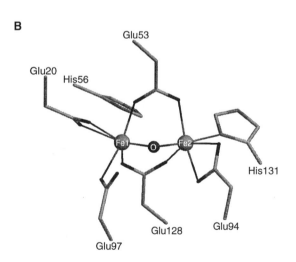

FIGURE 10.6. Structural features of *D. vulgaris* Rbr (deMaré et al. 1996). **A,** Rbr subunit with protein backbone and iron atoms (spheres). **B,** Diiron-oxo site with amino acid side chain ligands. Models generated via *RASMOL* (Sayle and Milner-White 1995) and coordinates from 1RYT in the Protein Databank.

Diiron-oxo sites similar to those in Rbr are known to activate dioxygen in other enzymes. However, the reaction of fully reduced (i.e., all-Fe(II)) Rbr with dioxygen was found to be relatively sluggish compared to those of the dioxygen-activating enzymes (Coulter et al. 2000b). Fully reduced Rbr is oxidized relatively rapidly by H_2O_2, but no catalase activity has been detected for Rbr or Ngr (Pierik et al. 1993). The specific SOD activities for several Rbrs were also found to be extremely low relative to those of known SODs (Coulter et al. 1999). A Possible in vivo connection to the relatively rapid reaction of fully reduced Rbr with H_2O_2 vs. oxygen was the observation that an H_2O_2-resistant mutant of the microaerophile *Spirillum volutans* showed constitutive overexpression of a protein whose molecular weight and partial amino acid sequence were homologous to those of Rbr (Alban et al. 1998). This *S. volutans* mutant also demonstrated high NADH peroxidase activity relative to the wild-type strain but retained the wild-type air sensitivity (Alban and Krieg 1998).

Coulter et al. (1999, 2000b) showed that, in vitro, *D. vulgaris* Rbr has the ability to function as the terminal component of an NADH peroxidase, catalyzing the reduction of hydrogen peroxide to water (Eq. 10.4).

$$NADH + H^+ + H_2O_2 \rightarrow NAD^+ + 2H_2O \qquad (10.4)$$

Ngr also exhibited this NADH peroxidase activity in vitro. For these studies, since the authors have not yet identified a native NADH:Rbr oxidoreductase, they substituted a nonnative iron-sulfur flavorotein reductase, BenC, from *Acinetobacter* sp. ADP1. Thus BenC and Rbr together catalyze Eq. 10.4, with estimated turnover numbers of ~2000 min^{-1} for Rbr homodimer.

Since a Δrbr strain of *D. vulgaris* was not available, Lumpio et al. (2001) tested the ability of Rbr and Ngr to protect appropriate *E. coli* strains against either superoxide–induced or hydrogen peroxide–induced oxidative stress. As shown Figure 10.7A, *D. vulgaris* Rbr was not able to rescue the SOD-deficient *E. coli* strain from superoxide stress, whereas *D. vulgaris* Rbo was able to do so. Conversely, Figure 10.7B shows that *D. vulgaris* Rbr and Ngr were able to increase viability of an *E. coli* calatase-deficient strain exposed to H_2O_2, whereas Rbo actually reduced the viability (Lumppio et al. 2001). It is important to note that the protection afforded the catalase-deficient strain by Ngr and Rbr presumably occurred in competition with at least some of the 30 or more endogenous *E. coli* proteins, whose expressions are induced in response to peroxide stress (Rocha and Smith 1998).

Rbo and Rbr as an Oxidative Stress Defense System in *Desulfovibrio vulgaris*

Although Rbr and Rbo are located on separate operons in *D. vulgaris*, genes encoding Rbo- and Rbr-like proteins are located within what appear to be the same operon or gene cluster in several other anaerobic bacteria

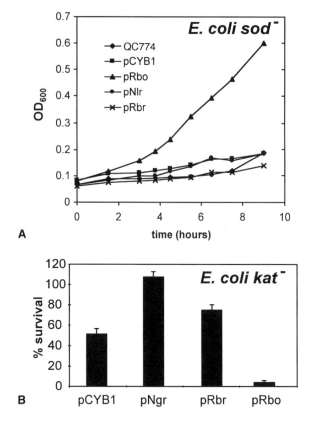

FIGURE 10.7. Tests of the ability of *D. vulgaris* Rbr, Ngr, and Rbo to protect *E. coli* against oxidative stress. **A,** Aerobic growth over time of an aerobically grown SOD-deficient *E. coli* strain (QC774) containing plasmids expressing either nothing (*pCYB1*), Rbo (*pRbo*), Rbr (*pRbr*), or Nlr (*pNlr*). **B,** Percent survival after 30 min exposure to 2.5 mM H_2O_2 of an aerobically grown catalase-deficient *E. coli* strain containing different plasmids. Reprinted with permission from Lumppio et al. (2001), copyright 2001 American Society for Microbiology.

and archaea (Das et al. 2001; Lumppio et al. 2001). This genetic proximity suggests a functional relationship between Rbo and Rbr. Given the results discussed above, one logical possibility is that Rbo and Rbr are complementary components of an alternative oxidative stress defense system (Fig. 10.8). During periods of air exposure, superoxide would be generated by adventitious oxidation of various redox enzymes. Both periplasmic and cytoplasmic compartments are shown in Figure 10.8, because superoxide is not membrane permeable at neutral pH. Lumppio et al. (2001) discovered a gene encoding a homolog to FeSODs in *D. vulgaris*; that encodes an N-terminal signal peptide, implying a periplasmic localization. The putative SOD is, therefore, suggested to remove superoxide generated within the periplasmic space (Fig. 10.8). No Rbo gene or protein, including that from *D. vulgaris*, shows any evidence for a signal peptide and is, therefore, suggested to catalyze reduction of cytoplasmically generated superoxide. The resulting hydrogen peroxide, plus other internally or externally generated hydrogen peroxide (because the uncharged H_2O_2 is membrane permeable), is then suggested to be catalytically reduced to water via the peroxidase activity of Rbr.

The immediate electron donor to both Rbo and Rbr is suggested to be the small electron-transfer protein rubredoxin, on the basis that the rubredoxin gene is co-transcribed with the Rbo gene in *D. vulgaris* and with Rbr genes in other anaerobes (Lumppio et al. 2001). The ultimate intracellular source of electrons in the SOR and peroxidase reactions of Rbo and Rbr, respectively, remain to be determined. *Desulfovibrio vulgaris* also contains a heme catalase, which, since it lacks a signal peptide, is presumed to be cytoplasmic. Because heme catalases typically have Michaelis constants for hydrogen peroxide in the millimolar to molar range (Zámocky and Koller 1999), the *D. vulgaris* catalase might complement the housekeeping function of Rbr by minimizing the effects of exposure to occasionally large bursts of extracellularly generated H_2O_2. The Rbr gene in *D. vulgaris* is co-transcribed with another gene, designated *fur* (Lumppio et al. 1997), which encodes a protein showing highest sequence homology to PerR, a peroxide-responsive regulatory protein in some other bacteria (Bsat et al. 1998; Zou et al. 1999). The product of the *fur*-like gene may, therefore, regulate the Rbo/Rbr oxidative stress response in *D. vulgaris*.

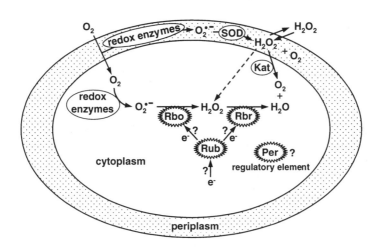

FIGURE 10.8. Proposed model for an Rbo/Rbr oxidative stress protection system in *D. vulgaris*. *Rub*, rubredoxin; *Kat*, heme catalase; *Per*, product of the *fur*-like gene. Reprinted with permission from Lumppio et al. (2001), copyright 2001 American Society for Microbiology.

Conclusion

Some aspects of the proposed Rbr/Rbo oxidative stress defense system in *D. vulgaris* resemble those recently suggested for oxidative stress protection in the anaerobic hyperthermophilic archaeon *Pyrococcus furiosus* (Jenney et al. 1999). *Pyrococcus furiosus* contains an Nlr-like protein with superoxide reductase activity as well as an Rbr, the genes for which are tandemly located. The microorganismic segregation of SOD/catalase between aerobes and anaerobes appears to be less distinct than for Rbo/Rbr, which, as noted above, have so far been found only in air-sensitive microbes (Kirschvink et al. 2000). The latter segregation suggests that the Rbo/Rbr oxidative stress protection system is well suited to protection of anaerobic life in an aerobic world.

Acknowledgments. Research in my laboratory referred to in this chapter was supported by NIH grant GM50388. I thank Eric Coulter, Heather Lumppio, Neeta Shenvi, Junjie Xiong, and Joseph Emerson for obtaining many of the results noted here and Gerrit Voordouw for his generosity in sharing *D. vulgaris* strains and his knowledge of *D. vulgaris* genetics.

References

Alban PS, Krieg NR. 1998. A hydrogen peroxide resistant mutant of *Spirillum volutans* has NADH peroxidase activity but no increased oxygen tolerance. Can J Microbiol 44:87–91.

Alban PS, Popham DL, Rippere KE, Krieg NR. 1998. Identification of a gene for a rubrerythrin/nigerythrin-like protein from *Sprillum volutans* by using amino acid sequence data from mass spectrometry and NH$_2$-terminal sequencing. J Appl Microbiol 85:875–82.

Bielski BHJ, Cabelli D. 1991. Highlights of current research involving superoxide and perhydroxyl radicals in aqueous solution. Int J Radiat Biol 59:291–319.

Brumlik MJ, Voordouw G. 1989. Analysis of the transcriptional unit encoding the genes for rubredoxin (*rub*) and a putative rubredoxin oxidoreductase (*rbo*) in *Desulfovibrio vulgaris* (Hildenborough). J Bacteriol 171:4996–5004.

Bsat N, Herbig A, Casillas-Martinez L, et al. 1998. *Bacillus subtilis* contains multiple Fur homologues: identification of the iron uptake (Fur) and peroxide regulon (PerR) repressors. Mol Microbiol 29:189–98.

Chung H-J, Choi J-H, Kim E-J, et al. 1999. Negative regulation of the gene for Fe-containing superoxide dismutase by an Ni-responsive factor in *Streptomyces coelicolor*. J Bacteriol 181:7381–4.

Coehlo AV, Matias P, Fülop V, et al. 1997. Desulfoferrodoxin structure determined by MAD phasing and refinement to 1.9-Å reveals a unique combination of a tetrahedral FeS$_4$ centre with a square pyramidal FeSN$_4$ centre. J Biol Inorg Chem 2:680–9.

Coulter ED, Emerson JP, Kurtz DM, Jr, Cabelli DE. 2000a. Superoxide reactivity of rubredoxin oxidoreductase (desulfoferrodoxin) from *Desulfovibrio vulgaris*: a pulse radiolysis study. J Am Chem Soc 122:11555–6.

Coulter ED, Shenvi NV, Beharry Z, et al. 2000b. Rubrerythrin-catalyzed substrate oxidation by dioxygen and hydrogen peroxide. Inorg Chim Acta 297:231–4.

Coulter ED, Shenvi NV, Kurtz DM, Jr. 1999. NADH peroxidase activity of rubrerythrin. Biochem Biophys Res Commun 255:317–23.

Cypionka H. 2000. Oxygen respiration by *Desulfovibrio* species. Annu Rev Microbiol 54:827–48.

Das A, Coulter ED, Kurtz DM Jr, Ljungdahl LG. 2001. Five-gene cluster in *Clostridium thermoaceticum* consisting of two divergent operons encoding rubre-doxin oxidoreductase-rubredoxin and rubrerythrin-type A flavoprotein-high-molecular-weight Rubredoxin. J Bacteriol 183:1560–7.

Deckers HM, Voordouw G. 1996. The *dcr* gene family of *Desulfovibrio*: implications from the sequence of *dcr*H and phylogenetic comparison with other *mcp* genes. Antonie Leeuwenhoek 70:21–9.

deMaré F, Kurtz DM Jr, Nordlund P. 1996. The structure of *Desulfovibrio vulgaris* rubrerythrin reveals a unique combination of rubredoxin-like FeS$_4$ and ferritin-like diiron domains. Nat Struct Biol 3:539–46.

Dos Santos WG, Pacheco I, Liu MY, et al. 2000. Purification and characterization of an iron superoxide dismutase and a catalase from the sulfate-reducing bacterium *Desulfovibrio gigas*. J Bacteriol 182:796–804.

Fu R, Voordouw G. 1997. Targeted gene-replacement mutagenesis of *dcr*A, encoding an oxygen sensor of the sulfate-reducing bacterium *Desulfovibrio vulgaris* Hildenborough. Microbiology 143:1815–26.

Frazão C, Silva G, Gomes CM, et al. 2000. Structure of a dioxygen reduction enzyme from *Desulfovibrio gigas*. Nat Struct Biol 7:1041–5.

Gort AS, Imlay JA. 1998. Balance between endogenous superoxide stress and antioxidant defenses. J Bacteriol 180:1402–10.

Hatchikian CE, LeGall J, Bell GR. 1977. Significance of superoxide dismutase and catalase activities in the strict anaerobes, sulfate-reducing bacteria. In: Michael AM, McCord JM, Fridovich I, editors. Superoxide and superoxide dismutase. New York: Academic Press. p 159–72.

Imlay KRC, Imlay JA. 1996. Cloning and analysis of *sodC*, encoding the copper-zinc superoxide dismutase of *Escherichia coli*. J Bacteriol 178:2564–71.

Jenney FE Jr, Verhagen MJM, Cui X, Adams MWW. 1999. Anaerobic microbes: oxygen detoxification without superoxide dismutase Science. 286:306–9.

Johnson MS, Zhulin IB, Gâpuzan ME, Taylor BL. 1997. Oxygen-dependent growth of the obligate anaerobe *Desulfovibrio vulgaris* Hildenborough. J Bacteriol 179: 5598–601.

Jurica MS, Stoddard BL. 1998. Mind your B's and R's: bacterial chemotaxis, signal transduction and protein recognition. Structure 6:809–13.

Kirschvink JL, Gaidos EJ, Bertani LE, et al. 2000. Paleoproterozoic snowball earth: extreme climatic and geochemical global change and its biological consequences. Proc Nat Acad Sci USA 97:1400–5.

Liochev SI, Fridovich I. 1997. A mechanism for complementation of the *sodA sodB* defect in *Escherichia coli* by overproduction of the *rbo* gene product (desulfo-ferrodoxin) from *Desulfoarculus baarsii*. J Biol Chem 272:25573–5.

Loewen PC, Klotz MG, Hassett DJ. 2000. Catalase—an "old" enzyme that contin-ues to surprise us. ASM News 66:76–82.

Lombard M, Fontecave M, Touati D, Niviere V. 2000. Reaction of the desulfoferro-doxin from *Desulfoarculus baarsii* with superoxide anion. Evidence for a super-oxide reductase activity. J Biol Chem 275:115–21.

Lumppio HL, Shenvi NV, Garg RP, et al. 1997. A rubrerythrin operon and nigery-thrin gene in *Desulfovibrio vulgaris* (Hildenborough). J Bacteriol 179:4607–15.

Lumppio HL, Shenvi NV, Summers AO, et al. 2001. Rubrerythrin and rubredoxin oxidoreductase in *Desulfovibrio vulgaris*. A novel oxidative stress protection system. J Bacteriol 183:101–8.

Marschall C, Frenzel P, Cypionka H. 1993. Influence of oxygen on sulfate reduction and growth of sulfate-reducing bacteria. Arch Microbiol 159:168–73.

Moura I, Tavares P, Moura JJ, et al. 1990. Purification and characterization of desul-foferrodoxin. A novel protein from *Desulfovibrio desulfuricans* (ATCC 27774) and from *Desulfovibrio vulgaris* (strain Hildenborough) that contains a distorted rubredoxin center and a mononuclear ferrous center. J Biol Chem 265:21596–602.

Mowbray SL, Sandgren MO. 1998. Chemotaxis receptors: a progress report on struc-ture and function. J Struct Biol 124:257–75.

Pianzzola MJ, Soubes M, Touati D. 1996. Overproduction of the *rbo* gene product from *Desulfovibrio* species suppresses all deleterious effects of lack of superox-ide dismutase in *Escherichia coli*. J Bacteriol 178:6736–42.

Pierik AJ, Wolbert RBG, Portier GL, et al. 1993. Nigerythrin and rubrerythrin from *Desulfovibrio vulgaris* each contain two mononuclear iron centers and two dinu-clear iron clusters. Eur J Biochem 212:237–45.

Rocha ER, Smith CJ. 1998. Characterization of a peroxide-resistant mutant of the anaerobic bacterium *Bacteroides fragilis*. J Bacteriol 180:5906–12.

Sayle R, Milner-White EJ. 1995. RASMOL—Biomolecular graphics for all. Trends Biochem Sci 20:374–6.

Silva G, Oliveira S, Gomes CM, et al. 1999. *Desulfovibrio gigas* neelaredoxin. A novel superoxide dismutase integrated in a putative oxygen sensory operon of an anaerobe. Eur J Biochem 259:235–43.

Stenkamp RE. 1994. Dioxygen and hemerythrin. Chem Rev 94:715–26.

Storz G, Imlay JA. 1999. Oxidative stress. Curr Opin Microbiol 2:188–94.

Touati D. 2000. Iron and oxidative stress in bacteria. Arch Biochem Biophys 373:1–6.

van Niel EWJ, Gottschal JC. 1998. Oxygen consumption by *Desulfovibrio* strains with and without polyglucose. Appl Environ Microbiol 64:1034–9.

Voordouw JK, Voordouw G. 1998. Deletion of the *rbo* gene increases the oxygen sensitivity of the sulfate-reducing bacterium *Desulfovibrio vulgaris* Hildenbor-ough. Appl Environ Microbiol 64:2882–7.

Xiong J, Kurtz DM Jr, Ai J, Sanders-Loehr J. 2000. A hemerythrin-like domain in a bacterial chemotaxis protein. Biochemistry 39:5117–25.

Zámocky M, Koller F. 1999. Understanding the structure and function of catalases: clues from molecular evolution and in vitro mutagenesis. Prog Biophys Mol Biol 72:19–66.

Zou P-J, Borovok I, Ortiz de Orué Lucana D, et al. 1999. The mycelium-associated *Streptomyces reticuli* catalase-peroxidase, its gene and regulation by FurS. Microbiology 145:549–59.

Suggested Reading

Deckers HM, Voordouw G. 1994. Identification of a large family of genes for puta-tive chemoreceptor proteins in an ordered library of the *Desulfovibrio vulgaris* Hildenborough genome. J Bacteriol 176:351–8.

11
One-Carbon Metabolism in Methanogenic Anaerobes

JAMES G. FERRY

Anaerobic microbes convert complex organic matter to the one-carbon compounds CH_4 and CO_2, a process essential to the global carbon cycle. The process requires a consortium of at least three interacting metabolic groups, the first two of which are primarily from the Bacteria domain and convert the organic matter to hydrogen, CO_2, formate, and acetate. The third group, the methanoarchaea, is found exclusively in the Archaea domain and converts the metabolic products of the first two groups to CH_4. One-carbon reactions play important roles in the metabolism of all three metabolic groups, but these reactions are dominant in the methanoarchaea. About two thirds of the nearly 1 billion tons of biologically produced CH_4 generated each year originates from reductive demethylation of acetate in the acetate fermentation pathway. The remaining one third is contributed by reduction of CO_2, with electrons originating from formate or hydrogen in the CO_2-reduction pathway. This chapter presents general features of the two pathways with a focus on recent developments regarding the enzymes catalyzing one-carbon reactions. The reader is referred to review articles that provide a more detailed background for these enzymes (Ferry 1999; Thauer 1998) or emphasize other aspects of the methanogenic pathways, including the bioenergetics (Schafer et al. 1999).

Cofactors

The cofactors involved in the methanogenic pathways are shown in Figure 11.1. Except for molybdopterin, the cofactors are largely confined to the methanoarchaea and *Archaeoglobus fulgidus*, which is a sulfate-reducing anaerobe from the Archaea domain. Coenzyme M, methanofuran, and factor F_{430} are unique structures first discovered in the methanoarchaea. Factor III and H_4MPT resemble vitamin B_{12} and tetrahydrofolate (H_4F), respectively, the latter two of which are commonly found in the Bacteria domain. Although H_4MPT closely resembles H_4F at the corresponding positions where C-1 fragments bind, thermodynamic differences confer differ-

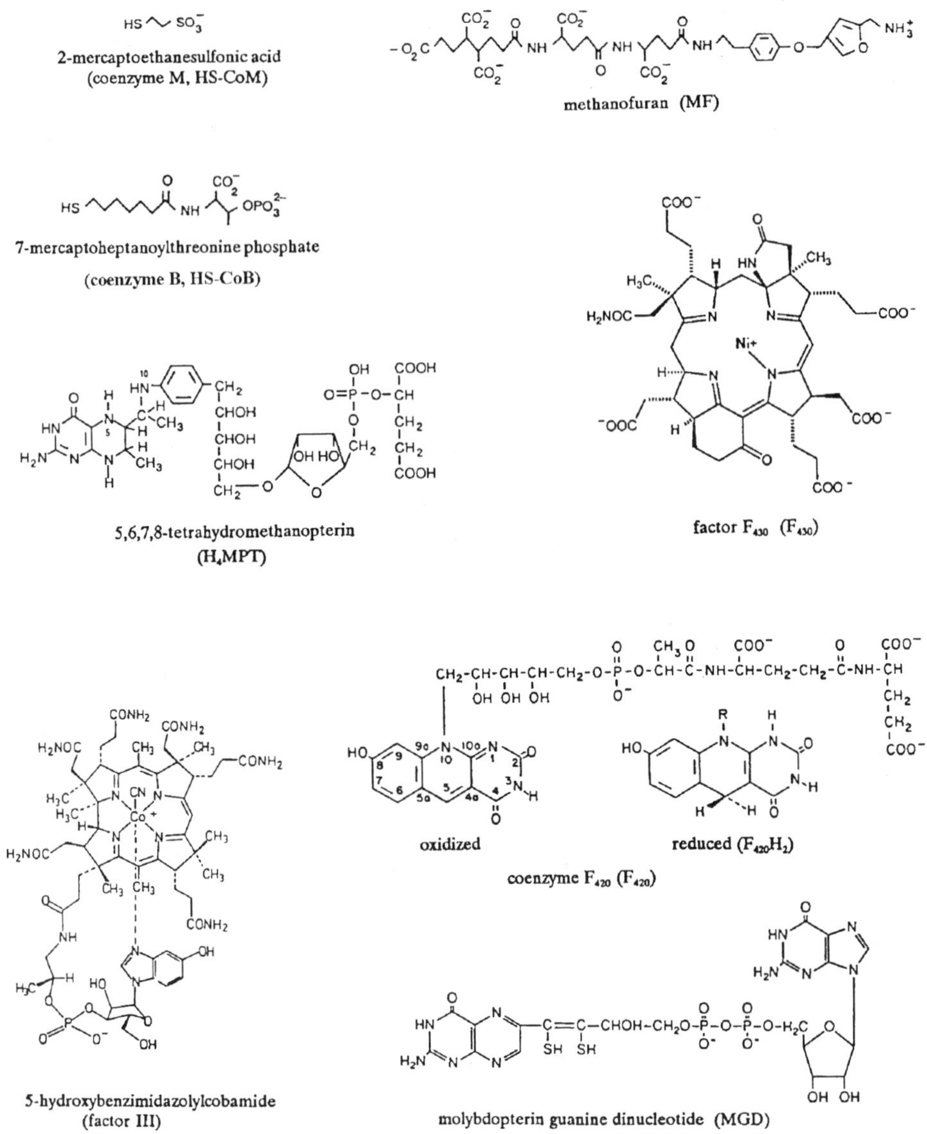

HS~SO₃⁻

2-mercaptoethanesulfonic acid
(coenzyme M, HS-CoM)

methanofuran (MF)

7-mercaptoheptanoylthreonine phosphate

(coenzyme B, HS-CoB)

5,6,7,8-tetrahydromethanopterin
(H₄MPT)

factor F₄₃₀ (F₄₃₀)

5-hydroxybenzimidazolylcobamide
(factor III)

oxidized

reduced (F₄₂₀H₂)

coenzyme F₄₂₀ (F₄₂₀)

molybdopterin guanine dinucleotide (MGD)

FIGURE 11.1. Structures of cofactors required for catalysis of one-carbon reactions in methanogenic pathways.

ent properties (Maden 2000). Methyl groups are more readily oxidized on H₄MPT than on H₄F, and the redox reactions for H₄MPT-dependent enzymes are coupled to more negative reductants than the pyridine nucleotides that are generally used in H₄F-dependent pathways. These differences are the result of the properties of the arylamine nitrogen N-10 on the two carriers. In H_4F, N-10 is subject to electron withdrawal by the carbonyl group of p-aminobenzoate; whereas in H_4MPT, an electron-donating methylene group is present in this position. Although H_4MPT is the carrier of C-1 fragments between formyl and methyl oxidation levels in the methanoarchaea, H_4F serves this function in the Bacteria domain. The only known exceptions are the aerobic proteobacterial methylotrophs (Vorholt et al. 1999). The methylotroph *Methylobacterium extorquens* AM1 contains a pathway for the oxidation of formaldehyde involving a methenyl-H_4MPT cyclohydrolase (Pomper et al. 1999) and a NADP-dependent methylene-H_4MPT dehydrogenase (Hagemeier et al. 2000).

A phylogenetic analysis of H_4MPT-dependent enzymes reveals distinct archaeal and bacterial branches of the tree that are clearly related; however, the results still do not resolve whether a common ancestor contained H_4MPT-dependent enzymes or the methylotrophs acquired genes from the Archaea domain. Nonetheless, analysis of the genomic sequences from the Bacteria domain indicate that only the methylotrophic proteobacteria harbor H_4MPT-dependent enzymes are consistent with lateral transfer from archaeal donors (Vorholt et al. 1999).

Coenzyme M was shown to function as the central cofactor of aliphatic epoxide carboxylation in *Xanthobacter* strain Py2, an aerobe from the Bacteria domain (Allen et al. 1999). The organism metabolizes short-chain aliphatic alkenes via oxidation to epoxyalkanes, followed by carboxylation to β-ketoacids. An enzyme in the pathway catalyzes the addition of coenzyme M to epoxypropane to form 2-(2-hydroxypropylthio)ethanesulfonate. This intermediate is oxidized to 2-(2-ketopropylthio)ethanesulfonate, followed by a NADPH-dependent cleavage and carboxylation of the β-ketothioether to form acetoacetate and coenzyme M. This is the only known function for coenzyme M outside the methanoarchaea.

Reactions Common to Both Pathways

The two major methanogenic pathways have in common the transfer of a methyl group from methyl-H_4MPT or methyl-H_4SPT (where H_4SPT is a close analog of H_4MPT) to coenzyme M (HS-CoM) and reductive demethylation of the CH_3-S-CoM to methane (Fig. 11.2; Table 11.1). The transfer reaction (AC1) is coupled to the generation of a sodium gradient (high outside) catalyzed by a membrane-bound methyltransferase (Mtr). Methyl-CoM methylreductase (Mcr) catalyzes the demethylation reaction (AC2) with HS-CoB as the electron donor. The heterodisulfide CoM-S-S-CoB is formed as a consequence of the demethylation reaction. Reduction of the heterodisulfide to the corresponding sulfhydryl forms (AC3) of the cofactors is catalyzed by heterodisulfide reductase (Hdr). Although the demethylation reaction has a large negative change in free energy, it is not coupled to energy generation; however, a proton gradient is generated by passage

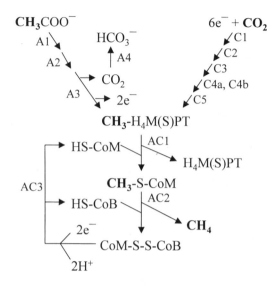

FIGURE 11.2. General features of the CO_2-reduction and acetate fermentation pathways showing reactions unique and common to each.

of electrons through a membrane-bound electron-transport chain terminating in reduction of the heterodisulfide. Thus reduction of the heterodisulfide CoM-S-S-CoB, which is common to all pathways, drives the phosphorylation of ADP by way of a chemiosmotic mechanism that provides ATP for growth.

Methyl-H_4MPT:coenzyme M methyltransfrase (Mtr) is an integral membrane-bound complex made up of eight subunits. MtrA contains factor III, the methylation of which is an intermediate in the reaction. It is hypothesized that sodium ion translocation is coupled to a conformational change in MtrA when Co^{1+} is methylated and histidine binds as a lower axial ligand (Gottschalk and Thauer 2001). Hippler and Thauer (1999) showed that MtrH is the subunit that transfers the methyl group from Methyl-H_4MPT to MtrA.

TABLE 11.1. Reactions common to the CO_2-reduction and acetate fermentation pathways for methane formation and enzymes that catalyze them.

Reaction	Enzyme	Abbreviation
5-methyl-H_4MPT + HS-CoM → CH_3-S-CoM + H_4MPT	N^5-methyltetrahydromethanopterin: coenzyme M methyltransferase	AC1
CH_3-S-CoM + HS-CoB → CH_4 + CoM-S-S-CoB	Methyl-coenzyme M methylreductase	AC2
CoM-S-S-CoB + $2e^-$ + $2H^+$ → HS-CoM + HS-CoM	Heterodisulfide reductase	AC3

Mcr contains two coenzyme F_{430} prosthetic groups. A high-resolution structure of the enzyme from *Methanobacterium thermoautotrophicum*, complexed with either HS-CoM plus HS-CoB or CoM-S-S-CoB, lead to a proposed reaction mechanism in which the nickel atom of F_{430} is methylated with CH_3-S-CoM followed by protonolysis to methane (Ermler et al. 1997). The structure also revealed five modified amino acids at or near the active site, four of which are methylated and the other is a thioglycine forming a thiopeptide bond. It was shown that the methyl groups derive from the methyl group of methionine, and not by a fortuitous methylation with CH_3-S-CoM, which suggests a functional role for the modified residues (Selmer et al. 2000). The presence of modified residues in the crystal structure of the *Methanosarcina barkeri* enzyme further supports this proposal (Grabarse et al. 2000). The recombinant enzyme has not been produced precluding site-directed replacement of residues to test the proposed functions.

The Carbon Dioxide Reduction Pathway

The one-carbon reactions and enzymes unique to the carbon dioxide reduction pathway are shown in Figure 11.2 and Table 11.2. Electrons required for the reductive reactions are derived from the oxidation of either hydrogen or formate, catalyzed by hydrogenase or formate dehydrogenase.

$$4H_2 + CO_2 \rightarrow CH_4 + 2H_2O \tag{11.1}$$

TABLE 11.2. Reactions unique to the CO_2-reduction pathway and enzymes that catalyze them.

Reaction	Enzyme	Abbreviation
CO_2 + MF + $2e^-$ + $2H^+ \rightarrow$ formyl-MF	Formylmethanofuran dehydrogenase	C1
Formyl-MF + H_4MPT $\rightarrow$ 5-formyl-H_4MPT + MF	Formylmethanofuran: tetrahydromethanopterin formyltransferase	C2
5-formyl-H_4MPT + $H^+ \rightarrow$ 5,10-methenyl-H_4MPT^+ + H_2O	N^5,N^{10}-methenyltetrahydromethanopterin cyclohydrolase	C3
5,10-methenyl-H_4MPT^+ + $F_{420}H_2 \rightarrow$ 5,10 methylene-H_4MPT +F_{420} + H^+	N^5,N^{10}-methylenetetrahydromethanopterin dehydrogenase; F_{420} dependent	C4a
5,10-methenyl-H_4MPT^+ + $H_2 \rightarrow$ 5,10-methylene-H_4MPT + H^+	N^5,N^{10}-methylenetetrahydromethanopterin dehydrogenase; hydrogen dependent	C4b
5,10-methylene-H_4MPT + $F_{420}H_2 \rightarrow$ 5-methyl-H_4MPT + F_{420}	N^5,N^{10}-methylenetetrahydromethanopterin reductase	C5

$$HCOO^- + 4H^+ \rightarrow CH_4 + 3CO_2 + 2H_2O \qquad (11.2)$$

In the first reductive step, the molybdopterin-containing formyl-methanofuran dehydrogenase (Fmf) functions reversibly to reduce CO_2 to the formyl level; however, the electron donor is unknown. In *M. barkeri*, the genes encoding the five subunits form a transcription unit and are cotranscribed with a gene (*fmdF*) predicted to encode a polyferredoxin, which is a candidate for the electron donor (Vorholt et al. 1996). FmdB has a deduced sequence similar to the formate dehydrogenase of *Methanobacterium formicicum*, previously shown to contain molybdopterin (Johnson et al. 1991), a result suggesting FmdB is the catalytic subunit containing the cofactor. *N*-carboxymethanofuran (carbamate) formation from unprotonated methanofuran and CO_2 is the first step in the reduction. The formation of the carbamate proceeds spontaneously, and under physiologic conditions, the rate is similar to the maximal rate of methane formation (Bartoschek et al. 2000). Furthermore, the rate of carbamate formation is not enhanced by purified formylmethanofuran dehydrogenase. Thus it appears that an enzyme-catalyzed carbamate formation is not required.

The crystal structure of the homotrimeric cyclohydrolase (Mch) from *Methanopyrus kandleri* (Grabarse et al. 1999) reveals a new α/β fold distinct from the human H_4F-dependent bifunctional dehydrogenase/cyclohydrolase. However, the archaeal and eucaryal structures are related by both having two α/β domains (A and B) with a large hydrophobic pocket between them, which is postulated to be the active site. Unfortunately, the archaeal structure was not obtained with the substrate bound, precluding meaningful predictions of the mechanism, although a binding site was hypothesized based primarily on sequence conservation and two clusters of basic residues that each bind a phosphate molecule. The basic residues are proposed to interact with the negatively charged α-hydroxyglutarylphosphate of the substrate.

Two genetically distinct methylene-H_4MPT dehydrogenases were described: one coenzyme F_{420} dependent (Mtd), catalyzing reaction C4a, and another hydrogen dependent (Hmd), catalyzing reaction C4b. Coenzyme F_{420} is an obligate two-electron carrier ($E_m = -350\,mV$) that donates or accepts a hydride ion. Characterization of the hydrogen-dependent dehydrogenases from several strains indicates that the enzymes contain no iron or nickel, nor do they have other characteristics of classical hydrogenases; however, evidence has been presented for a tightly bound thermolabile cofactor of <1000 Da (Buurman et al. 2000). The genomic sequences of *M. thermoautotrophicum* (Smith et al. 1997) and *Methanococcus jannaschii* (Bult et al. 1996) predict two additional putative *hmd* genes (*hmdII* and *hmdIII*). When grown under hydrogen-limiting conditions, synthesis of Hmd in *Methanothermobacter marburgensis* (formerly *M. thermoautotrophicum* strain Marburg) is constant, whereas Mtd is higher (Afting et al. 2000). In addition, a classical nickel-containing F_{420}-dependent hydrogenase (Frh) with high affinity for hydrogen is induced. Under nickel limitation, the synthesis of Frh is repressed; however, Mtd and Hmd (with a 10-fold lower affinity for hydrogen than Frh) are induced. It was also shown that the regulation of all three enzymes is at the transcriptional level, although additional translational or posttranslational mechanisms of regulation cannot be ruled out.

These results support the previous hypothesis that Frh and Mtd function to reduce methylene-H_4MPT when hydrogen is limiting, whereas Hmd provides this function when hydrogen is not limiting (Afting et al. 1998). These results also suggest that under nickel limitation, Hmd and Mtd function in tandem to oxidize hydrogen and reduce F_{420}, consistent with the observation that the two enzymes catalyze the conversion in vitro. Afting et al. (2000) reported that the transcription of *hmdII* is up-regulated and *hmdIII* is down-regulated under hydrogen limitation, suggesting a role in hydrogen metabolism, although neither enzyme has hydrogenase activity.

The crystal structures of the methylene-H_4MPT reductases from *M. kandleri* (kMer) and *M. thermoautotrophicum* (tMer) have been determined (Shima et al. 2000) in the homodimeric (kMer and tMer) and homotetrameric (kMer) states, revealing a core TIM-barrel fold with three insertion regions for the subunit from each species. It is proposed that the oligomeric structures are physiologically relevant. Although neither enzyme was crystalized with the substrates, binding sites were proposed for F_{420} and H_4MPT in a cleft formed by insertions 2 and 3, plus the C-terminal end of the TIM-barrel. A most interesting feature is an unusual cis-peptide bond between Gly 64/Ile 65 (kMer) and Gly 61/Val62 (tMer) in the hypothetical F_{420} binding site, which is of postulated importance based on the high conservation of this region in all reductases sequenced to date. Production of the recombinant enzyme and a crystal structure with the substrates bound will be necessary to determine the mechanism. Although methylene-H_4F reductase has functional similarity to Mer, the structures share no similarity, suggesting a distant relationship within the TIM-barrel family; however, the structures of Mer and luciferase have common features (including the nonprolyl cis-peptide bond), suggesting a common TIM-barrel ancestor.

Carbonic anhydrases, which catalyze the reversible hydration of CO_2 ($CO_2 + H_2O \rightarrow HCO3^- + H^+$), are widely distributed in the Bacteria and Archaea domains, suggesting the enzymes have diverse functions in procaryotes (Smith et al. 1999). The genomes of *M. jannaschii and M. thermoautotrophicum* contain open reading frames (ORFs) with deduced sequences having high identity to carbonic anhydrases from both the β- and γ-classes (Smith et al. 1999). Both species are autotrophic and, therefore, have a high CO_2 requirement for both methanogenesis and cell carbon synthesis; thus it has been proposed that carbonic anhydrases may play a role in the acquisition of CO_2 (analogous to the role proposed in plants) and/or interconversion between CO_2 and HCO_3^- to supply the appropriate substrate to enzymes in methanogenic and CO_2 fixation pathways. The β-class carbonic anhydrase from *M. thermoautotrophicum* (Cab) has been produced in *Escherichia coli* as a homotetramer with one zinc per subunit (Smith and Ferry 1999; Smith et al. 2000). Kinetic analyses indicate a zinc-hydroxide-catalytic mechanism (Smith et al. 2000). The overall enzyme-catalyzed reaction occurs in two distinct half reactions. The first half reaction is the interconversion of CO_2 and HCO_3^- (Eqs. 11.3 and 11.4) and involves nucleophilic attack of the zinc hydroxyl ion on CO_2. This is followed by exchange of zinc-bound HCO_3^- with water. The second half reaction corresponds to intramolecular proton transfer steps leading to protonation of solvent buffer (Eqs. 11.5 and 11.6), which regenerates the zinc-hydroxide at the active site.

$$E\text{-}Zn^{2+}\text{-}OH^- + CO \rightleftarrows E\text{-}Zn^{2+}\text{-}HCO_3^- \tag{11.3}$$

$$E\text{-}Zn^{2+}\text{-}HCO_3^- + H_2O \rightleftarrows E\text{-}Zn^{2+}\text{-}H_2O + HCO_3^- \tag{11.4}$$

$$E\text{-}Zn^{2+}\text{-}H_2O \rightleftarrows {}^+H\text{-}E\text{-}Zn^{2+}\text{-}OH^- \tag{11.5}$$

$${}^+H\text{-}E\text{-}Zn^{2+}\text{-}OH^- + B \rightleftarrows E\text{-}Zn^{2+}\text{-}OH^- + BH^+ \tag{11.6}$$

where E is the enzyme and B is the buffer.

The crystal structure (Strop et al. 2001) reveals a homodimer with the zinc atom ligated by the sulfur atoms of two cysteines (Cys 32 and Cys 90) and the nitrogen atom of a histidine (His 87), as is the case for the plant-type enzyme (Fig. 11.3). The active site contains an HEPES buffer molecule in a position that implicates involvement of Asp 34 in the transport of protons after ionization of the zinc-bound water.

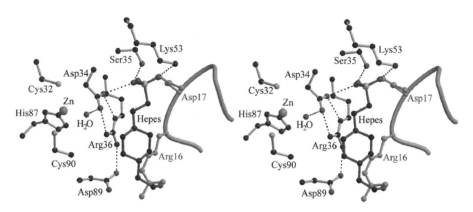

FIGURE 11.3. The active site of Cab. Reprinted with permission from Strop et al. (2001), copyright 2001 American Society for Biochemistry and Molecular Biology.

Fermentation of Acetate

Approximately two thirds of the methane produced in nature derives from the methyl group of acetate.

$$CH_3CO_2^- + H_2O \rightarrow CH_4 + HCO_3^- \tag{11.7}$$

The reactions unique to the pathway for *Methanosarcina thermophila* are shown in Figure 11.2 and Table 11.3. In the pathway, the carbon-carbon bond of acetate is cleaved, followed by reduction of the methyl group to methane with electrons originating from oxidation of the carbonyl group to carbon dioxide; thus the pathway is a true fermentation.

Acetate kinase (Ack) and phosphotransacetylase (Pta) function together to activate acetate to acetyl-CoA before cleavage by the CO dehydrogenase/acetyl-CoA synthase complex (Cdh). Both enzymes have been hyperproduced in *E. coli* in an highly active form, leading to a crystal structure for Ack and site-specific replacement of residues to investigate the catalytic mechanisms. The crystal structure (Buss et al. 2001) of Ack is an α_2-homodimer, consistent with the enzyme purified from *M. thermophila*. Features of the overall fold suggest Ack is the newest member of the sugar kinase/Hsc70/actin superfamily of phosphotranferases, and possibly the common ancestor to all kinases. The active site cleft is identified by bound ATP, although only the α- and β-phosphates are observable (Fig. 11.4). The β-phosphate points away from the active site, possibly the consequence of charge repulsion by a sulfate ion in the active site, resulting from the crystallization conditions. It is proposed that the sulfate ion occupies the site of the γ-phosphate of ATP. Site-directed mutagenesis was used to replace active-site residues individually to probe the catalytic mechanism. The results of kinetic and biochemical analyses of each variant (Singh-Wissmann et al. 1998; Ingram-Smith et al. 2000; Singh-Wissmann et al. 2000) suggests a direct in-line mechanism in which the transition state phosphate is stabilized by Arg 91 and Arg 241 (Fig. 11.4).

The CO dehydrogenase/acetyl-CoA synthase (Cdh) complex cleaves the C-C and C-S bonds in the acetyl moiety of acetyl-CoA, oxidizes the car-

TABLE 11.3. Reactions unique to the acetate fermentation pathway and enzymes that catalyze them.

Reaction	Enzyme	Abbreviation
$CH_3COO^- + ATP \rightarrow CH_3CO_2PO_3^{-2} + ADP$	Acetate kinase	A1
$CH_3CO_2PO_3^{-2} + HS\text{-}CoA \rightarrow CH_3COSCoA + Pi$	Phosphotransacetylase	A2
$CH_3COSCoA + H_4SPT + H_2O \rightarrow CH_3\text{-}H_4SPT + 2e^- + 2H^+ + CO_2 + HS\text{-}CoA$	CO dehydrogenase/acetyl-CoA synthase	A3
$CO_2 + H_2O \rightarrow HCO_3^- + H^+$	Carbonic anhydrase	A4

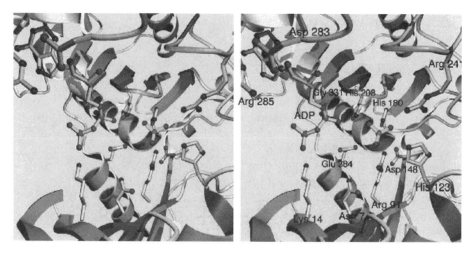

FIGURE 11.4. The active site of acetate kinase. Reprinted with permission from Buss et al. (2001), copyright 2001 American Society for Microbiology.

bonyl group to CO_2 (CO dehydrogenase activity), and transfers the methyl group to H_4SPT (reaction A3). The five-subunit complex contains three enzyme components. The nickel/iron-sulfur (Ni/Fe-S) component contains the CdhA and CdhB subunits. The corrinoid/iron-sulfur (Co/Fe-S) component contains the CdhD and CdhE subunits. The third component is the CdhC subunit. There are three metal clusters (A, B, and C) in the Ni/Fe-S component. Cluster A (Ni-X-$[Fe_4S_4]$, where atom X is unknown) is the proposed site for cleavage of the C-C and/or C-S bonds of acetyl-CoA based on analogy with the well-characterized function for cluster A in the acetogenic clostridia. The methyl group is transferred to the factor III corrinoid of the Co/Fe-S component, where H_4SPT is methylated. The carbonyl group is thought to be oxidized on cluster C, which also has the composition Ni-X-$[Fe_4S_4]$. Cluster B is a conventional Fe_4S_4 center proposed to shuttle electrons from cluster C to a ferredoxin.

The deduced sequence of *cdhC* identifies a cysteine motif identical to that in the well-characterized clostridial CO dehydrogenase/acetyl-CoA synthase proposed to bind cluster A. Murakami and Ragsdale (2000) showed that dissociation of CdhC from the complex by protease treatment results in loss of an electron paramagnetic resonance (EPR) signal diagnostic of cluster A, confirming the proposed cluster A in CdhC. The decrease in the EPR signal correlates with a loss in the CO/acetyl-CoA exchange reaction, implicating a function for CdhC in cleavage of the C-C and/or C-S bonds. Thus the questions are raised as to why two A clusters are present in the complex and why an association of CdhC is necessary for C-C and C-S bond cleavage. One explanation, consistent with these results, is that C-S bond cleavage occurs on cluster A in CdhC with transfer of the acetyl group to cluster A on CdhA where the C-C bond is cleaved.

The carbonic anhydrase (Cam) in *M. thermophila* cells is elevated several fold when the energy source is shifted to acetate, suggesting a role for this enzyme in the acetate-fermentation pathway. It is proposed that Cam functions outside the cell membrane to convert CO_2 to a charged species (reaction A4) thereby facilitating removal of product from the cytoplasm. Cam is the prototype of a new class (γ) of carbonic anhydrases, independently evolved from the other two classes (α and β). The crystal structure of Cam reveals a novel left-handed parallel β-helix fold (Kisker et al. 1996). Apart from the histidines ligating zinc, the activesite residues of Cam have no recognizable analogs in the active sites of the α- and β-classes. Kinetic analyses establish that the enzyme has a zinc-hydroxide mechanism similar to that of Cab (Alber et al. 1999).

The crystal structure of the cobalt-substituted enzyme was obtained with bicarbonate bound to the metal (Iverson et al. 2000). The structure shows Asn 202 and Gln75 hydrogen bonded to the metal-bound bicarbonate, suggestzing potential roles for these residues in either transition-state stabilization or orientation and polarization of CO_2 for attack from the zinc-hydroxyl (Fig. 11.5). The crystal structure also shows three discrete conformations for Glu 84, suggesting a role for this residue in the transfer of protons out of the active site; indeed, kinetic analyses of Glu 84 variants combined with chemical rescue experiments establish this residue as critical for proton transfer (Tripp and Ferry 2000). The location of Glu 62 adjacent to Glu 84 suggests a potential role in proton transfer as well. Although kinetic analyses of site-specific variants establish an essential role for Glu 62 in the CO_2 hydration steps (Eqs. 11.3 and 11.4), the results were inconclusive regarding an additional role in proton transfer (Eqs. 11.5 and 11.6).

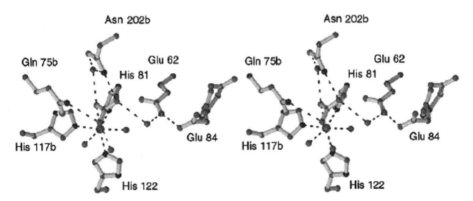

FIGURE 11.5. The active site of cobalt-substituted Cam complexed with bicarbonate. Reprinted with permission from Iverson et al. (2000), copyright 2000 American Chemical Society.

Conclusions

One-carbon reactions play large roles in the two major pathways for methane formation in nature. Several of the enzymes catalyzing these reactions have novel features, contributing to a larger understanding of biochemistry and microbial biology. The crystal structures for many of these enzymes are known, providing insight into the reaction mechanisms. The entire sequence of the genomes of several CO_2-reducing and acetate-fermenting methanoarchaea have been published or are nearing completion, an advance that will surely lead to the identification of still more novel proteins and enzymes involved in these pathways.

Acknowledgments. Research at The Pennsylvania State University was supported by grants DE-FG02-95ER20198 from the Department of Energy, and GM44661 from the National Institutes of Health.

References

Afting C, Hochheimer A, Thauer RK. 1998. Function of H_2-forming methylenetetrahydromethanopterin dehydrogenase from *Methanobacterium thermoautotrophicum* in coenzyme F_{420} reduction with H_2. Arch Microbiol 169:206–10.

Afting C, Kremmer E, Brucker C, et al. 2000. Regulation of the synthesis of H_2-forming methylenetetrahydromethanopterin dehydrogenase (Hmd) and of HmdII and HmdIII in *Methanothermobacter marburgensis*. Arch Microbiol 174: 225–32.

Alber BE, Colangelo CM, Dong J, et al. 1999. Kinetic and spectroscopic characterization of the gamma carbonic anhydrase from the methanoarchaeon *Methanosarcina thermophila*. Biochemistry 38:13119–28.

Allen JR, Clark DD, Krum JG, Ensign SA. 1999. A role for coenzyme M (2-mercaptoethanesulfonic acid) in a bacterial pathway of aliphatic epoxide carboxylation. Proc Natl Acad Sci USA 96:8432–7.

Bartoschek S, Vorholt JA, Thauer RK, et al. 2000. N-carboxymethanofuran (carbamate) formation from methanofuran and CO_2 in methanogenic archaea. Thermodynamics and kinetics of the spontaneous reaction. Eur J Biochem 267: 3130–8.

Bult CJ, White O, Olsen GJ, et al. 1996. Complete genome sequence of the methanogenic archaeon, *Methanococcus jannaschii*. Science 273:1058–73.

Buss KA, Cooper DR, Ingram-Smith C, et al. 2001. Urkinase: structure of acetate kinase, a member of the ASKHA superfamily of phosphotransferases. J Bacteriol 183:680–6.

Buurman G, Shima S, Thauer RK. 2000. The metal-free hydrogenase from methanogenic archaea: evidence for a bound cofactor. FEBS Lett 485:200–4.

Ermler U, Grabarse W, Shima S, et al. 1997. Crystal structure of methyl-coenzyme M reductase: the key enzyme of biological methane formation. Science 278: 1457–62.

Ferry JG. 1999. Enzymology of one-carbon metabolism in methanogenic pathways. FEMS Microbiol Rev 23:13–38.

Gottschalk G, Thauer RK. 2001. The Na^+ translocating methyltransferase complex from methanogenic. Archaea Biochim Biophys Acta 1505:28–36.

Grabarse WG, Mahlert F, Shima S, et al. 2000. Comparison of three methyl-coenzyme M reductases from phylogenetically distant organisms: unusual amino acid modification, conservation and adaptation. J Mol Biol 303:329–44.

Grabarse W, Vaupel M, Vorholt JA, et al. 1999. The crystal structure of methenyl-tetrahydromethanopterin cyclohydrolase from the hyperthermophilic archaeon *Methanopyrus kandleri*. Structure Fold Des 7:1257–68.

Hippler B, Thauer RK. 1999. The energy conserving methyltetrahydrometha-nopterin:coenzyme M methyltransferase complex from methanogenic archaea: function of the subunit MtrH. FEBS Lett 449:165–8.

Ingram-Smith C, Barber RD, Ferry JG. 2000. The role of histidines in the acetate kinase from *Methanosarcina thermophila*. J Biol Chem 275:33765–70.

Iverson TM, Alber BE, Kisker C, et al. 2000. A closer look at the active site of gamma-carbonic anhydrases: high resolution crystallographic studies of the carbonic anhydrase from *Methanosarcina thermophila*. Biochemistry 39:9222–31.

Johnson JL, Bastian NR, Schauer NL, et al. 1991. Identification of molybdopterin guanine dinucleotide in formate dehydrogenase from *Methanobacterium formicicum*. FEMS Microbiol Lett 77:2–3.

Kisker C, Schindelin H, Alber BE, et al. 1996. A left-handed beta-helix revealed by the crystal structure of a carbonic anhydrase from the archaeon *Methanosarcina thermophila*. EMBO J 15:2323–30.

Maden BEH. 2000. Tetrahydrofolate and tetrahydromethanopterin compared: functionally distinct carriers in C-1 metabolism. Biochem J 350:609–29.

Murakami E, Ragsdale SW. 2000. Evidence for intersubunit communication during acetyl-CoA cleavage by the multienzyme CO dehydrogenase/acetyl-CoA synthase complex from *Methanosarcina thermophila*. Evidence that the beta subunit catalyzes C-C and C-S bond cleavage. J Biol Chem 275:4699–707.

Schafer G, Engelhard M, Muller V. 1999. Bioenergetics of the Archaea. Microbiol Mol Biol Rev 63:570–620.

Selmer T, Kahnt J, Goubeaud M, et al. 2000. The biosynthesis of methylated amino acids in the active site region of methyl-coenzyme M reductase. J Biol Chem 275: 3755–60.

Shima S, Warkentin E, Grabarse W, et al. 2000. Structure of coenzyme F_{420} dependent methylenetetrahydromethanopterin reductase from two methanogenic Archaea. J Mol Biol 300:935–50.

Singh-Wissmann K, Ingram-Smith C, Miles RD, Ferry JG. 1998. Identification of essential glutamates in the acetate kinase from *Methanosarcina thermophila*. J Bacteriol 180:1129–34.

Singh-Wissmann K, Miles RD, Ingram-Smith C, Ferry JG. 2000. Identification of essential arginines in the acetate kinase from *Methanosarcina thermophila*. Biochemistry 39:3671–7.

Smith D, Doucette-Stamm RLA, Deloughery C, et al. 1997. Complete genome sequence of *Methanobacterium thermoautotrophicum* ΔH: functional analysis and comparative genomics. J Bacteriol 179:7135–55.

Smith KS, Ferry JG. 1999. A plant-type (beta-class) carbonic anhydrase in the thermophilic methanoarchaeon *Methanobacterium thermoautotrophicum*. J Bacteriol 181:6247–53.

Smith KS, Cosper NJ, Stalhandske C, et al. 2000. Structural and kinetic characterization of an archaeal beta-class carbonic anhydrase. J Bacteriol 182:6605–13.

Smith KS, Jakubzick C, Whittam TC, Ferry JG. 1999. Carbonic anhydrase is an ancient enzyme widespread in prokaryotes. Proc Natl Acad Sci USA 96:15184–9.

Strop P, Smith KS, Iverson TM, et al. 2001. Crystal structure of the "cab"-type beta class carbonic anhydrase from the archaeon *Methanobacterium thermoautotrophicum*. J Biol Chem 276:10299–305.

Thauer RK. 1998. Biochemistry of methanogenesis: a tribute to Marjory Stephenson Microbiology 144:2377–406.

Tripp BC, Ferry JG. 2000. A structure-function study of a proton transport pathway in a novel gamma-class carbonic anhydrase from *Methanosarcina thermophila*. Biochemistry 39:9232–40.

Vorholt JA, Chistoserdova L, Stolyar SM, et al. 1999. Distribution of tetrahydromethanopterin-dependent enzymes in methylotrophic bacteria and phylogeny of methenyl tetrahydromethanopterin cyclohydrolases. J Bacteriol 181:5750–7.

Vorholt JA, Vaupel M, Thauer RK. 1996. A polyferredoxin with eight [4Fe-4S] clusters as a subunit of molybdenum formylmethanofuran dehydrogenase from *Methanosarcina barkeri*. Eur J Biochem 236:309–17.

Suggested Reading

Hagemeier CH, Chistoserdova L, Lidstrom ME, et al. 2000. Characterization of a second methylene tetrahydromethanopterin dehydrogenase from *Methylobacterium extorquens* AM1. Eur J Biochem 267:3762–9.

Pomper BK, Vorholt JA, Chistoserdova L, et al. 1999. A methenyl tetrahydromethanopterin cyclohydrolase and a methenyl tetrahydrofolate cyclohydrolase in *Methylobacterium extorquens* AM1. Eur J Biochem 261:475–80.

12
Selenium-Dependent Enzymes from *Clostridia*

William T. Self

Stickland (1935) initially described the reductive deamination of glycine to ammonia and acetate in cultures of *Clostridium sporogenes*. During the purification of the enzyme responsible for this reaction—glycine reductase (GR) from *Clostridium sticklandii*—it was noted that GR activity was barely detectable in the later stages of growth of batch cultures in a defined medium (Turner and Stadtman 1973). The addition of selenium in the form of selenite overcame this loss in activity and suggested selenium might be required for GR. To investigate this possibility, cultures of *C. sticklandii* were supplemented with selenium in a radiolabeled form, ^{75}Se-selenite. Protein A, one of the three chromatographically separable components of GR, was purified and, in fact, was labeled with [75]Se (Turner and Stadtman 1973). This was the first report of a purified enzyme that contained selenium and began the work to define both the nature of selenium in proteins and the mechanism of incorporation of selenium into these selenoproteins.

A report by Flohe et al. (1973) revealed selenium was also present in bovine glutathione peroxidase. Soon after, Andreesen and Ljungdahl (1973) also detected selenium in formate dehydrogenase (FDH) from *Clostridium thermoaceticum* (renamed *Moorella thermoacetica*). These seminal reports of selenium-requiring enzymes stimulated interest in defining the biological role for selenium in proteins. In the years since these initial studies were published, the pathway for selenium incorporation into proteins has been well defined in *Escherichia coli* (Böck et al. 1991). However, the mechanism and regulation of incorporation of selenium into proteins in mammals is only now beginning to be understood at the biochemical level. Much of the groundwork for the current research in this area has been laid by numerous investigators studying selenoproteins from the genus *Clostridium*.

Table 12.1 summarizes the selenoenzymes isolated from clostridia, which fall into three major categories. The first group is the amino acid reductases, such as GR, sarcosine reductase (SR), betaine reductase (BR), and proline reductase (PR). The second class of selenoenzymes includes FDH. Although a number of clostridial species are suspected to contain a

TABLE 12.1. Selenium-dependent enzymes from Clostridiaceae.

Enzyme	Organism(s)	Reference
Glycine reductase protein A	*C. sticklandii*	Turner and Stadtman (1973)
	C. purinolyticum	Garcia and Stadtman (1991)
	C. litorale	Kreimer and Andreesen (1995)
Glycine reductase protein B	*E. acidaminophilum*	Wagner et al. (1999)
	C. litorale	Kreimer and Andreesen (1995)
Proline reductase	*C. sticklandii*	Kabisch et al. (1999)
Formate dehydrogenase	*C. thermoaceticum*	Yamamoto et al. (1983)
Nicotinic acid hydroxylase	*C. barkeri*	Dilworth (1982)
Xanthine dehydrogenase	*C. acidi-urici*	Wagner et al. (1984)
	E. barkeri	Schräder et al. (1999)
	C. purinolyticum	Self and Stadtman (2000)
Purine hydroxylase	*C. purinolyticum*	Self and Stadtman (2000)

selenium-dependent FDH, only the NADP-dependent tungsten-containing enzyme from *C. thermoaceticum* has been purified and characterized (Yamamoto et al. 1983). A third class of selenoenzymes in clostridia is selenium-dependent molybdenum hydroxylases. These hydroxylases are key enzymes in the metabolism of purinolytic clostridia and specialize in the fermentation of purines, such as adenine, xanthine, and uric acid. These selenoenzymes are unique in that those studied to date do not contain selenium in the form of selenocysteine but in the form of a labile cofactor, which has yet to be fully characterized. Although a number of other enzymes have been studied from clostridia that are involved in the metabolism of selenium, this chapter is a brief summary of the selenium-dependent enzymes listed in Table 12.1.

Glycine Reductase

Glycine reductase, which catalyzes the reductive deamination of glycine to ammonia and acetate, was shown to produce one equivalent of ATP from ADP in extracts of *C. sticklandii* (Stadtman 1966). GR consists of three chromatographically separable fractions, termed A, B and C. Figure 12.1 shows the requirement for selenium supplementation for optimal GR activity in large-scale cultures of *C. sticklandii*. The determination that protein A was labeled with ^{75}Se demonstrated the need for added selenium in a so-called rich culture medium. Selenium-deficient extracts of *C. sticklandii* could be complemented by purified selenoprotein A with restoration of GR activity. This complementation confirmed the requirement for selenium for protein A activity in the overall reaction of GR. The remaining components of GR (fractions B and C) were found to be associated with cell membranes during fractionation, and thus the use of osmotic shock was employed to obtain an enriched pool for separation of these proteins. Studies using puri-

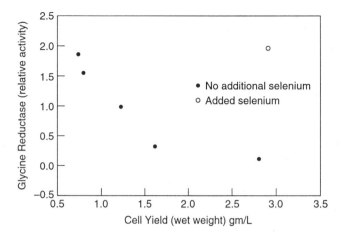

FIGURE 12.1. Requirement for selenium in glycine reductase activity in *C. sticklandii* when cultured in a defined medium. Data from Turner and Stadtman (1973).

fied components of the B and C fractions determined that protein B contained a pyruvate cofactor (Arkowitz and Abeles 1991) and that protein C was involved in the production of acetyl-phosphate (Arkowitz and Abeles 1989). Thus the overall reaction of GR is as follows:

$$NH_3CH_2COO^- + Pi + 2e^- \rightarrow CH_3COOPO_3^{2-} + NH_4^+ \qquad (12.1)$$

Dithiothreitol can be used as the source of electrons for the reaction in vitro, although it is widely held that the reduction of the GR in vivo is linked to the pool of reduced nucleotides (NADH or NADPH). A more detailed description of the characterization of each component of the GR is given below.

Selenoprotein A

Selenoprotein A is a small, acidic protein that contains selenium in the form of selenocysteine (Cone et al. 1976). The determination that selenium is present in the polypeptide in the form of selenocysteine was the first chemical identification of the selenium moiety in any selenoprotein. Numerous selenoenzymes have been studied since this initial report, and in almost all cases (with the exception of the selenium-dependent molybdenum hydroxylases) selenium is found in the form of selenocysteine. The cloning and sequencing of the gene for selenoprotein A revealed, as expected, a TGA codon within the coding sequence (Garcia and Stadtman 1991). The occurrence of a TGA to encode selenocysteine was described earlier during the analysis of the genes encoding selenium-dependent FDH-H of *E. coli* (Zinoni et al. 1986).

Selenoprotein A is remarkably heat stable, as seen by the loss of only 20% of activity on boiling at pH 8.0 for 10 min (Turner and Stadtman 1973). Although selenoprotein A contains one tyrosine and no tryptophan residues, it contains six phenylalanine residues and thus has an unusual absorbance spectrum (Cone et al. 1977). Upon reduction, a unique absorption peak emerges at 238 nm, presumably due to the ionized selenol of selenocysteine, which is not present in the oxidized enzyme. The activity of selenoprotein A was initially measured as its ability to complement fractions B and C for production of acetate from glycine, in the presence of reducing equivalents (e.g., dithiothreitol). Numerous purification schemes were adopted for isolation of selenoprotein A, all of which employed the use of an anion exchange column to exploit the strongly acidic character of the protein.

In addition to *C. sticklandii*, GR selenoprotein A was purified from other clostridial species, including *C. purinolyticum* (Sliwkowski and Stadtman 1988) and *Clostridium litoralis* (Dietrichs et al. 1991) as well as the closely related species *Eubacterium acidaminophilum* (Dietrichs et al. 1991). Each of these proteins is similar in size and composition of amino acids, and current evidence suggests these contain selenium in the form of selenocysteine. Selenoprotein A from both *C. litoralis* and *E. acidaminophilum* was purified using its ability to further stimulate a thioredoxin reductase (TrxR)-like flavoprotein activity (Dietrichs et al. 1991). This enhancement of the TrxR-like activity implicated a TrxR-like enzyme as a possible link to the pool of reduced nucleotides. In support of this theory, genes encoding components of GR of *Clostridium litorale* were reported to be flanked by genes encoding both thioredoxin (Trx) and TrxR (Kreimer and Andreesen 1995). Although selenoprotein A is the best characterized component of GR, its catalytic role is better discussed in the context of the role for proteins isolated from fractions B and C.

Glycine Reductase Protein B

The initial report that GR protein A was a selenoprotein also included the separation of the remaining components of GR into two fractions using DE-32 ion exchange chromatography (Turner and Stadtman 1973), termed fractions B and C. Protein B was partially purified in and was noted to be a large molecule, based on its mobility in nondenaturing polyacrylamide gel electrophoresis (PAGE), with a molecular weight of 200,000–250,000. It was found to be sensitive to borohydride treatment and hydroxylamine, both known to inhibit carbonyl-dependent enzymes. The sensitivity to these reagents was reduced in the presence of glycine (Tanaka and Stadtman 1979), indicating protein B may be interacting with the substrate glycine. Moreover, protein B directed a slow exchange of tritium from tritium-labeled glycine into water, suggesting a Schiff's base intermediate was being formed. This early work on the role for protein B not only demonstrated that glycine was interacting with protein B, but also indicated that fraction C was required for ATP generation (discussed below).

Eubacterium acidaminophilum not only reductively deaminates glycine to ammonia and acetate but also expresses enzymes capable of reductively deaminating sarcosine and betaine when cells are cultured in the presence of formate (Hormann and Andreesen 1989). One would expect that the enzymes catalyzing the latter two reactions might be similar to GR. In fact, the substrate-specific protein B for sarcosine reductase was purified and found to be similar to GR protein B from *C. sticklandii* (Meyer et al. 1995). Apparently, this organism has evolved means to use different amino acids as electron acceptors and preferentially expresses each in response to conditions in the environment.

In addition to SR, components of GR from *E. acidaminophilum* have also been isolated. In a study outlining the purification of the substrate-specific protein of GR, the genes encoding GR proteins from *E. acidaminophilum* were also sequenced. Unexpectedly, a TGA codon was uncovered in a gene encoding a subunit of protein B as well as the genes encoding selenoprotein A, indicating selenocysteine is present in both A and B proteins of GR (Wagner et al. 1999). In agreement with the sequence, purified GR protein B obtained from cells grown in medium containing ^{75}Se was labeled with selenium. The labeled protein, approximately 47 kDa, is apparently present in a complex with two other proteins, approximately 22 kDa and 25 kDa, to make up the components of protein B. The latter two smaller proteins are encoded by one gene, *grdE*, which is subsequently cleaved after translation (Wagner et al. 1999). The genes encoding protein B components for sarcosine and betaine reductase have also been cloned from this organism, and sequence analysis revealed a gene encoding a protein similar to the 47 kDa protein from GR, which also contains a TGA codon (Andreesen et al. 1999). A reaction mechanism to explain the need for selenocysteine in a substrate-specific protein B had not previously been considered but has been proposed (Andreesen et al. 1999). Further biochemical analysis of purified components of GR and its close relatives (sarcosine and betaine reductases) from this organism are needed to elucidate the role for selenocysteine in these subunits and determine more specifically the catalytic mechanism of these enzymes.

Glycine Reductase Protein C

Early attempts at purification of the protein C component of the GR complex focused on the reconstitution of all three components to generate acetate in the presence of glycine. However, a critical discovery was made in regard to the overall reaction mechanism of GR by Arkowitz and Abeles (1989) when it was determined that acetyl phosphate was generated by GR in the presence of inorganic phosphate. This result explained the nature of the one equivalent of ATP generated (through acetate kinase) and significantly altered the methods employed for purification of protein C. Stadt-man (1989) showed that acetyl phosphate, normally generated by reaction of the proposed acetyl-thiol ester intermediate on protein C with phosphate, could be decomposed in the reversible back reaction with arsenate to form acetyl arsenate. Spontaneous decomposition of the unstable arsenate ester to acetate resulted in net hydrolysis of acetyl phosphate. Inhibition of protein C activity by alkylation supported a role of a thiol group in the reaction. In studies using the $Se[^{14}C]$carboxymethyl derivative of selenoprotein A as substrate, Stadtman and Davis (1991) showed that it was decomposed to $[^{14}C]$acetate by protein C in the presence of arsenate and confirmed that prior alkylation of protein C in the absence of acetyl phosphate inhibited the reaction.

The critical discovery that acetyl phosphate is generated and the information gained from several studies of each of the components of GR allowed an enzyme mechanism to be proposed (Arkowitz and Abeles 1991). However, with the current knowledge that one of the subunits of protein B also contains selenium, further work is needed to characterize the intermediates of the reaction and to explain the role of an additional selenocysteine residue. Whether this additional selenocysteine residue in protein B might serve as a direct reductant of the postulated thioselenide derivative of selenoprotein A and possibly serve as a link to the Trx-TrxR system is unknown. It should also be noted that the selenium-limited cultures that were initially studied during analysis of selenoprotein A (Turner and Stadtman 1973) apparently contained active fractions of proteins B and C, suggesting the role for selenium in protein B may not prove to be absolutely necessary for enzyme catalysis.

Proline Reductase

Clostridium sticklandii also expresses a proline reductase that can reductively cleave proline to δ-aminovalerate (Seto and Stadtman 1976). PR was first purified by Seto and Stadtman (1976) by following the decomposition of proline in the presence of dithiothreitol or NADH. They found PR to have a denatured mass of approximately 30 kDa (sodium dodecyl sulfate–polyacrylomide gel electrophoresis; SDS-PAGE) and a native size of approximately 300 kDa. The addition of selenite to the growth medium of *C. sticklandii* did increase the specific activity of PR in extracts by threefold; however, no selenium was detected in the purified enzyme. It should be noted that this purified enzyme had lost the ability to couple reduction of proline to NADH and thus probably was missing one or more components of the complete enzyme complex.

In a more recent report, Kabisch et al. (1999) noted the purification of PR, also from *C. sticklandii*. The purified enzyme contained three subunits (SDS-PAGE), two of which were labeled with fluorescein thiosemicarbazide, indicating the presence of pyruvoyl groups. Previously, a pyruvate-containing peptide isolated from the N-terminal region of PR (Seto 1980) after mild alkaline hydrolysis was generated by cleavage between the Ser and Glu residues. The genes encoding the subunits of proline reductase were also cloned and sequenced and found to encode a selenoprotein. The native size of the proline reductase complex was determined to be 870 kDa, in contrast to the smaller size determined previously (300 kDa). Despite the apparent disparities in the two reports, a simple analysis of the assay of enzyme activity may reveal the differences. In the earlier work, the decomposition of proline was monitored instead of production of δ-aminovalerate, and cell extracts were treated with detergents before chromatographic separation. In the more recent work no detergent was used, and the production of δ-aminovalerate was followed. However, stoichiometric conversion of proline to δ-aminovalerate had been established earlier (Stadtman and Elliott 1957). Nevertheless, PR is now known to be a complex of proteins, including a selenoprotein, which greatly resembles the GR enzyme complex. Further study of this enzyme in regard to the detailed reaction mechanism is needed to define a role for selenium in the reaction.

Formate Dehydrogenase

Soon after the report that the GR of *C. sticklandii* contained selenium, Andreesen and Ljungdahl (1973) reported the partial purification of a NADP-dependent FDH from the thermophilic anaerobe *C. thermoaceticum*. Determination that FDH activity increased with the addition of selenium during growth of this organism in part sparked the experiments to test the effect of selenium on GR (Turner and Stadtman 1973). FDH activity of *C. thermoaceticum* not only responded positively to the addition of selenium to the culture medium during growth but also increased on addition of tungsten and molybdenum. These findings confirmed the observations made many years before (Pinsent 1954) that selenite and molybdate were required for formation of FDH by *E. coli* and related organisms. Selenium, in the form of ^{75}Se, cochromatographed with FDH activity in this initial study. A more detailed study describing the purification of FDH (FDH is exquisitely sensitive to oxygen inactivation) confirmed the enzyme contained both selenium and tungsten in stoichiometric amounts (Yamamoto et al. 1983). Molybdenum was not present in stoichiometric amounts, but the level of FDH activity did increase significantly in extracts when molybdate was added to the medium. This suggested that expression of FDH may be regulated by the presence of molybdenum, although it is not needed for enzyme activity.

SDS-PAGE analysis of purified FDH revealed two subunits and, in combination with native PAGE, suggested a $\alpha_2\beta_2$ structure. Selenium was not released from the protein during denaturing gel electrophoresis and was found exclusively in the α-subunit. A recent entry in the Genbank database reveals that the gene encoding the α-subunit has been cloned and sequenced and contains a TGA codon encoding selenocysteine (accession number 1658401). Thus the selenium-dependent FDH from *C. thermoaceticum* is similar to other selenium-dependent FDH enzymes studied in *E. coli* (Böck and Sawers 1996), with the exception that a tungsten cofactor is present instead of a molybdenum cofactor. In *E. coli* FDH-H, selenium was determined to be directly coordinated to the Mo atom in the molybdenum cofactor molybdopterin guanine dinucleotide, both by spectral analysis (Gladyshev et al. 1994b) and by structural analysis (Boyington et al. 1997). Most likely the selenium atom in FDH from *C. thermoaceticum* is closely associated with the tungsten molecule in this enzyme. Perhaps future studies of selenium- and tungsten-dependent FDH enzymes from clostridia, such as that described above, could further probe the differences between tungsten- and molybdenum-dependent FDH isoenzymes.

Selenium-Dependent Molybdenum Hydroxylases

Xanthine oxidase (XO) was the first enzyme studied from the family of enzymes now known as the molybdenum hydroxylases (Hille 1999). XO, which catalyzes the hydroxylation of xanthine to uric acid is abundant in cow's milk and contains several cofactors, including FAD, two Fe-S centers, and a molybdenum cofactor, all of which are required for activity (Massey and Harris 1997). Purified XO has been shown to use xanthine, hypoxanthine, and several aldehydes as substrates in the reduction of methylene blue (Booth 1938), used as an electron acceptor. Early studies also noted that cyanide was inhibitory but could only inactivate XO during preincubation, not during the reaction with xanthine (Dixon 1927). The target of cyanide inactivation was identified to be a labile sulfur atom, termed the cyanolyzable sulfur (Wahl and Rajagopalan 1982), which is also required for enzyme activity.

A group of clostridia that efficiently ferments purines, such as xanthine and uric acid, were initially studied in Barker's laboratory (Barker and Beck 1941). Xanthine dehydrogenase (XDH) was purified from one of these organisms (Bradshaw and Barker 1960) and found to be similar to bovine XO. During the study of XDH from other strains of purinolytic clostridia it was reported that XDH activity responded positively to selenium addition to the culture medium (Wagner and Andreesen 1979), suggesting selenium might be required for XDH activity. In addition to XDH, other related molybdenum hydroxylases have been investigated from these purinolytic clostridia, and a brief overview of the need for selenium in each is given below.

Xanthine Dehydrogenase

Clostridium acidi-urici and *Clostridium cylindrosporum* were characterized in regard to their ability to ferment purines to ammonia, CO_2, and acetic acid (Barker and Beck 1941). A critical enzyme in the pathway for use of purines as carbon and energy sources in these organisms is XDH. Purified from *C. cylindrosporum* by Bradshaw and Barker (1960), XDH contained significant iron, molybdenum, and flavin as cofactors and used xanthine and other purine derivatives as substrates. *C. cylindrosporum* XDH was similar to the well-studied bovine XO (Massey and Harris 1997), and these authors likely never suspected their purified enzyme contained selenium. However, later analysis of the growth of *C. acidi-urici* and *C. cylindrosporum* and activity of XDH in extracts of these organisms (Wagner and Andreesen 1979) indicated that inclusion of selenite resulted in a 14-fold increase in XDH activity, suggesting that XDH enzymes from both organisms required selenium for activity.

The first description of a selenium-containing XDH was reported in an analysis of the purification of XDH from *C. acidi-urici*. In that report, XDH was shown to contain 0.13 mole of selenium per mole of enzyme (Wagner et al. 1984). Electron spin resonance of a dithionite-reduced form of XDH revealed a spectrum similar to the analogous desulfo-XO prepared by incubation with cyanide. These results suggested the selenium atom that occurs in XDH is analogous to the cyanolyzable sulfur previously described in XO (Wahl and Rajagopalan 1982). In addition, these results demonstrated that selenium is required for activity and, as the authors suggest, significant amounts of a labile form of selenium may have been lost during purification. A more recent report on the purification of XDH from *Eubacterium barkeri*, a nicotinic acid fermentor originally described by Stadtman et al. (1972), revealed selenium was present in a molar ratio of 0.24:1 with respect to each set of three unique subunits (Schräder et al. 1999). Selenium was removed from the native enzyme by treatment with cyanide, and the loss of selenium resulted in concomitant loss in activity. XDH activity could be restored by incubation with selenide and a reducing agent, similar to the technique used to restore activity of XO using sulfide (Wahl and Rajagopalan 1982).

Another study (Self and Stadtman 2000) described the purification of XDH from *C. purinolyticum*, a purine-fermenting strain originally isolated as an adenine-fermentor (Dürre et al. 1981). Selenium was labile (cyanolyzable) and required for XDH activity. Similar to XDH from *E. barkeri*, XDH from *C. purinolyticum* consisted of three subunits (determined by SDS-PAGE). However, XDH from *C. purinolyticum* was significantly stable when isolated under aerobic conditions. Although these reports solidify the previously speculative data that selenium is present as a labile cofactor in clostridial XDH, the exact nature of selenium and the molecules to which this selenium is bound within the active site of XDH remain unknown and the focus of current investigation.

Nicotinic Acid Hydroxylase

Another selenium-containing molybdenum hydroxylase that has been isolated from *Clostridium barkeri* (identical to *Eubacterium barkeri*) is nicotinic acid hydroxylase (NAH). *Clostridium barkeri* was isolated initially as a fermentor of nicotinic acid and thus NAH is a key enzyme in the efficient fermentation of nicotinic acid as a source of carbon and energy. NAH contained selenium when purified from cells labeled with ^{75}Se-selenite. However, this label was lost during denaturing gel electrophoresis and also on heating of the enzyme (Dilworth 1982). Exhaustive analysis of selenium-labeled alkylation products of NAH under various conditions revealed selenium was bound as a labile cofactor (Dilworth 1982), and not as selenocysteine. This report was the first to describe a selenium-dependent enzyme that did not contain selenium in the form of selenocysteine.

NAH is composed of four subunits (SDS-PAGE) and contains a molybdenum cofactor (Dilworth 1983). Analysis of the electron paramagnetic resonance (EPR) spectra of the molybdenum center of NAH revealed a coordination of molybdenum to selenium (Gladyshev et al. 1994b). Apparently NAH is much like other selenium-dependent molybdenum hydroxylases such as XDH from *C. barkeri* and other purinolytic clostridia. Whether or not the selenium is present as a ligand of molybdenum or is coordinated to molybdenum while being bound to another molecule (e.g., sulfur of cysteine) is still not known. The nature of the selenium cofactor and the mechanism of its incorporation into NAH are most likely similar to XDH and thus also require more study.

Purine Hydroxylase

In addition to the molybdenum hydroxylases mentioned above, a new selenium-dependent hydroxylase with specificity for purine and hypoxanthine as substrates, termed purine hydroxylase, was uncovered during purification of XDH from *C. purinolyticum* (Self and Stadtman 2000). Purified PH was labeled with ^{75}Se and was not reduced in the presence of xanthine as a substrate. As with other selenium-dependent molybdenum hydroxylases, selenium was removed by treatment with cyanide with parallel loss in catalytic activity. Selenium was also efficiently removed in the presence of low ionic strength buffer during final dialysis of PH, indicating that ionic strength affects the stability of the labile selenium cofactor in this enzyme.

The subunit structure of PH, an $\alpha_4\beta_4\gamma_4\delta_4$ (SDS-PAGE and gel permeation chromatography), is similar to that of the NAH (Gladyshev et al. 1994a). Comparative kinetic analysis of PH and XDH from *C. puri-*

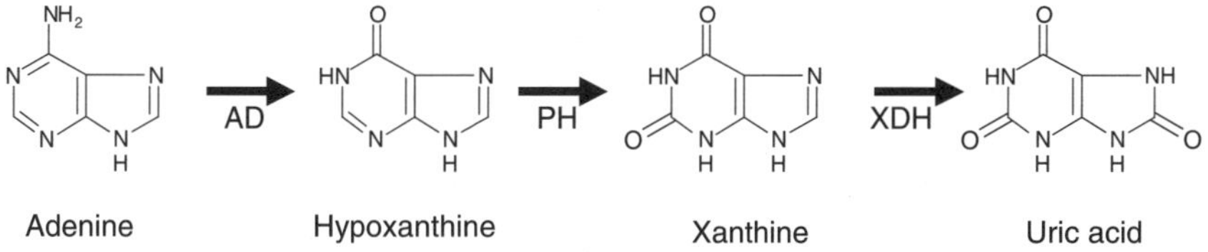

FIGURE 12.2. Pathway for adenine fermentation in *C. purinolyticum*. *AD*, adenine deaminase.

nolyticum revealed that XDH was specific for xanthine, and PH acted most efficiently on either purine or hypoxanthine. The products of the reaction of purine with PH were hypoxanthine and xanthine, based on high-performance liquid chromatography (HPLC) analysis, demonstrating that PH can efficiently hydroxylate either the two or the six positions of the purine ring (Fig. 12.2). A revised pathway for hydroxylation of purines in *C. purinolyticum* to include PH is shown in Figure 12.2. It should be noted that this organism will not grow in the absence of adenine (Dürre et al. 1981). Thus the production of hypoxanthine from adenine and therefore the hydroxylation of hypoxanthine may be critical to this organism's niche in the metabolism of purines. It is possible that other purinolytic anaerobes, some eukaryotes or members of the archea may possess an enzyme similar to PH with specificity for hypoxanthine in the process of purine degradation. This possibility as well as the further study of the nature of and delivery of selenium to PH is currently under investigation.

Conclusion

This brief description of selenium-dependent enzymes, studied from both amino acid and purine-fermenting members of the genus *Clostridium*, emphasizes the significant effect these investigations have had on selenium biochemistry. The study of basic metabolic pathways in bacteria can lead to a significant base of knowledge, which is drawn on by cell biologists and biochemists alike to further their research in mammals. Both the nature of selenium within selenoenzymes as well as the pathway for incorporation of selenium have been well established in *E. coli* and related organisms, and the information gained in these studies helps guide selenium research in all areas, even today. It is hoped, even in this era of biological science dominated by DNA sequence data, that the biochemical characterization of enzymes and pathways from bacterial sources will continue to lead basic science in the future.

References

Andreesen JR, Ljungdahl LG. 1973. Formate dehydrogenase of *Clostridium thermoaceticum*: incorporation of selenium-75, and the effects of selenite, molybdate, and tungstate on the enzyme. J Bacteriol 116:867–73.

Andreesen JR, Wagner M, Sonntag D, et al. 1999. Various functions of selenols and thiols in anaerobic Gram-positive, amino acids-utilizing bacteria. Biofactors 10: 263–70.

Arkowitz RA, Abeles RH. 1989. Identification of acetyl phosphate as the product of clostridial glycine reductase. Evidence for an acyl enzyme intermediate. Biochemistry 28:4639–44.

Arkowitz RA, Abeles RH. 1991. Mechanism of action of clostridial glycine reductase—isolation and characterization of a covalent acetyl enzyme intermediate. Biochemistry 30:4090–7.

Barker HA, Beck JV. 1941. The fermentative decomposition of purines by *Clostridium acidi-urici* and *Clostridium cylindrosporum*. J Biol Chem 141:3–27.

Böck A, Sawers G. 1996. Fermentation. In: Neidhardt FC, Curtiss R III, Ingraham JL, et al., editors. *Escherichia coli* and *Salmonella*: cellular and molecular biology. 2nd ed. Washington, DC: ASM Press. p 262–82.

Böck A, Forchhammer K, Heider J, et al. 1991. Selenocysteine: the 21st amino acid. Mol Microbiol 5:515–20.

Booth VH. 1938. On the specificity of xanthine oxidase. Biochem J 32:494–502.

Boyington JC, Gladyshev VN, Khangulov SV, et al. 1997. Crystal structure of formate dehydrogenase H: catalysis involving Mo, molybdopterin, selenocysteine and an Fe_4S_4 cluster. Science 275:1305–8.

Bradshaw WH, Barker HA. 1960. Purification and properties of xanthine dehydrogenase from *Clostridium cylindrosporum*. J Biol Chem 235:3620–9.

Cone JE, del Rio RM, Davis JN, Stadtman TC. 1976. Chemical characterization of the selenoprotein component of clostridial glycine reductase: identification of selenocysteine as the organoselenium moiety. Proc Natl Acad Sci USA 73: 2659–63.

Cone JE, del Rio RM, Stadtman TC. 1977. Clostridial glycine reductase complex. Purification and characterization of the selenoprotein component. J Biol Chem 252:5337–44.

Dietrichs D, Meyer M, Rieth M, Andreesen JR. 1991. Interaction of selenoprotein PA and the thioredoxin system, components of the NADPH-dependent reduction of glycine in *Eubacterium acidaminophilum* and *Clostridium litorale*. J Bacteriol 173:5983–91.

Dilworth GL. 1982. Properties of the selenium-containing moiety of nicotinic acid hydroxylase from *Clostridium barkeri*. Arch Biochem Biophys 219:30–8.

Dilworth GL. 1983. Occurrence of molybdenum in the nicotinic acid hydroxylase from *Clostridium barkeri*. Arch Biochem Biophys 221:565–9.

Dixon M. 1927. The effect of cyanide on the Schardinger enzyme. Biochem J 21:840.

Dürre P, Andersch W, Andreesen JR. 1981. Isolation and characterization of an adenine-utilizing, anaerobic sporeformer, *Clostridium purinolyticum* sp. nov. Int J Syst Bacteriol 31:184–94.

Flohe L, Gunzler WA, Schock HH. 1973. Glutathione peroxidase. A selenoenzyme. FEBS Lett 32:132–4.

Garcia GE, Stadtman TC. 1991. Selenoprotein A component of the glycine reductase complex from *Clostridium purinolyticum*: nucleotide sequence of the gene shows that selenocysteine is encoded by UGA. J Bacteriol 173: 2093–8.

Gladyshev VN, Khangulov SV, Axley MJ, Stadtman TC. 1994b. Coordination of selenium to molybdenum in formate dehydrogenase H from *Escherichia coli*. Proc Natl Acad Sci USA 91:7708–11.

Gladyshev VN, Khangulov SV, Stadtman TC. 1994a. Nicotinic acid hydroxylase from *Clostridium barkeri*: electron paramagnetic resonance studies show that selenium is coordinated with molybdenum in the catalytically active selenium-dependent enzyme. Proc Natl Acad Sci USA 91:232–6.

Hormann K, Andreesen JR. 1989. Reductive cleavage of sarcosine and betaine by *Eubacterium acidaminophilum* via enzyme systems different from glycine reductase. Arch Microbiol 153:50–9.

Hille R. 1999. Molybdenum enzymes. Essays Biochem 34:125–37.

Kabisch UC, Grantzdorffer A, Schierhorn A, et al. 1999. Identification of D-proline reductase from as a selenoenzyme and indications for a catalytically active pyruvoyl group derived from a cysteine residue by cleavage of a proprotein. J Biol Chem 274:8445–54.

Kreimer S, Andreesen JR. 1995. Glycine reductase of *Clostridium litorale*. Cloning, sequencing, and molecular analysis of the *grdAB* operon that contains two in-frame TGA codons for selenium incorporation. Eur J Biochem 234: 192–9.

Massey V, Harris CM. 1997. Milk xanthine oxidoreductase: the first one hundred years. Biochem Soc Trans 25:750–5.

Meyer M, Granderath K, Andreesen JR. 1995. Purification and characterization of protein PB of betaine reductase and its relationship to the corresponding proteins glycine reductase and sarcosine reductase from *Eubacterium acidaminophilum*. Eur J Biochem 234:184–91.

Pinsent J. 1954. The need for selenite and molybdate in the formation of formic dehydrogenase by members of the coli-aerogenes group of bacteria. Biochem J 57:10–16.

Schräder T, Rienhofer A, Andreesen JR. 1999. Selenium-containing xanthine dehydrogenase from *Eubacterium barkeri*. Eur J Biochem 264:862–71.

Self WT, Stadtman TC. 2000. Selenium-dependent metabolism of purines: a selenium-dependent purine hydroxylase and xanthine dehydrogenase were purified from *clostridium purinolyticum* and characterized. Proc Natl Acad Sci USA 97:7208–13.

Seto B. 1980. Chemical characterization of an alkali-labile bond in the polypeptide of proline reductase from *Clostridium sticklandii*. J Biol Chem 255:5004–6.

Seto B, Stadtman TC. 1976. Purification and properties of proline reductase from *Clostridium sticklandii*. J Biol Chem 251:2435–9.

Sliwkowski MX, Stadtman TC. 1988. Selenium-dependent glycine reductase: differences in physicochemical properties and biological activities of selenoprotein A components isolated from *Clostridium sticklandii* and *Clostridium purinolyticum*. Biofactors 1:293–6.

Stadtman TC. 1966. Glycine reduction to acetate and ammonia: identification of ferredoxin and another low molecular weight acidic protein as components of the reductase system. Arch Biochem Biophys 113:9–19.

Stadtman TC. 1989. Clostridial glycine reductase: protein C, the acetyl group acceptor, catalyzes the arsenate-dependent decomposition of acetyl phosphate. Proc Natl Acad Sci USA 86:7853–6.

Stadtman TC, Davis JN. 1991. Glycine reductase protein C. Properties and characterization of its role in the reductive cleavage of Se-carboxymethyl-selenoprotein A. J Biol Chem 266:22147–53.

Stadtman TC, Elliott P. 1957. Studies on the Enzymatic Reduction of amino acids: II. Purification and properties of a_D-proline reductase and a proline racemase from *Clostridium sticklandii*. J Biol Chem 228:983–97.

Stadtman ER, Stadtman TC, Pastan I, Smith LD. 1972. *Clostridium barkeri* sp. n. J Bacteriol 110:758–60.

Stickland LH. 1935. Studies in the metabolism of the strict anaerobes (genus *Clostridium*). Part IV. The reduction of glycine by *Cl. sporogenes*. Biochem J 29:889–98.

Tanaka H, Stadtman TC. 1979. Selenium-dependent clostridial glycine reductase. Purification and characterization of the two membrane-associated protein components. J Biol Chem 254:447–52.

Turner DC, Stadtman TC. 1973. Purification of protein components of the clostridial glycine reductase system and characterization of protein A as a selenoprotein. Arch Biochem Biophys 154:366–81.

Wagner R, Andreesen JR. 1979. Selenium requirement for active xanthine dehydrogenase from *Clostridium acidiurici* and *Clostridium cylindrosporum*. Arch Microbiol 121:255–60.

Wagner R, Cammack R, Andreesen JR. 1984. Purification and characterization of xanthine dehydrogenase from *Clostridium acidiurici* grown in the presence of selenium. Biochim Biophys Acta 791:63–74.

Wagner M, Sonntag D, Grimm R, et al. 1999. Substrate-specific selenoprotein B of glycine reductase from *Eubacterium acidaminophilum*. Biochemical and molecular analysis. Eur J Biochem 260:38–49.

Wahl RC, Rajagopalan KV. 1982. Evidence for the inorganic nature of the cyanolyzable sulfur of molybdenum hydroxylases. J Biol Chem 257:1354–9.

Yamamoto I, Saiki T, Liu SM, Ljungdahl LG. 1983. Purification and properties of NADP-dependent formate dehydrogenase from *Clostridium thermoaceticum*, a tungsten-selenium-iron protein. J Biol Chem 258:1826–32.

Zinoni F, Birkmann A, Stadtman TC, Böck A. 1986. Nucleotide sequence and expression of the selenocysteine-containing polypeptide of formate dehydrogenase (formate-hydrogen-lyase-linked) from *Escherichia coli*. Proc Natl Acad Sci USA 83:4650–4.

13
How the Diverse Physiologic Potentials of Acetogens Determine Their In Situ Realities

Harold L. Drake and Kirsten Küsel

The in situ activity of a cell is determined by which physiologic traits are expressed under in situ conditions. Thus the ecological impact of acetogens is determined by the in situ manifestation of their physiologic potentials. Since their physiologic capabilities are diverse, it must be anticipated that the in situ activities of acetogens are likewise diverse. Indeed, because acetogens constitute a phylogenetically diverse bacteriologic group, it should be anticipated that they can inhabit, and have an effect on, diverse habitats. The intent of this chapter is to outline recent work that highlights what is known about the realities of acetogens in both test tubes and ecosystems.

Acetogenic Bacteria and the Acetyl-CoA Pathway

Acetogens are obligately anaerobic bacteria capable of reducing CO_2 to acetate via the acetyl-CoA pathway (also referred to as the Wood-Ljungdahl pathway) (Fig. 13.1). The acetyl-CoA pathway consists of two reductive branches that together reduce CO_2 to acetyl-CoA, is a terminal electron-accepting process, conserves energy, and can serve as a source of usable carbon for the synthesis of biomass (Wood and Ljungdahl 1991; Drake 1994; Ljungdahl 1994). The overall reduction of CO_2 to acetate with hydrogen has a free energy change of approximately −100 kJ per mole acetate synthesized. Energy is conserved by generating either proton- or sodium-dependent chemiosmotic potentials that can be coupled to the synthesis of ATP by ATPases or used for energy-dependent processes, such as active transport (Das and Ljungdahl 2000; Müller et al. 2001). As noted below, use of this pathway is not the only option acetogens have relative to the conservation of energy.

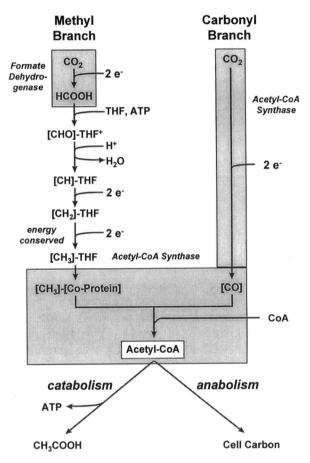

FIGURE 13.1. The acetyl-CoA pathway showing reactions central to the initial reduction of CO_2 and the synthesis of acetyl-CoA (*shaded*). *THF*, tetrahydrofolate; *Co-protein*, corrinoid protein; *brackets*, the C_1 unit is bound to a cofactor or structurally associated with an enyzme.

Phylogenetic Diversity: Who Has the Acetyl-CoA Pathway?

To date, there are 19 bacterial genera that can use the acetyl-CoA pathway and be termed acetogens: *Acetitomaculum, Acetoanaerobium, Acetobacterium, Acetohalobium, Acetonema, Caloramator, Clostridium, Eubacterium, Holophaga, Moorella, Natroniella, Natronoincola, Oxobacter, Ruminococcus, Sporomusa, Syntrophococcus, Thermoacetogenium, Thermoanaerobacter,* and *Treponema* (Drake et al. 2003). Based on their 16S rRNA gene sequence similarities, these genera are not tightly clustered phylogenetically but are widely dispersed throughout the domain Bacteria. Together with taxonomically unclassified isolates, the total number of known acetogenic species approximates 100. Acetate-producing organisms that do not use the acetyl-CoA pathway are sometimes incorrectly referred to as acetogens; thus one should take special note of how this term is being applied relative to the description of an organism. Two additional less obvious points should also be noted when considering whether an organism is an acetogen: Bacteria not previously identified as acetogens (i.e., true users of the acetyl-CoA pathway) might nonetheless be acetogens, and the in situ acetogenic capacity of a bacterium can be lost when the organism is cultivated in the laboratory (Küsel et al. 2000, 2001).

Use of the acetyl-CoA pathway by nonacetogens and members of the domain Archaea for biomass synthesis or the oxidation of acetate demonstrates that this pathway is a fundamental process of the Prokaryotes and is widely dispersed (Wood and Ljungdahl 1991; Drake 1994). It has been postulated that the pathway was the first autotrophic process of the planet, and it is estimated that acetogenesis yields trillions of kilograms of acetate globally each year. Thus it is now recognized that the acetyl-CoA pathway is important to the global carbon cycle. These more recent considerations of the acetyl-CoA pathway are in marked contrast to earlier times when acetogens and acetogenesis were considered little more than a biological curiosity.

Metabolic Versatility: What Acetogens Do in Test Tubes

What are the realities of acetogens when cultivated in test tubes, i.e., when cultivated under our terms in the laboratory? Historically important answers to these questions were found when the first two acetogens *Clostridium aceticum* (Wieringa 1936) and *Moorella thermoacetica* (originally *Clostridium thermoaceticum*) (Fontaine et al. 1942) were isolated. Two unique physiologic features were observed with these organisms: the stoichiometric conversion of glucose to acetate (Eq. 13.1) and the hydrogen-dependent reduction of CO_2 to acetate (Eq. 13.2).

$$C_6H_{12}O_6 \rightarrow 3CH_3COOH \tag{13.1}$$

$$4H_2 + 2CO_2 \rightarrow CH_3COOH + 2H_2O \tag{13.2}$$

Reactions 13.1 and 13.2 are now recognized as the hallmark signatures of acetogens. During many decades of research, these two reactions were emphasized in evaluating the realities of acetogens in test tubes and in studies that ultimately resolved the enzymology of the acetyl-CoA pathway (Drake 1994; Ljungdahl 1994). In retrospect, the high phylogenetic diversity of acetogens strongly suggests that acetogens harbor diverse physiologic potentials that are not indicated by Eqs. 13.1 and 13.2. Indeed, they do.

Sources of Reductant

In Eq. 13.1, glucose is oxidized and decarboxylated via glycolysis and pyruvate ferredoxin oxidoreductase, and the resulting eight reducing equivalents are used to synthesize acetate via the acetyl-CoA pathway (Fig. 13.1). Six reducing equivalents are used in the methyl branch of the pathway to reduce CO_2 to the methyl level, and two reducing equivalents are used in the carbonyl branch of the pathway to reduce CO_2 to the carbonyl level. In Eq. 13.2, the eight reducing equivalents required to reduce CO_2 to acetate are derived from hydrogen via hydrogenase.

Work in more recent years has demonstrated that acetogens are capable of using diverse substrates as a source of reductant. It is beyond the scope of this chapter to delineate fully the wide range of substrates that can be utilized by acetogens. Oxidizable substrates include CO, hydrogen, carbohydrates, alcohols, carboxylic acids (including dicarboxylic acids), aldehydes, substituent groups of aromatic compounds, and numerous other organic and halogenated substrates (Drake 1994; Schink 1994). Although acetogens can use a wide range of substrates and participate in complex trophic linkages with other organisms, acetogens do not appear to be able to use complex polymers, such as lignin or cellulose.

Certain one-carbon substrates, such as CO and the methoxyl group of aromatic compounds, are particularly well suited for use by acetogens because they are readily incorporated into either the carbonyl or the methyl branches of the acetyl-CoA pathway. In this regard, acetogens appear to be specialized in the use of substituent groups of aromatic compounds. Use of such substituent groups can have regulatory effects on the flow of carbon and reductant in the cell. For example, use of the aldehyde group of certain lignin-derived aromatic compounds by *Clostridium formicoaceticum* inhibits the cell's ability to use fructose (Frank et al. 1998). In the presence of both fructose and a usable hydroxybenzaldehyde, reductant for the synthesis of acetate from CO_2 is preferentially derived from the aldehyde group of hydroxybenzaldehyde (Fig. 13.2).

Use of Diverse Terminal Electron Acceptors

Acetogens can also use a diverse number of terminal electron acceptors substrates (Drake 1994; Drake et al. 1997); these acceptors and the resulting reduced end products include:

$$\text{carbon dioxide} \rightarrow \text{acetate} \tag{13.3}$$

$$\text{fumarate} \rightarrow \text{succinate} \tag{13.4}$$

$$\text{methoxylated phenylacrylates} \rightarrow \text{methoxylated phenylpropionates} \tag{13.5}$$

$$\text{nitrate} \rightarrow \text{nitrite} \tag{13.6}$$

$$\text{nitrite} \rightarrow \text{ammonium} \tag{13.7}$$

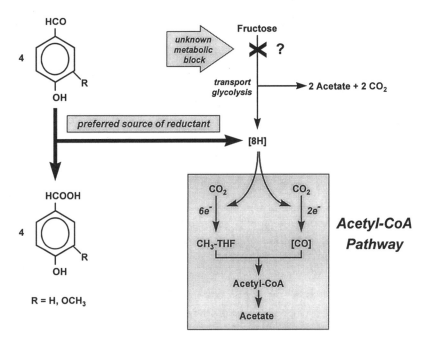

FIGURE 13.2. The preferred flow of reductant from aromatic aldehydes to the acetyl-CoA pathway by the acetogen *C. formicoaceticum*. *THF*, tetrahydrofolate; *brackets*, the C_1 unit is bound to a cofactor or structurally associated with an enzyme.

$$\text{thiosulfate} \rightarrow \text{sulfide} \qquad (13.8)$$

$$\text{dimethylsulfoxide} \rightarrow \text{dimethylsulfide} \qquad (13.9)$$

$$\text{pyruvate} \rightarrow \text{lactate} \qquad (13.10)$$

$$\text{acetaldehyde} \rightarrow \text{ethanol} \qquad (13.11)$$

$$\text{protons} \rightarrow \text{hydrogen gas} \qquad (13.12)$$

Each acetogen displays a different propensity relative to the use of alternative electron acceptors. However, a few generalizations can be made.

First, most if not all acetogens can produce more than acetate as their sole reduced end product. Thus referring to acetogens as homoacetogens is usually a misnomer. Acetogens can facilitate the homoacetogenic conversion of a substrate to acetate, but doing so depends on the condition of growth. For example, the classic homoacetogen *M. thermoacetica* preferentially reduces nitrate instead of CO_2 by repressing a membranous cytochrome *b* that is essential to electron flow on the methyl branch of the acetyl-CoA pathway (Fröstl et al. 1996); thus, in the presence of nitrate, acetate is not synthesized via the acetyl-CoA pathway (Fig. 13.3). Hydro-

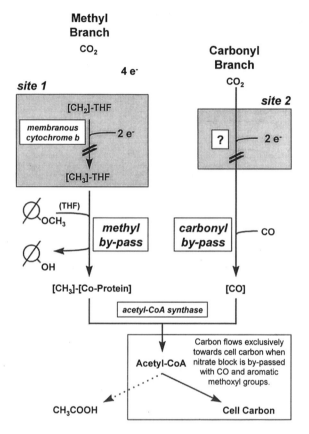

FIGURE 13.3. The nitrate-dependent blocks (*shaded*) in the flow of reductant and carbon when *M. thermoacetica* is cultivated under nitrate-dissimilating conditions. *THF*, tetrahydrofolate; *Co-protein*, corrinoid protein; *brackets*, the C_1 unit is bound to a cofactor or structurally associated with an enzyme.

genase activity is also strongly affected by nitrate. Although the function of hydrogenase(s) during heterotrophic growth on sugars is unknown, it is likely that at least one function is as an oxidoreductase in the intracellular management of sugar-derived reductant rather than as a classic hydrogen-producing or -consuming hydrogenase. Nitrate might also have other regulatory effects on the acetyl-CoA pathway (Arendsen et al. 1999). Another example is *Peptostreptococcus productus*. This co-called homoacetogen shifts reductant flow away from the acetyl-CoA pathway to classic fermentation and the production of lactate, succinate, ethanol, and formate when grown with high concentrations of fructose or conditions that limit the use of extracellular CO_2 (Misoph and Drake 1996).

Second, the availability of CO_2 can influence the ability of an acetogen to engage an alternative terminal electron-accepting process. However, even in the presence of CO_2, alternative electron acceptors can be used, and in many cases simultaneously to the use of CO_2. For example, *C. formicoaceticum* can simultaneously reduce CO_2 to acetate and phenyl-acrylates to phenylpropionates (Misoph et al. 1996).

Third, thermodynamically, the ability of acetogens to use diverse electron acceptors indicates that they can accommodate a wide range of redox conditions. For example, the ability of *M. thermoacetica* to use nitrate indicates that this organism does not require stringently reduced conditions (−300 mV) to grow and might be found in habitats with fluxuating redox conditions. Indeed, this organism is easily isolated from aerated soils (Goessner et al. 1998).

Finally, the reduction of certain electron acceptors might not always be coupled to the conservation of energy.

Life in the Presence of Oxygen

The diverse physiologic capabilities of acetogens in the laboratory indicate that their trophic relationships and adaptation strategies in the real world are not exclusively determined by their ability to use the acetyl-CoA pathway. Because acetogens colonize habitats with unstable redox conditions, such as well-aerated soils, it is likely that they are, by one means or another, able to cope with oxygen. Indeed, acetogens have the ability to consume trace levels of oxygen (Küsel et al. 2001; Karnholz et al. 2002). Acetogens such as *M. thermoacetica*, *Acetobacterium woodii*, *Clostridium magnum*, and *Sporomusa silvacetica* can tolerate and consume 0.5–2% oxygen in the headspace of culture tubes. The ability of acetogens to reduce oxygen is dependent on numerous enzymes and does not appear to be coupled to the conservation of energy; rather this physiologic capacity is likely used to reduce trace levels of oxygen that are inhibitory to growth in their respective habitats (Das et al. 2001; Drake et al. 2002).

It is obvious enough that acetogens, like methanogens, can form important trophic links to other anaerobes under strictly anoxic conditions; such information has contributed significantly to our understanding of how acetogens influence the flow of carbon in anoxic habitats, such as gastrointestinal tracts of animals and water-saturated soils and sediments (discussed below). However, how acetogens orient themselves with other bacteria in soils or other habitats subject to fluxuations in oxygen is not well understood. Because acetogens would likely prefer anoxic niches as habitats, they might be members of microcommunities (microcolonies) that contain nonacetogens that are capable of rapidly consuming oxygen but do not compete with the acetogens for the same substrate. Indeed, given the type of microzone in which an acetogen would likely prefer in a habitat like soil, the best nonacetogenic partner would be a microaerophile. Such a trophic relationship was documented for two soil bacteria that were isolated together in mixed culture (Gößner et al. 1999). One organism (*Thermicanus*

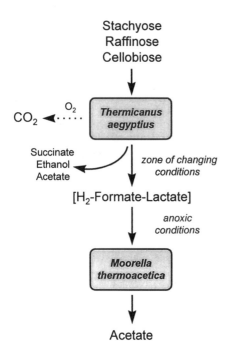

FIGURE 13.4. Postulated trophic inter-action of the microaerophile *T. aegyptius* and the acetogen *M. thermoacetica* (used with permission; Gößner et al. 1999).

aegyptius) is a microaerophile that converts small-chain carbohydrates to products that are subsequently used by the acetogen *M. thermoacetica* (Fig. 13.4). This trophic relationship would be of obvious advantage to an acetogen in aerated soils.

Diverse Habitats and Environmental Realities

Acetogens can seemingly be isolated from all habitats. Most isolates have been obtained from obligately anoxic habitats, such as freshwater or marine sediments, sewage sludge, and gastrointestinal tracts (Drake 1994; Drake et al. 2003). Although acetogens appear to be present in practically all strictly anoxic environments, we know relative-little about their diverse in situ activities in these habitats. In contrast to methanogens, which produce a gas that can be easily evaluated to assess their in situ activities, the activities of acetogens are far more difficult to assess under field conditions. Acetate, the main characteristic product of acetogens, is also produced by a variety of facultative and obligate anaerobes and is subject to complex turnover dynamics in the presence of electron acceptors.

In anoxic environments with low amounts of sulfate or usable Fe(III), the microbial degradation of organic matter occurs via a complex network of trophic links that collectively terminate in the production of methane

(McInerney and Bryant 1981). The main processes in which acetogens participate are highlighted in Figure 13.5. Acetogens compete with primary fermentors for monomeric compounds and with secondary fermentors for typical fermentation products, such as lactate, ethanol, and hydrogen (Schink 1994). In the case of hydrogen, acetogens compete with Fe(III) reducers, sulfate reducers, and methanogens.

The standard free energy change for methanogenesis from hydrogen and CO_2 is more exergonic than that of acetogenesis ($\Delta G'_0 = -135.6$ kJ per mole methane and $\Delta G'_0 = -104.6$ kJ per mole acetate, respectively). However, certain conditions compromise or inhibit methanogenesis (Fig. 13.5). In a complex ecosystem, the metabolic interactions of the anaerobic populations

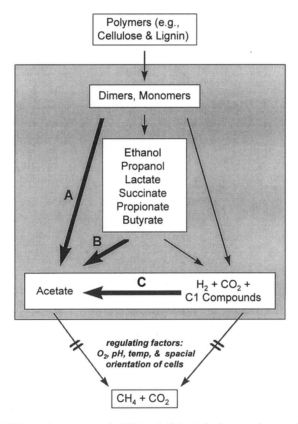

FIGURE 13.5. Flow of carbon in habitats deficient in inorganic terminal electron acceptors (e.g., sulfate and iron). Reaction C is catalzyed exclusively by acetogens. *Shaded*, reactions facilitated by acetogens and other fermentative microorganisms; *thick arrows*, reactions **A–C** in which acetogens participate; Arrow size does not correlate quantitatively with carbon flow. Based in part on the 3-stage model of McInerney and Bryant (1981).

are influenced by physical and chemical parameters (e.g., pH, temperature, periodic oxygenation, spatial arrangements, and different microbial population sizes) that can favor acetogens. For example, in a spatially heterogenic system, acetogens would be more competitive for hydrogen by positioning themselves near hydrogen-producing cells. Thus, under certain circumstances, the acetogenic conversion of hydrogen and one-carbon substrates to acetate can be a significant process (Fig. 13.5; reaction C).

Little information about the spatial distribution of acetogens in ecosytems has been published. Although community structure can been assessed with molecular probes, the phylogenetic diversity of acetogens makes such an assessment difficult (Lovell and Hui 1991). Nonetheless, in the past several years, the presence and in situ activities of acetogens have been detected in hypersaline sediments, deep aquifers, oxic soils, and plant roots, demonstrating that acetogenesis is an underlying process in ecosystems previously unrecognized as habitats of acetogens.

Gastrointestinal Tracts

Acetogenesis is an ongoing process in the gastrointestinal tract of humans and other animals. However, the extent to which acetogenesis from H_2-CO_2 or from organic substrates contributes to the overall turnover of organic matter differs among host species. In ruminants, the production of CH_4 is the predominant terminal electron sink during the microbial anaerobic degradation of organic matter in the rumen (Mackie and Bryant 1994). The molar proportions of short-chain fatty acids produced in the rumen approximate 63% acetic, 21% propionic, 14% butyric, and 2% higher acids. Between 60 and 80% of the daily metabolizable energy intake in ruminant animals is derived from short-chain fatty acids by absorption. Methanogenic bacteria are the predominant hydrogen users in the rumen (Hungate 1976; Bryant 1979). Theoretically, it would be beneficial to the host animal if hydrogen and CO_2 were converted to acetate rather than to CH_4, as this would minimize the loss of usable energy from the host. Some studies indicate that acetogenic bacteria capable of using hydrogen grow preferentially on organic substrates in the rumen (Leedle and Greening 1988). Depending on the host animal investigated and the substrate used, the numbers of acetogenic bacteria in the rumen vary between 10^6 and 10^9 acetogens per gram ruminal content, whereas the numbers of methanogens in the rumen approximate 10^8–10^9 methanogens per gram ruminal content (Leedle and Greening 1988). Several acetogenic hydrogen-using acetogenic strains were isolated from the rumen (Rieu-Lesme et al. 1996). However, to date, information on the diversity and ecology of acetogens in the rumen is inadequate for the successful manipulation of this ecosystem and the selective rerouting of reductant to acetate rather than CH_4.

In humans, dietary components not absorbed in the upper digestive tract reach the colon where they are fermented to acetate, propionate, butyrate, hydrogen, and CO_2. It has been assumed that 95% of the short-chain fatty acids produced are absorbed and used by the host. The daily intestinal production of 10–30 g acetate may be partly attributed to the activity of acetogens. The formation of [1,2-^{14}C]acetate from [3,4-^{14}C]glucose by fecal microbes supports the hypothesis that acetogenesis is an ongoing process in the human colon (Miller and Wolin 1996). Between 30 and 50% of Europeans harbor large populations of methanogens, and methanogenesis is probably the main hydrogenotrophic pathway in this population; in contrast, the reduction of CO_2 to acetate is a major colonic process of non-CH_4 excreting humans (Bernalier et al. 1996). The capacity of acetogens of the human gut to synthesize [^{13}C]acetate from [13]CO_2 was demonstrated (Lajoie et al. 1988). In non-CH_4 excreting humans, fecal populations of hydrogen-using acetogens are higher than in CH_4-excreting humans. Thus hydrogen-dependent acetogenesis appears to be quantitatively important in the presence of low numbers of methanogens (Leclerc et al. 1997).

Termites have a diverse and dense hindgut microbial community that aids in digestion and is the source of fermentation products such as acetate, hydrogen, and CH_4. The symbiotic hindgut microflora of the wood-feeding lower termites includes cellolytic protozoa and bacteria that effect an acetogenic fermentation of the wood polysaccharides ingested by the termite (Brauman et al. 1992). In the extracellular hindgut fluid of *Reticulitermes flavipes*, acetate constitutes 94–99% of the total short-chain fatty acid pool. Cellulose is fermented by the protozoa in the central region of the hindgut to acetate, hydrogen, and CO_2; acetogens then convert the hydrogen and CO_2 to acetate (Breznak 1994).

The in situ spatial distribution of acetogens appears to contribute to their success in consuming hydrogen in termite gut ecosystems. Attachment of hydrogen-using acetogenic spirochetes of the genus *Treponema* to termite gut protozoa likely facilitates the interspecies transfer of hydrogen; under in situ conditions, hydrogen concentrations are likely in excess of the hydrogen threshold values (i.e., the lowest concentration of hydrogen that acetogens can use for the reductive synthesis of acetate from CO_2) for acetogens (Leadbetter et al. 1999). Thus acetogenesis outcompetes methanogenesis for reductant in wood- and in grass-feeding termites. In contrast, acetogenesis from hydrogen and CO_2 appears to be of little significance in fungus-growing and soil-feeding termites, which evolve CH_4 (Brauman et al. 1992). In the soil-feeding termite *Cubitermes* spp., methanogens and acetogens coexist within the posterior hindgut, which has low hydrogen partial pressures in situ; in contrast, the anterior gut regions accumulate hydrogen. Although acetogens are not able to outcompete methanogens for hydrogen in termite gut homogenates, under in situ conditions, substrates other than hydrogen might support the growth of acetogens in the hindgut, or a cross-epithelial hydrogen transfer from anterior gut regions might create microniches favorable for hydrogen-dependent acetogenesis (Tholen and Brune 1999).

Anoxic Sediments, Water-Logged Soils, and Aquifers

Acetogenic bacteria are ubiquitous in anoxic aquatic habitats such as freshwater, estuarine, and marine sediments. In some freshwater sediments, parameters like pH and temperature seem to favor acetogenesis over methanogenesis. Acetogens can outcompete methanogens for hydrogen at the in situ pH of 6.2 and more acidic pH values in lake sediments (Phelps and Zeikus 1984). At low pH, hydrogen-consuming methanogens appear to be more restricted than hydrogen-consuming acetogens, and the flow of carbon and reductant goes primarily through acetate to CH_4. Similarly, in permanently cold sediments, acetogenesis precedes acetotrophic methanogenesis (i.e., the production of CH_4 from acetate). At low in situ temperatures, acetogens successfully compete with methanogens for hydrogen in certain lake sediments and in sediments polluted with papermill waste water. Hydrogenotrophic methanogens can be activated only at higher incubation temperatures (Conrad et al. 1989; Nozhevnikova et al. 1994). The isolation of psychrophilic or psychrotrophic acetogens from those habitats (Kotsyurbenko et al. 1996) underlines the potential importance of acetogenesis in cold habitats.

Flooded rice paddy soils and tundra wetland soils can also have low temperatures that improve the in situ competitiveness of acetogens (Conrad et al. 1989; Nozhevnikova et al. 1994). In paddy soils, approximately 80% of the CH_4 formed is derived from acetate, and acetogens appear to be important in the turnover of acetate in such soils (Chin and Conrad 1995). Acetogenic bacteria in the soil of laboratory microcosms containing rice plants approximate 10^3–10^5 cells per gram of dry soil, and *Sporomusa* might be an important acetogenic genera in such soil (Rosencrantz et al. 1999).

In some marine sediments, the sulfate concentration in pore waters can vary owing to seasonal thermal stratification, and the flow of carbon and reductant can shift from the reduction of sulfate to methanogenesis (Hoehler et al. 1999). During this transition period, the concentration of acetate increases, which might be due to a temporary decoupling of acetate-producing and -consuming processes (Sansone and Martens 1982). Hydrogen concentrations of the sediment are elevated at the beginning of this transition period, making the acetogenic reduction of CO_2 more favorable. Thus acetogenesis might be an important transient process in marine sediments and in other ecosytems that undergo such geochemical fluctuations (Hoehler et al. 1999).

Carbon-limited, low-temperature environments (e.g., deep subsurface aquifers) might also be favorable habitats for acetogens. Acetate accumulates and sulfate is reduced in microcosms containing subsurface sandstones and shales (Krumholz et al. 1997). The acetogen *Acetobacterium psammolithicum* and the sulfate reducer *Desulfomicrobium hypogeium* were isolated from such subsurface material by enrichment with hydrogen. Experimental results suggest that sulfate reduction is the dominant hydrogen-consuming process and acetogens produce acetate via the degradation of low molecular weight organic compounds in these subsurface habitats (Krumholz et al. 1999).

In deep granitic groundwater containing hydrogen and CH_4, acetogens and acetotrophic methanogens dominate the viable cell counts of various physiologic groups (Kotelnikova and Petersen 1997). Acetogens approximate up to 10^4 cells per milliliter of groundwater. In radiotracer experiments, $^{14}CH_4$ and $[^{14}C]$-acetate were formed from $^{14}CO_2$, and $[^{14}C]$-acetate was converted to $^{14}CH_4$. Thus acetogenesis appears to be an ongoing process in this hydrogen-based autotrophic biosphere.

Hypersaline Environments

The anaerobic decomposition of organic matter in the sediments of hypersaline ecosystems (e.g., inland lakes or marine salterns) has received relatively little attention. However, short-chain fatty acids and hydrogen accumulate in such sediments (Oren 1988), indicating that interspecies hydrogen transfer is inhibited or limited during decomposition. Thus hydrogen-dependent acetogenesis might be more favorable when the concentration of hydrogen is elevated. To maintain cell turgor at high salt concentrations, some bacteria accumulate organic osmolytes like betaine. Halophilic, hydrogen-using acetogens (e.g., *Acetohalobium arabaticum*) can convert betaine to trimethylamine and acetate (Zavarzin et al. 1994). Two new haloalkaliphilic acetogenic bacteria, *Natroniella acetigena* and *Natronoincola histidinovorans* (Zhilina et al. 1998), were isolated from soda lakes; both isolates use glutamate, which is also accumulated by moderate halophiles. The capacity of halophilic acetogens to use betaine and glutamate indicates that acetogens are important in the turnover dynamics of osmoprotective compounds in hypersaline habitats.

Seagrass Roots

Biogeochemical processes occur at accelerated rates in the rhizosphere. Based on phospholipid fatty acid and most probable number analyses, the seagrass rhizosphere of certain marine sediments contain significantly higher numbers of sulfate-reducing and acetogenic bacteria than do adjacent, unvegetated sediments (Küsel et al. 1999b). Acetogenic O-demethylation activity (i.e., the capacity to use the methoxyl group of certain aromatic compounds) is tightly associated with seagrass roots. Hybridization of root thin sections with ^{33}P-labeled nucleic acid probes for *Acetobacterium* and *Eubacterium limosum* reveal intracellular bacteria that hybridize with these probes. The bacteria that hybridize with the acetogen probe form clusters and occur mostly in the rhizoplane and outermost cortex (Küsel et al. 1999b).

Root homogenates of the seagrass *Halodule wrightii* contain approximatlely 10^4 culturable hydrogen-using acetogens per gram fresh-weight root. The hydrogen-using acetogen RD1 was isolated from the highest growth-positive dilution of a root homogenate (Küsel et al. 2001). Phylogenetically, RD1 is closely related to *Clostridium glycolicum*, a saccharolytic fermentor not previously known to be an acetogen. Although seagrasses are rooted in highly reduced, anoxic habitats, an oxygen gradient is generated around the roots owing to the transport of oxygen that is produced by leaf photosynthesis during the day. RD1 tolerates up to 6% oxygen in the head space of liquid cultures (Küsel et al. 2001). During the glucose-dependent growth of RD1 in the presence of oxygen, ethanol and hydrogen become significant end products, indicating that, under higher redox conditions, carbon flow switches from acetogenesis to classic fermentation. Thus seagrass rhizosphere and roots appear to be colonized by acetogens that have a high tolerance to transient, root-generated oxygen gradients.

Aerated and Well-Drained Soils

Although acetogens are obligate anaerobes, the environmental roles of acetogens are not restricted to highly reduced, permanently anoxic habitats. Anoxic sites occur in terrestrial soils when the consumption of oxygen exceeds its supply, thus yielding oxygen-free microzones. Anoxic microzones occur in soil aggregates and litter (Sextone et al. 1985; Van der Lee 1999). Litter and soil have tremendous spontaneous capacities to form acetate from endogenous matter under anoxic conditions, indicating that a subcommunity of the microflora can respond rapidly to anoxic conditions (Küsel and Drake 1995, 1996; Peters and Conrad 1996; Wagner et al. 1996). In soils and litter, supplemental hydrogen, CO, and ethanol are converted to acetate in stoichiometries, which approximate those associated with hydrogen-, CO-, and ethanol-dependent acetogenesis, demonstrating that acetogens are members of the acetate-forming microflora of soils and litter.

The number of cultured anaerobes from both forest mineral soil and litter is identical with the number of cultured acetate-producing anaerobes. In soil or litter that has a relatively neutral pH, hydrogen-using acetogens dominate the cultured obligate anaerobes and, for both soil and litter, range from 8×10^3 to 1×10^5 hydrogen-using acetogens per gram dry material (Küsel et al. 1999a) Acidic forest soils have similar ranges for hydrogen-using acetogens (Peters and Conrad 1995). The actual number of soil acetogens is likely much higher than these values indicate, because many acetogens grow poorly or not at all on H_2-CO_2. Nonetheless, it is significant to note that only 1% of the hydrogen-using acetogens in soil and litter are detected after pasteurization. This finding indicates that a large percentage of the spore-forming, hydrogen-using acetogens are in a vegetative, active state or that the dominant hydrogen-using acetogens present in soil and litter are not spore formers (Küsel et al. 1999a).

At in situ temperatures, acetate is a stable end product in anoxic microcosms, and acetotrophic methanogenesis is induced only after extended incubation periods (Küsel and Drake 1995). Acetogenic activities are relatively stable when soils are subjected to oxic drying or varying fluxes of oxygen (Wagner et al. 1996). The capacity of the forest soil acetogen *S. silvacetica* (Kuhner et al. 1997) to grow in the presence of low concentrations of oxygen indicates that it is well adapted to aerated soils (Karnholz et al. 2002). The capacity of soils to form acetate from H_2-CO_2 is enhanced by high temperatures (Küsel and Drake, 1995; Wagner et al. 1996), suggesting that soils that are subject to elevated temperatures might also harbor thermophilic acetogens. The isolation of different strains of *M. thermoacetica* from high-temperature Kansas and Egyptian soils (Goessner et al. 1998) demonstrates that this classic, thermophilic acetogen is geographically wide spread in aerated soils.

The temporal and spatial variability of oxygen and other terminal electron acceptors in soils suggests that the consumption of acetate might be linked to oxidative processes. Oxygen and nitrate are rapidly consumed when added to preincubated anoxic litter or soil microcosms (Küsel and Drake 1995; Wagner et al. 1996). The consumption of oxygen and acetate is concomitant to the production of CO_2, according to the following stoichiometry: $CH_3COOH + 2O_2 \rightarrow 2CO_2 + 2H_2O$. The rate of acetate consumption exceeds the rate of formation under anoxic conditions, indicating a high turnover dynamic under in situ conditions. The oxidation of acetate in soil might also be linked to other anaerobic processes, such as denitrification, the dissimilation of nitrate to ammonium, or the reduction of Fe(III) to Fe(II) (Küsel and Drake, 1995; Wagner et al. 1996; Küsel et al. 2002). For example, Hawaiian soils of volcanic origin contain large amounts of reducible Fe(III), and the acetate formed during the consumption of hydrogen appears to be anaerobically oxidized to CO_2 via the reduction of Fe(III) to Fe(II). Theoretically, the acetate formed in anoxic microzones of aerated soils is subject to a rapid turnover by virtue of the diffusion of oxygen into formerly anoxic zones or the transport of acetate with the soil solution into zones in which electron acceptors (e.g., nitrate, Fe(III), or oxygen) are present (Fig. 13.6). As such, acetate constitutes an important trophic link between different anaerobic and aerobic microbial populations that are collectively involved in the breakdown of organic matter in aerated soils.

Conclusion

For many decades, the capacity of acetogens to reduce CO_2 to acetate and the thermodynamics of the acetyl-CoA pathway casts the acetogens as somewhat obscure specialists and thermodynamically disadvantaged

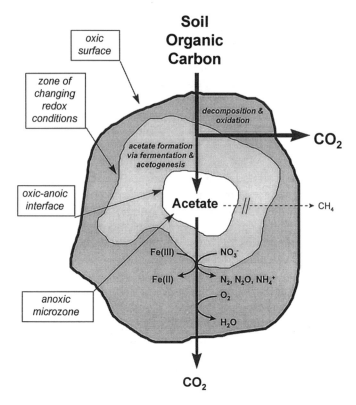

FIGURE 13.6. Model of the trophic relationships between acetate-forming and -consuming processes in a soil aggregate that contains nitrate, reducible Fe(III), and oxygen (modified from Drake et al. 1997).

relative to other obligately anaerobic microbes, such as methanogens and sulfate-reducing bacteria. However, extensive laboratory observations indicate that acetogens are highly robust and capable of engaging numerous metabolic processes that make them versatile and competitive under a wide range of in situ conditions. Thus they exist in a wide range of habitats. Although it seems fair to conclude that the activities of acetogens impinge on more processes than originally conceived, it is obvious that more work is needed to fully resolve all of their in situ activities.

Aknowledgments. Support for work in the authors' laboratory was provided by the German Ministry of Education, Science, Research, and Technology (PT BEO 51-0339476C).

References

Arendsen AF, Soliman MQ, Ragsdale SW. 1999. Nitrate-dependent regulation of acetate biosynthesis and nitrate respiration by *Clostridium thermoaceticum*. J Bacteriol 181:1489–95.

Bernalier A, Lelait M, Rochet V, et al. 1996. Acetogenesis from H_2 and CO_2 by methane- and non-methane-producing human colonic bacterial communities. FEMS Microbiol Ecol 19:193–202.

Brauman A, Kane MD, Labat M, Breznak JA. 1992. Genesis of acetate and methane by gut bacteria of nutritionally diverse termites. Science 257:1384–7.

Breznak JA. 1994. Acetogenesis from carbon dioxide in termite guts. In: Drake HL, editor. Acetogenesis. New York: Chapman & Hall. p 303–30.

Bryant MP. 1979. Microbial methane production—theoretical aspects. J Anim Sci 48:193–201.

Chin KJ, Conrad R. 1995. Intermediary metabolism in methanogenic paddy soil and the influence of temperature. FEMS Microbiol Ecol 18:85–102.

Conrad R, Bak F, Seitz HJ, et al. 1989. Hydrogen turnover by psychrotrophic homoacetogenic and mesophilic methanogenic bacteria in anoxic paddy soil and lake sediment. FEMS Microbiol Ecol 62:285–94.

Das A, Ljungdahl LG. 2000. Acetogenesis and acetogenic bacteria. In: Lederberg J, editor. Volume 1, Encyclopedia of microbiology. San Diego, CA: Academic Press. p 18–27.

Das A, Coulter ED, Kurtz DM, Jr, Ljungdahl LG. 2001. Five-gene cluster in *Clostridium thermoaceticum* consisting of two divergent operons encoding rubredoxin oxidoreductase—rubredoxin and rubrerythrin—type A flavoprotein—high-molecular-weight rubredoxin. J Bacteriol 183:1560–7.

Drake HL. 1994. Acetogenesis, acetogenic bacteria, and the acetyl-CoA "Wood/Ljungdahl" pathway: past and current perspectives. In: Drake HL, editor. Acetogenesis. New York: Chapman & Hall.

Drake HL, Daniel SL, Küsel K, et al. 1997. Acetogenic bacteria: what are the *in situ* consequences of their diverse metabolic versatilities? Biofactors 6:13–24.

Drake HL, Küsel K, Matthies C. 2002. Ecological consequences of the phylogenetic and physiological diversities of acetogens. Ant V Leeuwenhoek 81:203–13.

Drake HL, Küsel K, Matthies C. 2003. Acetogenic prokaryotes. In: Dworkin M, Falkow S, Rosenberg E, Schleifer K-H, and Stackebrandt E, editors. *The Prokaryotes*, 2nd ed., New York: Springer-Verlag (in press).

Fontaine FE, Peterson WH, McCoy E, et al. 1942. A new type of glucose fermentation by *Clostridium thermoaceticum* n. sp. J Bacteriol 43:701–15.

Frank C, Schwarz U, Matthies C, Drake HL. 1998. Metabolism of aromatic aldehydes as cosubstrates by the acetogen *Clostridium formicoaceticum*. Arch Microbiol 170:427–34.

Fröstl JM, Seifritz C, Drake HL. 1996. Effect of nitrate on the autotrophic metabolism of the acetogens *Clostridium thermoautotrophicum* and *Clostridium thermoaceticum*. J Bacteriol 178:4597–603.

Gößner A, Devereux R, Ohnemüller N, et al. 1999. *Thermicanus aegyptius* gen. nov., sp. nov., isolated from oxic soil, a facultative microaerophile that grows commensally with the thermophilic acetogen *Moorella thermoacetica*. Appl Environ Microbiol 65:5124–33.

Goessner AS, Kuesel K, Devereux R, Drake HL. 1998. Occurrence of thermophilic acetogens in Egyptian soils. p 366. Abstract N-1 of the 98th General Meeting of the American Society for Microbiology.

Hoehler TM, Albert DB, Alperin MJ, Martens CS. 1999. Acetogenesis from CO_2 in an anoxic marine sediment. Limnol Oceanogr 44:662–7.

Hungate RE. 1976. The rumen fermentation. In: Schlegel HG, Gottschalk G, Pfennig N, editors. Microbial production and utilization of Gases. Göttingen: Goltze. p 119–24

Karnholz A, Küsel K, Göner A, Schramm A, Drake HL. 2002. Tolerance and metabolic response of acetogenic bacteria toward oxygen. Appl Environ Microbiol 68:1005–9.

Kotelnikova S, Pedersen K. 1997. Evidence for methanogenic Archaea and homoacetogenic Bacteria in deep granitic rock aquifers. FEMS Microbiol Rev 20:339–49.

Kotsyurbenko OR, Nozhevnikova AN, Soloviova TI, Zavarin GA. 1996. Methanogenesis at low temperatures by microflora of tundra wetland soil. Ant V Leeuwenhoek 69:75–86.

Krumholz LR, Harris SH, Tay ST, Suflita JM. 1999. Characterization of two subsurface H_2-utilizing bacteria, *Desulfomicrobium hypogeium* sp. nov. and *Acetobacterium psammolithicum* sp. nov., and their ecological roles. Appl Environ Microbiol 65:2300–6.

Krumholz LR, McKinley JP, Ulrich GA, Suflita JM. 1997. Confined subsurface microbial communities in Cretaceous rock. Nature 386:64–6.

Kuhner CH, Frank C, Grießhammer A, et al. 1997. *Sporomusa silvacetica* sp. nov., an acetogenic bacterium isolated from aggregated forest soil. Int J Syst Bacteriol 47:352–8.

Küsel K, Drake HL. 1995. Effects of environmental parameters on the formation and turnover of acetate by forest soils. Appl Environ Microbiol 61:3667–75.

Küsel K, Drake HL. 1996. Anaerobic capacities of leaf litter. Appl Environ Microbiol 62:4216–9.

Küsel K, Dorsch T, Acker G, et al. 2000. *Clostridium scatologenes* strain SL1 isolated as an acetogenic bacterium from acidic sediments. Int J Syst Evol Microbiol 20:537–46.

Küsel K, Karnholz A, Trinkwalter T, et al. 2001. Physiological ecology of *Clostridium glycolicum* RD-l, an aerotolerant acetogen isolated from sea grass roots. Appl Environ Microbiol 67:4737–41.

Küsel K, Pinkart HL, Drake HL, Devereux R. 1999b. Acetogenic and sulfate-reducing bacteria inhabiting the rhizoplane and deep cortex cells of the seagrass *Halodule wrightii*. Appl Environ Microbiol 65:5117–23.

Küsel K, Wagner C, Drake HL. 1999a. Enumeration and metabolic product profiles of the anaerobic microflora in the mineral soil and litter of a beech forest. FEMS Microb Ecol 29:91–103.

Küsel K, Wagner C, Trinkwalter T, et al. 2002. Microbial reduction of Fe(III) and turnover of acetate in Hawaiian soils. FEMS Microbiol. Ecology 40:73–81.

Lajoie SF, Bank S, Miller TL, Wolin MJ. 1988. Acetate production from hydrogen and [^{13}C]carbon dioxide by the microflora of human feces. Appl Environ Microbiol 54:2723–7.

Leadbetter JR, Schmidt TM, Graber JR, Breznak JA. 1999. Acetogenesis from H_2 plus CO_2 by spirochetes from termite guts. Science 29:686–9.

Leclerc M, Bernalier A, Donadille G, Lelait M. 1997. H_2/CO_2 metabolism in acetogenic bacteria isolated from the human colon. Anaerobe 3:307–15.

Leedle JAZ, Greening RC. 1988. Methanogenic and acidogenic bateria in the bovine rumen: postprandial changes after feeding high- or low-forage diets once daily. Appl Environ Microbiol 54:502–6.

Ljungdahl LG. 1994. The acetyl-CoA pathway and the chemiosmotic generation of ATP during acetogenesis. In: Drake HL, editor. Acetogenesis, New York: Chapman & Hall. p 63–87.

Lovell CR, Hui Y. 1991. Design and testing of a functional group-specific DNA probe for the study of natural populations of acetogenic bacteria. Appl Environ Microbiol 57:2602–9.

Mackie RI, Bryant MP. 1994. Acetogenesis and the rumen: synthrophic relationships. In: Drake HL, editor. Acetogenesis, New York: Chapman & Hall. p 331–56.

McInerney MJ, Bryant MP. 1981. Basic principles of bioconversions in anaerobic digestion and methanogenesis. In: Sofer SS, Zaborsky OR, editors. Biomass conversion processes for energy and fuels. New York: Plenum. p 277–96.

Miller T, Wolin MJ. 1996. Pathways of acetate, propionate, and butyrate formation by the human fecal microbial flora. Appl Environ Microbiol 62:1589–92.

Misoph M, Drake HL. 1996. Effect of CO_2 on the fermentation capacities of the acetogen *Peptostreptococcus productus* U-1. J Bacteriol 178:3140–5.

Misoph M, Daniel SL, Drake HL. 1996. Bidirectional usage of ferulate by the acetogen *Peptostreptococcus productus* U-1: CO_2 and aromatic acrylate groups as competing electron acceptors. Microbiology 142:1983–8.

Müller V, Aufurth S, Rahlfs S. 2001. The Na^+-cycle in *Acetobacterium woodii*: identification and characterization of a Na^+-translocating F_1F_0-ATPase with a mixed oligomer of 8 and 16-kDa proteolipids. Biochim Biophys Acta 1505:108–20.

Nozhevnikova A, Kotsyurbenko OR, Simankova MV. 1994. Acetogenesis at low temperature. In: Drake HL, editor. Acetogenesis. New York: Chapman & Hall. p 416–31.

Oren A. 1988. Anaerobic degradation of organic compounds at high salt concentrations. Ant V Leeuwenhoek 54:267–77.

Peters V, Conrad R. 1995. Methanogenic and other strictly anaerobic bacteria in desert soil and other oxic soils. Appl Environ Microbiol 61:1673–6.

Peters V, Conrad R. 1996. Sequential reduction processes and initiation of CH_4 production upon flooding of oxic upland soils. Soil Biol Biochem 28:371–82.

Phelps TJ, Zeikus JG. 1984. Influence of pH on terminal carbon metabolism in anoxic sediments from a mildly acidic lake. Appl Environ Microbiol 48:1088–95.

Rieu-Lesme F, Dauga C, Morvan B, et al. 1996. Acetogenic coccoid spore-forming bacteria isolated from the rumen. Res Microbiol 147:753–64.

Rosencrantz D, Rainey FA, Janssen PH. 1999. Culturable populations of *Sporomusa* spp. and *Desulfovibrio* spp. in the anoxic bulk soil of flooded rice microcosms. Appl Environ Microbiol 65:3526–33.

Sansone FJ, Martens CS. 1982. Volatile fatty acid cycling in organic-rich marine sediments. Geochim Cosmochim Acta 46:1575–89.

Schink B. 1994. Diversity, ecology, and isolation of acetogenic bacteria. In: Drake HL, editor. Acetogenesis. New York: Chapman & Hall. p 273–302.

Sexstone AJ, Revsbech NP, Parkin TB, Tiedje JM. 1985. Direct measurement of oxygen profiles and denitrification rates in soil aggregates. Soil Sci Soc Am J 49: 645–51.

Tholen A, Brune A. 1999. Localization and *in situ* activities of homoacetogenic bacteria in the highly compartmentalized hindgut of soil-feeding higher termites (*Cubitermes* spp.). Appl Environ Microbiol 65:4497–505.

Van der Lee GEM, de Winder B, Bouten W, Tietema A. 1998. Anoxic microsites in douglas fir litter. Soil Biol Biochem 31:1295–301.

Wagner C, Griesshammer A, Drake HL. 1996. Acetogenic capacities and the anaerobic turnover of carbon in a Kansas prairie soil. Appl Environ Microbiol 62:494–500.

Wieringa KT. 1936. Over het verdwijnen van waterstof en koolzuur onder anaerobe voorwaarden. Ant V Leeuwenhoek 3:263–73.

Wood HG, Ljungdahl LG. 1991. Autotrophic character of acetogenic bacteria. In: Shively JM, Barton LL, editors. Variations in autotrophic life. San Diego, CA: Academic Press. p 201–50.

Zavarzin GA, Zhilina TN, Pusheva MA. 1994. Halophilic acetogenic bacteria. In: Drake HL, editor. Acetogenesis. New York: Chapman & Hall. p 432–44.

Zhilina TN, Detkova EN, Rainey FA, et al. 1998. *Natronoincola histidinovorans* gen. nov., sp. nov., a new alkaliphilic acetogenic anaerobe. Curr Microbiol 37:177–85.

14
Electron-Transport System in Acetogens

AMARESH DAS and LARS G. LJUNGDAHL

Most anaerobic bacteria conserve energy by substrate-level phosphorylation from the metabolism of glucose via the glycolytic fermentation. Some anaerobic bacteria also possess the ability to conserve energy from the metabolism of nonfermentative substrates such as acetate, formate, methanol, and CO. The energy is generated via chemiosmosis or electron-transport (ET) coupled phosphorylation. In this process, a variety of terminal electron acceptors are used, e.g., metal ions, oxides of carbon, nitrogen, and sulfur. The reduction of the terminal electron acceptors by the ET chain results in the generation of a transmembrane gradient of H^+ or Na^+, which is subsequently used by a membrane-bound ATP synthase to synthesize ATP from ADP and inorganic phosphate (P_i) (Senior 1988). Based on the nature of the compounds being used as the terminal electron acceptor, the anaerobic bacteria are further classified into several groups. Acetogens and methanogens use CO_2 as a terminal acceptor, nitrate reducers use nitrate or nitrite, and sulfate reducers use sulfate or dimethyl sulfoxide as terminal acceptor.

The acetogenic bacteria *Moorella thermoacetica*, *Moorella thermoautotrophica*, and *Acetobacterium woodii* specialize in the use of C-1 compounds, e.g., CO_2/H_2, CO, methanol, or formate. The C-1 compounds are metabolized into acetate via the acetyl-CoA pathway, which is described in several reviews (Ljungdahl 1986; Ragsdale 1991; Wood and Ljungdahl 1991; Drake 1994) and previous chapter by Drake and Küsel. However, under certain growth conditions, such as in the presence of nitrate or dimethylsufoxide (DMSO), acetate is not produced as the metabolic end product (Seifritz et al. 1993; Ljungdahl 1994; Fröstl et al. 1996; Arendsen et al. 1999). Instead, ammonia and dimethyl sulfide are produced as the final end products. This chapter reviews various aspects of the energy metabolism associated with the use of CO_2, nitrate, and DMSO as terminal electron acceptors, along with some discoveries demonstrating that at least some acetogens are oxygen tolerant.

Use of C-1 Compounds and the Conservation of Energy via the Acetyl-CoA Pathway

From classical growth studies, it was established that acetogens conserve energy from CO_2 fixation via the acetyl-CoA pathway (Andreesen et al. 1973; Ljungdahl 1994). When acetogens grow autotrophically, no net ATP is generated at the substrate level via the acetyl-CoA pathway. Therefore, energy must be generated by a chemiosmotic mechanism. Calculations of Gibbs free energy ($\Delta G_0'$) for reactions of the acetyl-CoA pathway indicate a net gain of free energy of $-95 \, kJ/mol$ (Diekert and Wohlfarth 1994). Anaerobic bacteria require -60 to $-80 \, kJ$ per mol of free energy for the synthesis of 1 mol of ATP (Thauer et al. 1977). Therefore, at least 1 mol of ATP could be synthesized via chemiosmosis using the free energy generated from reactions of the acetyl-CoA pathway. It is unclear how the acetyl-CoA pathway is involved in the generation of chemiosmotic energy in acetogens. Some enzymes of the acetyl-CoA pathway are partially associated with the membrane, e.g., carbon monoxide dehydrogenase/acetyl-CoA synthase. (CODH/ACS) and methylene-H_4-reductase in *M. thermoautotrophica* (Hugenholtz et al. 1987) and methyl transferase in *A. woodii* (Müller and Gottschalk 1994). The bifunctional CODH/ACS is the key enzyme of the acetyl-CoA pathway. Its structure was recently elucidated by crystallographic analysis (Doukov et al. 2002). It is $\alpha_2\beta_2$ heterotetramer. The CODH-reaction is catalyzed by the β subunit and acetyl-CoA synthesis by the α subunit, which contains a Ni-Fe-Cu center. The CO was found to reduce cytochromes and menaquinone in membranes associated with CODH/ACS (Ivey 1986; Das et al. 1989) and generate a proton gradient (ΔpH) in membrane vesicles of *M. thermoautotrophica* that could drive amino acid uptake (Hugenholtz and Ljungdahl 1989, 1990). These findings provide the circumstantial evidence for the function of a membrane-bound ET chain in the generation of chemiosmotic energy.

However, the membrane of *A. woodii* lacks cytochromes or menaquinone. The chemiosmotic energy is most likely generated by the activities of a membrane-bound methyl transferase in this bacterium (Müller and Gottschalk 1994). The primary ion for the chemiosmotic energy in *M. thermoacetica* is H^+ (Ivey and Ljungdahl 1986) and that in *A. woodii* is Na^+ (Heise et al. 1992). The difference in the ionic specificity is reflected in the properties of their ATP synthases. Thus the ATP synthase from *M. thermoacetica* and *M. thermoautotrophica* pump protons (Das et al. 1997) while that from *A. woodii* is a Na^+-pump (Heise et al. 1992; Redlinger and Müller 1994).

Electron-Transport Pathway Linked to the Use of CO_2 as a Terminal Electron Acceptor in Acetogens

Membranes of *M. thermoacetica* contain two *b*-type cytochromes: cytochrome b_{560} (E_0' –200 mV) and cytochrome b_{554} (E_0' –48 mV) and a menaquinone MK7 (E_0' –74 mV) (Das et al. 1989). The reductions of cytochromes and menaquinone by CO via membrane-bound CODH/ACS (Ragsdale and Kumar 1996) suggest that the acetyl-CoA pathway could be linked to membrane ET in acetogenic bacteria. In the ET pathway, menaquinone is placed between the two cytochromes in the following order: cytochrome b_{560} → MK7 → cytochrome b_{554} (Das et al. 1989). Apparently, reducing equivalents generated from the oxidation of CO via the acetyl-CoA pathway provide electrons to the ET chain. The other physiologic electron donor is hydrogen, and the transfer of electrons from the oxidation of hydrogen is presumably catalyzed by a membrane-bound hydrogenase.

The mechanism of the transfer of electrons from the oxidation of CO or hydrogen to the ET chain is not clear. A ferredoxin (E_0' –365 mV) is most likely used as redox mediator between substrates, and the ET (Ljungdahl 1994; Ragsdale and Kumar 1996). A flavoprotein was partially characterized from membranes of *M. thermoacetica* (Ivey 1986). It has a low redox potential (E_0' –221 mV); therefore, it might be a redox mediator between the ferredoxin and the cytochrome b_{560}. Rubredoxins were proposed as the terminal electron acceptors for the ET chain (Das et al. 1989). *M. thermoacetica* contains two rubredoxins: Rd I and Rd II (Yang et al. 1980). They have redox potentials (Rd I: E_0' –27 mV; Rd II: E_0' +20 mV) that are high enough to accept electrons from cytochromes and menaquinone. Electron paramagnetic resonance (EPR) spectra of reduced membranes indicated the presence of a rubredoxin or rubredoxin-like protein in membranes of *M. thermoacetica* (Hugenholtz et al. 1989).

Several features of the components of ET chain found in acetogens have been reviewed (Ljungdahl 1994). Table 14.1 lists all known components of the ET chain and their subcellular locations. Some components were distributed in membranes and cytosols, e.g., CODH, methylene-H_4F-reductase, ferredoxin, and rubredoxin. They may provide important links between the soluble and membrane-bound ET systems. Figure 14.1 presents a model of the ET chain, slightly modified from the previously proposed model. Many features of this ET chain, including the roles of ferredoxin, flavoprotein, methylene tetrahydrofolate (MTHF), and rubredoxin, have not been tested experimentally. The possibility exists that other electron carriers are involved, and their genes could be clustered. This was further evidenced from our recent finding that the genes for oxidative stress protection proteins in *M. thermoacetica* are clustered (discussed below).

TABLE 14.1. Components of the ET chain and their subcellular locations in *M. thermoacetica*.

Membrane only	Cytosol only	Both membranes and cytosol
Menaquinone	Ferredoxin I	Ferredoxin II
Cytochrome b_{554}	Type A flavoprotein (FprA)	Flavoprotein/flavodoxin
F_1F_0 ATP synthase	Rubrerythrin	High molecular weight rubredoxin (Hrb)
	Rubredoxin oxidoreductase	CODH/ACS
	High molecular weight rubredoxin	Methanol dehydrogenase
	Formate dehydrogenase	Hydrogenase
		NADH dehydrogenase
		Cytochrome b_{560}
		Methylene-THF reductase
		Methyl transferase

Membrane

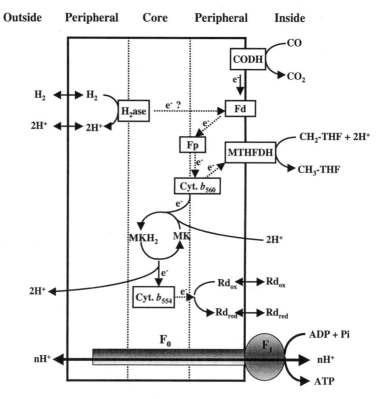

FIGURE 14.1. Proposed electron-transport pathway in the acetogenic bacteria *M. thermoacetica* and *M. thermoautotrophica*. *Cyt.*, cytochrome; *MTHFDH*, methylene tetrahydrofolate dehydrogenase; *Fd*, ferredoxin; *Fp*, flavoprotein; H$_2$ase, hydrogenase.

Oxidation of hydrogen or CO by membrane-bound dehydrogenases generates a proton gradient (ΔpH) in membrane vesicles of *M. thermoautotrophica*. Apparently, acetogens contain several hydrogenases. Hugenholtz and Ljungdahl (1989) reported a membrane-bound hydrogenase in *M. thermoautotrophica*. It oxidizes hydrogen with benzyl viologen as electron acceptor and generates a proton gradient. No membrane ET was involved in this process. Drake (1984) demonstrated a soluble hydrogenase in *M. thermoacetica*. The latter was shown to reduce *b*-type cytochromes in membranes, but whether this reaction is coupled to generation of a proton gradient has not been investigated.

Electron-Transport Pathway Linked to Use of Nitrate, DMSO, and Thiosulfate as Terminal Acceptors

Acetogenic bacteria could conserve energy-using nitrate, thiosulfate, and DMSO as terminal acceptors (Ljungdahl 1994; Fröstl et al. 1996). Cultures grown in the presence of these compounds produced negligible amount of acetate, even in the presence of CO_2. The Gibbs free energy ($\Delta G_0'$) for the reduction of CO_2, thiosulfate, DMSO, and nitrate are -12 kJ mol^{-1}, -18.9 kJ mol^{-1}, -93 kJ mol^{-1}, and -63 kJ mol^{-1}, respectively (Thauer et al. 1977). Therefore, from a thermodynamic standpoint thiosulfate, DMSO, and nitrate are preferred over CO_2 as terminal acceptors. Nitrate was shown to have a negative effect on the transcription of the genes encoding the acetyl-CoA pathway enzymes, resulting in 50–70% reduction at the protein levels of several key enzymes of the acetyl-CoA pathway (Arendsen et al. 1999). One of the distinct features of nitrate-grown cells is the absence of cytochromes in membrane (Seifritz et al. 1993). This is surprising, because nitrite reductase, which catalyzes the reduction of nitrite to ammonia in the final step of nitrate reduction, is a cytochrome *c*–dependent enzyme in many species, including anaerobic bacteria (McEwan et al. 1991; Lin and Stewart 1998). Apparently, *M. thermoacetica* has a novel type of nitrate-reduction system not common to other bacteria. Unlike nitrate-grown cells, DMSO-grown cells of *M. thermoacetica* contain cytochromes in membranes. The cytochrome content was estimated to be fourfold to fivefold higher in membranes of methanol-plus-DMSO-grown cells than in membranes of methanol-plus-CO_2-grown cells (Das and VanHoek, personal communication, 1996). Therefore, cytochromes are most likely involved in the reduction of DMSO in acetogens. The difference spectra of oxidized and dithionite-reduced cytochromes in membranes of *M. thermoacetica* were virtually identical whether cells were grown in the presence of DMSO or CO_2. This suggests that similar cytochromes are involved in both CO_2 and DMSO reduction. In the absence of cytochromes, ferredoxin, flavodoxin, or menaquinone is likely involved as a redox mediator in nitrate reduction. The involvement of menaquinone in the reduction of DMSO or nitrate was demonstrated in many bacteria, including anaerobic bacteria (McEwan et al. 1991; Lorenzen et al. 1994). Ferredoxin and flavoprotein were shown to be redox mediators in the assimilatory nitrate reduction in several species, including those from the genera *Klebsiella, Bacillus*, and *Synechococcus* (Lin and Stewart 1998).

Electron-Transport Pathway Linked to the Use of Oxygen in *Moorella thermoacetica*

Oxygen metabolism often generates O_2^- (superoxide) and H_2O_2 (hydrogen peroxide), which leads to oxidative stress in microorganisms. In aerobic bacteria, O_2^- and H_2O_2 are effectively removed by the enzymes superoxide dismutase (SOD) and catalase, respectively. These enzymes are often absent in anaerobic bacteria, and they have not been detected in *M. thermoacetica*. Studies indicate that the nonheme iron proteins such as rubrerythrin (Rbr) and rubredoxin oxidoreductase (Rbo) (also called desulfoferrodoxin or superoxide reductase, SOR) provide a novel oxidative stress protection system that is confined to air-sensitive bacteria and archaea (Alban et al. 1988; Chen et al. 1993; Jenney et al. 1999; Romao et al. 1999). Table 14.2 lists the functions of these proteins in several anaerobic microorganisms. Both Rbr and Rbo have been shown to restore aerobic growth in *sod*⁻ strains of *Escherichia coli* (Lehmann et al. 1996; Pianzolla et al. 1996). Furthermore, deletion of the *rbo* gene led to increased sensitivities toward oxygen and superoxide in *Desulfovibrio vulgaris* (Voodrouw and Voodrouw 1998).

We sequenced the genes encoding Rbr and Rbo from *M. thermoacetica*. They were present in a novel cluster containing three additional genes encoding rubredoxin (Rub), type A flavoprotein (FprA), and a high molecular weight rubredoxin (Hrb) (Das et al. 2001). The structural genes of these proteins are organized in two divergently oriented clusters: Cluster I carries two genes in the order *rbo-rub*, and cluster II has three genes in the order *rbr-fprA-hrb* (Fig. 14.2). Northern blot hybridization experiments

TABLE 14.2. Proposed functions of rubrerythrin, rubredoxin oxidoreductase, rubredoxin, high molecular weight rubredoxin, and type A flavoprotein in different bacteria.

Protein	Functions	Organism	References
Rbr	Catalase	*Spirillum volutans*	Alban et al. (1988)
	SOD	*Clostridium perfringens*	Lehman et al. (1996)
Rbo	SOD	*Desulfarculus baarsii*	Lombard et al. (2000)
	SOD	*Desulfovibrio vulgaris/E. coli*	Pianzolla et al. (1996)
Rub	Oxygen reduction	*Desulfovibrio gigas*	Chen et al. (1993)
FprA	Oxygen reduction	*D. gigas*	Gomes et al. (1997)
Hrb/FprA	Oxygen reduction	*Moorella thermoaceticum*	

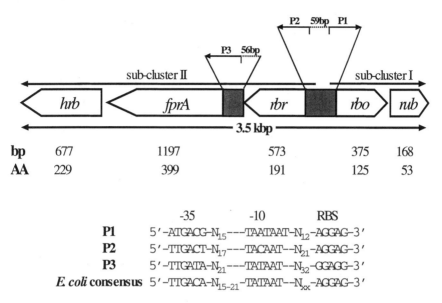

FIGURE 14.2. Organization of *rub, rbo, rbr, fprA*, and *hrb* (genes) in *M. thermoacetica* and of promoters P1, P2, and P3, as well as the relationship between the genes and their products.

revealed cluster I and cluster II as two separate operons (Fig. 14.2). Analysis of the deduced amino acid sequences of these proteins revealed the presence of highly conserved iron-binding motifs. They were identified as $CX_2CX_{29}CX_2C$ for the single $FeS(Cys)_4$ site in Rub; $EX_{30-34}EX_2HX_nEX_{30-34}$ EX_2H and $CX_2CX_{12}CX_2C$ for the di-iron and $FeS(Cys)_4$ sites, respectively, in Rbr; and $CX_2CX_{15}CC$ and $HX_{15}HX_9HX_{40}CX_2H$ for the $FeS(Cys)_4$ and $Fe(NHis)_4SCys$ centers, respectively, in Rbo. The Hrb contains two separate motifs: a rubredoxin motif $[FeS(Cys)_4]$ at the C-terminal end and a flavin motif at the NH_2-terminal end. The FprA also has two separate motifs: a flavin motif and a di-iron center.

Three other proteins with similar domain structure as that of FprA were reported in other bacteria (Wasserfallen et al. 1995; Gomes et al. 1997, 2000). The recombinant CthFprA and CthHrb, overexpressed in *E. coli*, were purified and characterized. Both FprA and Hrb were found to be present as dimers. Metal/cofactor analysis of the purified proteins revealed the presence of 2 mol each of iron and flavin (FMN) per mole dimer of Hrb and 4 mol of iron and 2 mol FMN per mole dimer of FprA. The EPR spectra of the purified proteins indicated that iron is present in a di-iron center in FprA and as a $Fe(Cys)_4$ cluster in Hrb.

FprA alone had a low level of NADH oxidase activity with oxygen as a terminal electron acceptor. Addition of Hrb stimulates the oxidase activity of FprA, and NADH supported 10-fold higher oxidase activity than

NADPH. The redox potentials of the purified Hrb and FprA were estimated to be -140 and $-60\,mV$, respectively. These values are consistent with the idea of Hrb being a redox mediator in the reduction of FprA by NADH and suggest the following electron transport pathway for the reduction of oxygen by the Hrb/FprA couple: NADH $\rightarrow$ Hrb $\rightarrow$ FprA $\rightarrow$ O_2. Apparent K_m and V_{max} values of the NADH oxidation reaction were determined to be $2.1\,\mu M$ and $1125\,\mu mol$ NADH oxidized per minute per micromol FprA monomer, respectively. When oxygen was measured under identical conditions, V_{max} was $527\,\mu mol$ oxygen consumed per minute per micromol monomer, i.e., the ratio of NADH oxidized per oxygen molecule reduced is $2:1$ and supports the following reaction, catalyzed by Hrb/FprA couple:

$$4e^- + 4H^+ + O_2 \rightarrow 2H_2O \tag{14.1}$$

Aerotolerant growth of *M. thermoacetica* in the absence of any reducing agent was reported (Karnholtz et al. 2002). We found that *M. thermoacetica* could reduce traces of oxygen, but the cultures did not grow until all dissolved oxygen was reduced. Levels of transcripts of *rbr*, *fprA*, and *hrb* were up-regulated in cells grown in the absence of reducing agent, indicating a possible role of the corresponding proteins in oxidative stress protection. However, no appreciable change in the protein concentrations of Rbr, FprA, and Hrb were observed in extracts of cells grown in the presence or absence of reducing agent. Rbo was not detectable in extracts of *M. thermoacetica*, whether cells were grown in the presence or absence of reducing agent. Several new proteins were specifically expressed when cells were grown in the absence of any reducing agent.

NADH oxidase activity was reported in extracts of *M. thermoacetica* (Karnholtz et al. 2002). We found that the cytosolic fractions of *M. thermoacetica* have significantly higher NADH oxidase activity than that of the membrane fractions. Consistent with these results, membrane and cytosolic fractions of *M. thermoacetica* were found to reduce oxygen. The reduction of oxygen by membranes was inhibited by the respiratory inhibitors rotenone and HQNO. This suggests the involvement of the ET chain in the reduction of oxygen. Analyses of the proteins in subcellular fractions of *M. thermoacetica* by western blotting experiments revealed the presence of Hrb in membranes while FprA and Rbr in the cytosol. This suggests a possible link of Hrb with the ET chain in the reduction of oxygen.

ATP Synthases of Gram-Positive Anaerobic Bacteria

ATP synthesis is the final step in the conservation of energy via chemiosmosis. This step is catalyzed by F_1F_0 ATP synthase, a multisubunit enzyme complex found exclusively in cytoplasmic membranes of bacteria (Senior 1988). It uses the chemiosmotic energy (ΔpH) generated from membrane ET to synthesize ATP from ADP and P_i. The most investigated bacterial

ATP synthase is that from *E. coli*. Its F_1 consists of five subunits (α, β, γ, δ, and ε), and its F_0 has three subunits (*a*, *b*, and *c*) (Foster and Fillingame 1979). The ATP synthases found in most other bacteria have a similar composition (Deckers-Hebestreit and Altendorf 1996). The *atp* operon encoding ATP synthase from most bacteria, including *M. thermoacetica* and *Clostridium pasteurianum* consists of 9 genes arranged in the order *atpI* (*i*), *atpB* (*a*), *atpE* (*c*), *atpF* (*b*), *atpH* (δ), *atpA* (α), *atpG* (γ), *atpD* (β), and *atpC* (ε) (Das and Ljungdahl 1997, 2000). The product of the first gene, the *i*-subunit, has never been found in ATP synthase purified from any source, and its function is not known. The *atp* operon of *A. woodii* contains 11 genes (Rahlfs et al. 1999). Two additional genes of the *A. woodii atp* operon are multiple copies of *atpE* encoding subunit *c2* and *c3* of the ATP synthase (Aufurth et al. 2000). The composition of the *atp* operons from different anaerobic and aerobic bacteria are compared in Figure 14.3. Northern blot analysis of total RNA isolated from *M. thermoacetica* and *C. pasteurianum* revealed transcription of all *atp* genes. From rather intensive studies in our laboratory we found that the purified F_0 moiety of the ATP synthase of the two *Moorella* species lacked the *a* and *b* subunits. It was nevertheless functional. How this is possible remains to be investigated.

The primary structure analysis of the ATP synthase subunits from *M. thermoacetica* and *C. pasteurianum* revealed unusual features. In the Mt*atp* operon, the 5′-end of the structural gene of the ATP synthase *a*-subunit, *atpB*, overlapped by 50 bp with the 3′-end of the structural gene of the *i*-subunit *atpI*. This could lead to tight co-regulation of *atpI* and *atpB*. The transcriptional analysis of *atpI* in *E. coli* revealed specific cleavage and shorter half-lives for *atpI*-transcripts. We reported similar findings for the transcripts of both *atpI* and *atpB* from *M. thermoacetica* (Das and Ljungdahl 1997). This suggests that the transcripts of *atpI* and *atpB* in the latter bacterium were subjected to posttranscriptional regulations, similar to that of the *E. coli atpI*. In addition, rare codons were present in the *a*-subunit gene of the *M. thermoacetica atp* operon. Rare codons reduce translational efficiency of genes. Thus they may have similar effect on *atpB* of *M. thermoacetica atp* operon (Mt*atp*). Apparently, posttranscriptional regulation could play an important role in the regulation of *atpB* in the Mt*atp* operon, resulting poor expression of the ATP synthase *a* subunit.

Conclusion

Acetogenic bacteria sustain nonfermentative autotrophic type of growth on C-1 compounds using the chemiosmotic energy generated from the membrane ET. The reducing equivalents generated from the metabolism of C-1 compounds via the acetyl-CoA pathway serve as electron donors to the ET chain. Two *b*-type cyochromes and a menaquinone are the major components of the ET chain in *M. thermoacetica* and *M. thermoautotrophica*.

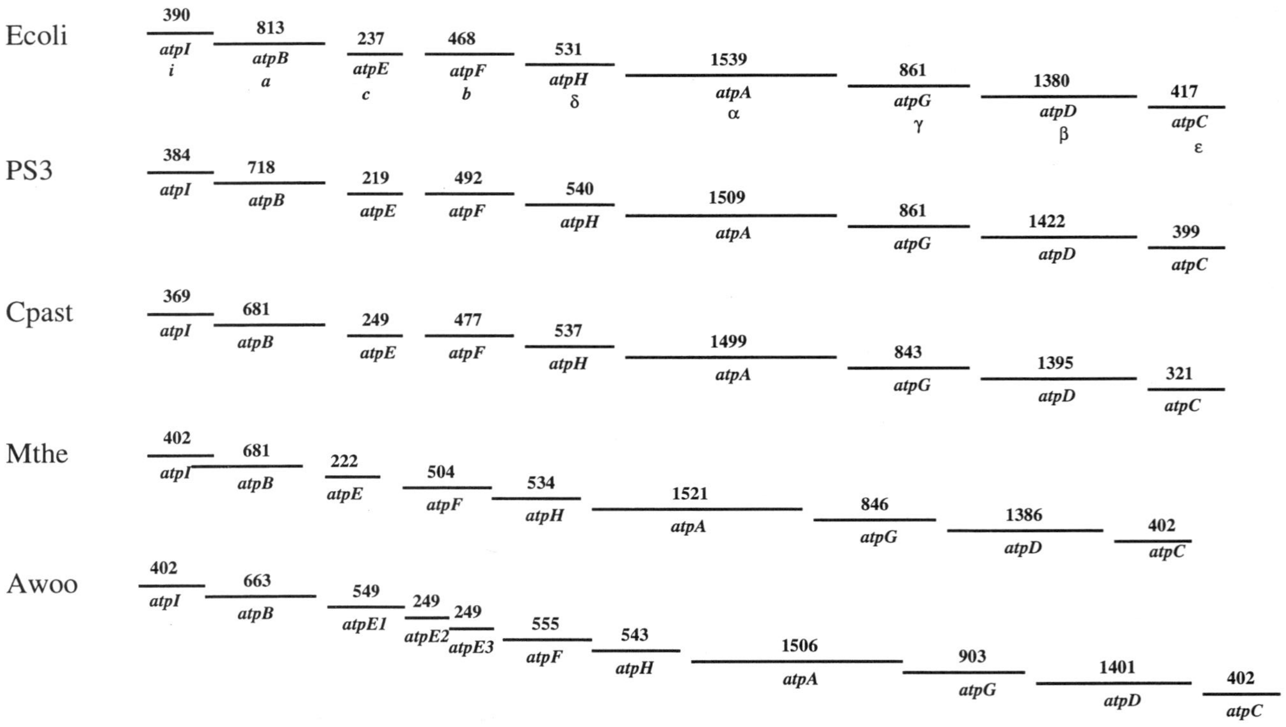

FIGURE 14.3. Composition and structural organization of *atp* genes in *E. coli* (*Ecoli*) (accession no. J01594), thermophilic bacterium PS3 (*PS3*) (accession no. X07804), *C. pasteurianum* (*Cpast*) (accession no. AF283808), *M. thermoacetica* (*Mthe*) (accession no. U64318), and *A. woodii* (*Awoo*) (U10505). The size of each gene (in base pairs) is also shown.

Other electron carriers are ferredoxins, flavoproteins, and rubredoxins. Acetogens use a variety of compounds as terminal electron acceptors, such as CO_2, DMSO, and nitrate. The use of DMSO or nitrate blocks acetogenesis from C-1 compounds via the acetyl-CoA pathway, and the function of the latter pathway is then limited only to supply cell carbon. The acetogenic bacteria exhibit aerotolerance in the absence of any reducing agent. Genes for the oxidative stress protection proteins in *M. thermoacetica* were present in a novel cluster of five genes encoding Rbr, Rbo, Rub, FprA, and Hrb. Membrane and cytosolic fractions of *M. thermoacetica* exhibit both NADH-oxidase and oxygen-uptake activities, which were inhibited by respiratory inhibitors (e.g., rotenone and HQNO), suggesting the involvement of membrane ET in the oxygen reduction. The purified recombinant Hrb and FprA were found to have significant NADH oxidase and oxygen-uptake activity when present together. The Hrb was found to be partially associated with membranes, suggesting its link to the ET chain in oxygen reduction. The *atp*-operon of *M. thermoacetica*, contains the genes for all subunits normally found in bacterial F_1F_0 ATP synthases, and they are transcribed as a part of the polycistronic operon. Analysis of the primary structure of the gene encoding the *a* subunit of MtheATP synthase and that of the deduced amino acid sequence of the protein revealed unusual features that could have adverse effects either on the posttranscriptional regulation of the gene or on the functions of the enzyme complex.

Acknowledgment. Support by grant DE-FG02-93ER20127 from the U.S. Department of Energy for work on acetogenic bacteria is gratefully acknowledged.

References

Alban PS, Popham DL, Rippere KE, Krieg NR. 1988. Identification of a gene for a rubrerythrin/nigerythrin-like protein in *Spirillum volutans* by using amino acid sequence data from mass spectrometry and NH_2-terminal sequencing. J Appl Microbiol 85:875–82.

Andreesen JR, Schaupp A, Neurauter C, et al. 1973. Fermentation of glucose, fructose, and xylose by *Clostridium thermoaceticum*: effect of metals on growth yield, enzymes, and synthesis of acetate from CO_2. J Bacteriol 114:743–51.

Arendsen AF, Soliman MQ, Ragsdale SW. 1999. Nitrate-dependent regulation of acetate biosynthesis and nitrate respiration by *Clostridium thermoaceticum*. J Bacteriol 181:1489–95.

Aufurth S, Schagger H, Müller V. 2000. Identification of subunits *a, b,* and *c*1 from *Acetobacterium woodii* Na^+-F_1F_0-ATPase. Subunits *c*1, *c*2, and *c*3 constitute a mixed *c*-oligomer. J Biol Chem 275:33297–301.

Chen L, Liu M-Y, LeGall J, et al. 1993. Purification and characterization of an NADH-rubredoxin oxidoreductase involved in the utilization of oxygen by *Desulfovibrio gigas*. Eur J Biochem 216:443–8.

Das A, Ljungdahl LG. 2000. The primary structure of the *atp* operon encoding the F$_1$F$_0$ ATP synthase from *Clostridium pasteurianum*. Growth and metabolism of acetogenic bacteria in the presence of oxygen. Abstracts of the 100th general meeting of the American Society for Microbiology. Washington, DC: American Society for Microbiology. p 374.

Das A, Coulter ED, Kurtz DM Jr, Ljungdahl LG. 2001. Five-gene cluster in *Clostridium thermoaceticum* consisting of two divergent operon encoding rubredoxin oxidoreductase-rubredoxin and rubrerythrin-type A flavoproteon-high molecular weight rubredoxin. J Bacteriol 183:1560–7.

Das A, Hugenholtz J, Van Halbeek H, Ljungdahl LG. 1989. Structure and function of a menaquinone involved in electron transport in membranes of *Clostridium thermoaceticum* and *Clostridium thermoautotrophicum*. J Bacteriol 171:5823–9.

Das A, Ivey DM, Ljungdahl LG. 1997. Purification and reconstitution into proteoliposomes of the F$_1$F$_0$ ATP synthase from obligately anaerobic bacterium *Clostridium thermoautotrophicum*. J Bacteriol 179:1714–20.

Das A, Ljungdahl LG. 1997. Composition and primary structure of the F$_1$F$_0$ ATP synthase from the obligately anaerobic bacterium *Clostridium thermoaceticum*. J Bacteriol 179:3746–55.

Deckers-Hebestreit G, Altendorf K. 1996. The F$_1$F$_0$-type ATP synthases of bacteria: structure and function of the F$_0$ complex. Annu Rev Microbiol 50:791–824.

Diekert G, Wohlfarth G. 1994. Energetics of acetogenesis from C$_1$ units. In: Drake HL, editor. Acetogenesis. New York: Chapman & Hall. p 157–79.

Doukov TI, Iverson TM, Seravalli J, et al. 2002. A Ni-Fe-Cu center in a bifunctional carbon monoxide/acetyl-CoA synthase. Science 298:567–72.

Drake HL. 1984. Demonstration of hydrogenase in extracts of the homoacetate-fermenting bacterium *Clostridium thermoaceticum*. J Bacteriol 150:702–9.

Drake HL. 1994. Acetogenesis, acetogenic bacteria, and the acetyl-CoA "Wood/Ljungdahl pathway": past and current perspectives. In: Drake HL, editor. Acetogenesis. New York: Chapman & Hall. p 3–60.

Foster DL, Fillingame RH. 1979. Energy-transducing H$^+$-ATPase of *Escherichia coli*: purification, reconstitution and subunit composition. J Biol Chem 254:8230–6.

Fröstl JM, Seifritz C, Drake HL. 1996. Effect of nitrate on the autotrophic metabolism of the acetogens *Clostridium thermoaceticum* and *Clostridium thermoautotrophicum*. J Bacteriol 178:4597–603.

Gomes CM, Silva G, Oliviera S, et al. 1997. Studies on redox centers of the terminal oxidase from *Desulfovibio gigas* and evidence for its interaction with rubredoxin. J Biol Chem 272:22502–8.

Gomes CM, Vicente JB, Wasserfallen A, Teixeira M. 2000. Spectroscopic studies and characterization of a novel electron-transfer chain from *Escherichia coli* involving a flavorubredoxin and its flavoprotein reductase partner. Biochemistry 39:16230–7.

Heise R, Müller V, Gottschalk G. 1992. Presence of a Na$^+$-translocating ATPase in membrane vesicles of the homoacetogenic bacterium *Acetobacterium woodii*. Eur J Biochem 206:553–7.

Hugenholtz J, Ljungdahl LG. 1989. Electron transport and electrochemical proton gradient in membrane vesicles of *Clostridium thermoautotrophicum*. J Bacteriol 171:2873–5.

Hugenholtz J, Ljungdahl LG. 1990. Amino acid transport in membrane vesicles of *Clostridium thermoautotrophicum*. FEMS Microbiol Lett 69:117–22.

Hugenholtz J, Ivey DM, Ljungdahl LG. 1987. Carbon monoxide driven electron transport in *Clostridium thermoautotrophicum* membranes. J Bacteriol 169: 5845–7.

Hugenholtz J, Morgan TV, Ljungdahl LG. 1989. EPR studies of electron transport in membranes of *Clostridium thermoautotrophicum*. Abstracts of the 89th general meeting of the American Society for Microbiology. Washington, DC: American Society for Microbiology. p 269.

Ivey DM. 1986. Generation of energy during CO_2 fixation in acetogenic bacteria [dissertation]. Athens: University of Georgia.

Ivey DM, Ljungdahl LG. 1986. Purification and characterization of the F_1-ATPase from *Clostridium thermoaceticum*. J Bacteriol 165:252–7.

Jenney FE Jr, Verhagen MF, Cui X, Adams MW. 1999. Anaerobic microbes: oxygen detoxification without superoxide dismutase. Science 286:306–9.

Karnholtz A, Küsel K, Gößner A, et al. 2002. Tolerance and metabolic response of acetogenic bacteria toward oxygen. Appl Env Microbiol 68:1005–9.

Lehmann Y, Meile L, Teuber M. 1996. Rubrerythrin from *Clostridium perfringens*: cloning of the gene, purification of the protein, and characterization of its superoxide dismutase function. J Bacteriol 178:7152–8.

Lin JT, Stewart V. 1998. Nitrate assimilation by bacteria. Adv Microbiol Physiol 39: 1–30.

Ljungdahl LG. 1986. The autotrophic pathway of acetate synthesis in acetogenic bacteria. Annu Rev Microbiol 40:415–50.

Ljungdahl LG. 1994. Biochemistry and energetics of acetogenesis and the acetyl-CoA pathway. In: Drake HL, editor. Acetogenesis. New York: Chapman & Hall. p 63–87.

Lombard M, Fontecave M, Touati D, Niviere V. 2000. Reaction of the desulfoferredoxin from *Desulfovibrio baarsii* with superoxide anion. Evidence for a superoxide reductase activity. J Biol Chem 275:115–21.

Lorenzen J, Steinwachs S, Unden G. 1994. DMSO respiration by the anaerobic bacterium *Wolinella succinogenes*. Arch Microbiol 162:277–81.

McEwan AG, Benson N, Bonnet TC, et al. 1991. Bacterial dimethyl sulfoxide reductases and nitrate reductases. Biochem Soc Trans 19:605–8.

Müller V, Gottschalk G. 1994. The sodium ion cycle in acetogenic and methanogenic bacteria: generation and utilization of a primary electrochemical sodium ion gradient. In: Drake HL, editor. Acetogenesis. New York: Chapman & Hall. p 127–56.

Pianzzola MJ, Soubes M, Touati D. 1996. Overproduction of the *rbo* gene product from *Desulfovibrio* species suppresses all deleterious effects of lack of superoxide dismutase in *Escherichia coli*. J Bacteriol 178:6736–42.

Ragsdale SW. 1991. Enzymology of acetyl-CoA pathway of CO_2 fixation. Crit Rev Biochem Mol Biol 26:261–300.

Ragsdale SW, Kumar M. 1996. Nickel-containing carbon monoxide dehydrogenase/acetyl-CoA synthase. Chem Rev 96:2515–39.

Rahlfs S, Aufurth S, Müller V. 1999. The Na^+-F_1F_0-ATPase operon from *Acetobacterium woodii*: operon structure and presence of multiple copies of *atpE*, which encode proteolipids of 8- and 18-kDa. J Biol Chem 274:33999–4004.

Redlinger J, Müller V. 1994. Purification of ATP synthase from *Acetobacterium woodii* and identification as a Na⁺-translocating F_1F_0 type enzyme. Eur J Biochem 223:275–83.

Romao CV, Liu MY, LeGall J, et al. 1999. The superoxide dismutase activity of desulfoferrodoxin from *Desulfovibrio desulfuricans* ATCC 27774. Eur J Biochem 261:438–43.

Seifritz C, Daniel SL, Gößner A, Drake HL. 1993. Nitrate as a preferred electron sink for the acetogen *Clostridium thermoaceticum*. J Bacteriol 175:8008–13.

Senior AE. 1988. ATP synthesis by oxidative phosphorylation. Physiol Rev 68: 177–231.

Thauer RK, Jungermann K, Decker K. 1977. Energy conservation in chemotrophic bacteria. Bacteriol Rev 41:100–80.

Voodrouw JK, Voodrouw G. 1998. Deletion of the rbo gene increases the oxygen sensitivity of the sulfate-reducing bacterium *Desulvovibrio vulgaris* Hildenborough. Appl Environ Microbiol 64:2882–7.

Wasserfallen A, Huber K, Leisinger T. 1995. Purification and structural characterization of a flavoprotein induced by iron limitation in *Methanobacterium thermoautotrophicum* Marburg. J Bacteriol 177:2436–41.

Wood HG, Ljungdahl LG. 1991. Autotrophic character of the acetogenic bacteria. In: Shively JM, Barton LL, editors. Variations of autotrophic life. New York: Academic Press. p 201–50.

Yang S-S, Ljungdahl LG, Dervartanian DV, Watt GD. 1980. Isolation of two rubredoxins from *Clostridium thermoaceticum*. Biochim Biophys Acta 590:24–33.

15
Microbial Inorganic Sulfur Oxidation: The APS Pathway

Donovan P. Kelly

Although this book is devoted primarily to anaerobic bacteria, the energy-generating oxidation of inorganic sulfur compounds is included because the work of Harry Peck in the 1960s showed that some biochemical features were common to both sulfate-reducing bacteria and aerobic thiosulfate-oxidizing thiobacilli. His work on thiosulfate metabolism by aerobic thiobacilli subsequently proved to be of fundamental importance to our understanding of bacterial sulfur compound oxidation. Some sulfur oxidizers are also facultative anaerobes, being able to couple inorganic sulfur oxidation to the respiratory reduction of nitrate to dinitrogen. The biochemistry, genetics, energetics, molecular biology, and taxonomy of the aerobic and facultatively anaerobic sulfur-oxidizing bacteria were the subjects of intense study for the whole of the twentieth century, and detailed consideration is beyond the scope of this chapter (Nathansohn 1902; Kelly 1982, 1988, 1989; Kelly et al. 1997; Friedrich 1998; Kelly 1999; Kelly and Wood 2000a, 2000b). My purpose is to summarize and assess the contribution of the work of Peck to the development of the understanding of inorganic sulfur oxidation from the latter half of the twentieth century to the present.

The View of Sulfur Compound Oxidation Before 1960

Various rod-shaped Gram-negative bacteria can obtain all the energy needed for their growth from inorganic sulfur compound oxidation and are known as chemolithoautotrophs. Sulfide, sulfur, thiosulfate, and tetrathionate are the favored substrates, and cell carbon is obtained by fixation of carbon dioxide, using the Calvin cycle. During the 1900s, many of these bacteria were classified as members of the genus *Thiobacillus*, which inspired an expectation of common oxidative pathways in the organisms studied. This expectation was ill-founded, as later work showed *Thiobacillus* to be a heterogeneous genus, and many of its species have now been reclassified into a range of new genera or assigned to other genera, including *Paracoccus*, *Acidiphilium*, and *Starkeya* (Kelly 1989; Moreira and Amils 1997;

Hiraishi et al. 1998; Rainey et al. 1999; Kelly and Wood 2000a, 2000b; Kelly et al. 2000, 2001). A corollary of this phylogenetic diversity is the absence of a single common biochemical pathway, as had already been suspected (Schlegel, 1975; Kelly 1982) and was later established beyond doubt (Kelly, 1988, 1989, 1990; Kelly et al. 1997; Friedrich 1998; Kelly 1999).

Up to the 1960s, most hypothetical published pathways for sulfide and thiosulfate oxidation involved polythionates ($S_nO_6^{2-}$) as intermediates and sulfite as the final intermediate before the production of sulfate (Nathansohn 1902; Tamiya et al. 1941; Vishniac and Santer 1957; London and Rittenberg 1964). Such schemes were supported by the common observation of polythionates, such as tetrathionate and trithionate, during thiosulfate oxidation, often arising sequentially as would be consistent with a pathway in which they were intermediates (Vishniac 1952; Kelly and Syrett 1966b). None of the pathways postulated up to 1960 explained in biochemical terms how the sulfane (S–) sulfur atom of thiosulfate was oxidized to the level of sulfite ($-SO_3^{2-}$) or how energy might be obtained from the oxidation of sulfite. Early observations of Nathansohn (1902) and Starkey (1935) led them to propose biochemical mechanisms whereby the thiosulfate ion was metabolized to produce sulfur and sulfate as approximately equimolar oxidation products:

$$Na_2S_2O_3 = O = Na_2SO_4 + S \qquad (15.1)$$

$$5S_2O_3^{2-} + 4O_2 + H_2O = 6SO_4^{2-} + 4S^0 + 2H^+ \qquad (15.2)$$

Although these equations held true for some strains of bacteria under some growth conditions, they did not help explain the commonly observed quantitative conversion of both sulfur atoms of thiosulfate to sulfate, rather than the liberation of the sulfane-sulfur mainly as elemental sulfur. During the 1960s, cyclic reactions of polythionates and other polysulfur compounds continued to be postulated as mechanisms for thiosulfate and polythionate metabolism (Trudinger 1967), but none of these was supported at the time by strong biochemical evidence. The time was opportune for a new approach to the problem of thiosulfate oxidation in thiobacilli.

1960: A Milestone Year in the Biochemistry of Sulfur Oxidation

In 1960, Peck published a paper in the *Proceedings of the National Academy of Sciences U.S.A.* that changed forever the perceptions of inorganic sulfur oxidation in bacteria. When Peck's paper appeared, I was a final-year undergraduate student being taught the biochemistry of autotrophic bacteria by Philip Syrett at University College London. Syrett perceived its significance immediately and summarized it in a lecture. That lecture was a prime stimulus to my desire to pursue a doctorate degree on the thiobacilli.

I would not have guessed then that I would still be fascinated by those bacteria, thanks to Peck and Syrett, more than 40 years later. So, what did that paper say and do?

Peck's first significant contribution was to look at *Thiobacillus thioparus* (the type species of the genus *Thiobacillus*) through the eyes of one who knew a lot about sulfate-reducing bacteria and about the enzymes involved in sulfate metabolism in yeast and mammalian tissues. This led him to think "maybe the same enzymes are involved in sulfur oxidation as in reduction." The seminal paper of 1960 showed that this was indeed the case.

Reductases and Sulfurylases as the Enzymes of Thiosulfate Oxidation

Using the type strain of *T. thioparus* (ATCC 5158) Peck (1960) showed that cell-free extracts of thiosulfate-grown bacteria oxidized thiosulfate when supplemented with reduced glutathione. The rate of oxygen uptake by the complete assay mixture in a Warburg flask was 92 μL oxygen per 20 min per 13 mg protein. Omitting thiosulfate lowered the rate to 34 μL, so thiosulfate oxidation (postulated in the assay to represent oxidation of sulfite from the sulfonate group of the thiosulfate) accounted for uptake of 58 μL oxygen per 20 min per 13 mg protein or 10 nmol oxygen per minute per milligram protein. Although this rate was low compared to the rate of oxidation by suspensions of intact bacteria, the activity was sufficient to enable assay of the enzymes responsible.

The extracts contained a reduced methylviologen-dependent thiosulfate reductase:

$$S_2O_3{}^{2-} + 4H^+ + 4e^- = 2SO_3{}^{2-} + 2H_2S \qquad (15.3)$$

and three enzymes known from sulfate reducing bacteria—adenylylsulfate (adenosine phosphosulfate; APS) reductase (EC 1.8.99.2), ADP sulfurylase, and ATP sulfurylase, as well as adenylate kinase. The specific activities of these have been recalculated from Peck's (1960) original data (Table 15.1), which enables a scheme for thiosulfate scission and oxidation to be deduced

TABLE 15.1. Activities of enzymes involved in thiosulfate metabolism as assayed in cell-free extracts of *T. thioparus*[a].

Enzyme assayed	Specific activity [nmol min^{-1} (mg protein)$^{-1}$]
ATP sulfurylase	305
ADP sulfurylase	485
APS reductase (methyl viologen)	10.7
Thiosulfate reductase (sulfide formation)	0.8
Adenylate kinase	28.2

[a] Recaculated from Peck's (1960) data.

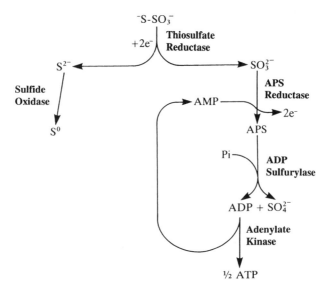

FIGURE 15.1. Thiosulfate oxidation by the APS pathway in *T. thioparus*. Data from Peck (1960).

(Fig. 15.1). The extracts also contained a particulate sulfide oxidase (0.5 mol oxygen consumed per mole sulfide, indicating elemental sulfur to be the product), at an activity of about 2 nmol oxygen per minute per milligram protein.

The scheme (Fig. 15.1) thus explained the production of both sulfate and sulfur in equimolar amounts from thiosulfate oxidation. In showing adenylylsulfate as an intermediate, it also provided a feasible route for the conservation of energy from sulfite oxidation by a substrate-level phosphorylation mechanism, in which ADP sulfurylase and adenylate kinase give rise to ATP:

$$2SO_3^{2-} + 2AMP = 2APS + 4e^- \qquad \text{(APS reductase)} \qquad (15.4)$$

$$2APS + 2P_i = 2ADP + 2SO_4^{2-} \qquad \text{(ADP sulfurylase)} \qquad (15.5)$$

$$2ADP = AMP + ATP \qquad \text{(adenylate kinase)} \qquad (15.6)$$

$$2SO_3^{2-} + AMP + P_i = 2SO_4^{2-} + ATP + 4e^- \qquad \text{(overall process)} \qquad (15.7)$$

Considering only the enzyme data, a weakness of the scheme could have been in the mechanism and activity of the thiosulfate reductase. This reductase activity was demonstrable only with methylviologen as the reductant, and no activity was seen with NADPH (Peck 1960). This showed that the reductase required an extremely electronegative electron donor: the E_0' for methylviologen is $-446\,mV$, whereas that for NADPH is only $-320\,mV$. This requirement for reduced methylviologen is consistent with the E_0' value for the couple $S_2O_3^{2-}/HS^- + HSO_3^- = -402\,mV$ (Thauer et al. 1977; Kelly 1999). A physiologic reductant for this reaction was not known and would need to be more electronegative than $-402\,mV$. The generation of such a reductant from the overall less electronegative process of sulfide and sulfite oxidation would be energetically unfavorable. Although the SO_4^{2-}/HSO_3^- couple is quite electronegative ($E_0' = -516\,mV$; Kelly 1999), the physiologic acceptor of electrons in sulfite oxidation is typically a c-type cytochrome, and there is no evidence for a couple as electronegative as $-402\,mV$. Clearly, electrons released by the APS reductase reaction (Eq. 15.4) could not couple to thiosulfate reductase, because the E_0' of the APS/AMP + HSO_3^- couple is only $-60\,mV$.

The activities of thiosulfate reductase originally reported were low ($\sim 0.8\,nmol$ sulfide per minute per milligram protein), when assayed at pH 7.0 by the Kaji and McElroy (1959) method. Possibly this method underestimated the thiosulfate reductase activity, measured as sulfide production (Peck 1960), because sulfide oxidase in the extract ($2\,nmol$ oxygen per minute per milligram protein) could have removed some of the sulfide product during the assay. Subsequently, thiosulfate reductase assays at pH 8.0 did demonstrate specific activities of $5–10\,nmol$ sulfide per minute per milligram protein (Peck and Fisher 1962), which would be sufficient to support the rates at which crude extracts oxidized thiosulfate and were similar to the activities of APS reductase in the same extracts.

A Retrospective View of Thiosulfate Cleavage and the Role of APS Reductase

It now seems probable that cleavage of thiosulfate in *T. thioparus* (and *Thiobacillus denitrificans*) depends not on a reductase as originally perceived by Peck (1960) but on a sulfur transferase of the rhodanese type (Peck 1968). Rhodanese is usually detected by its ability to transfer the sulfane-sulfur of thiosulfate to the nonphysiologic acceptor cyanide, producing thiocyanate and liberating sulfite:

$$^-S - SO_3 + CN^- = SCN^- + SO_3^{2-} \qquad (15.8)$$

In the cell, the acceptor for the sulfane-sulfur is probably a dithiol, such as lipoic acid thioredoxin (Nandi and Westley 1998), which can show considerably higher affinities for the sulfane-sulfur than does cyanide (Prieto et al. 1997). Such reactions were shown for the rhodanese of another thiosulfate-oxidizing bacterium, *Paracoccus versutus*, which transferred the sulfane-sulfur of thiosulfate to dihydrolipoate or dihydrolipoamide to form the corresponding persulfide (Silver and Kelly 1976a, 1976b; Silver et al. 1976). Cleavage by rhodanese bypasses the requirement for a reductase reaction with an E_0' below $-402\,mV$ and only requires a system whereby the reduced lipoate acceptor is regenerated. In support of this interpretation, the thiosulfate reductase and rhodanese activities of *Chlamydomonas* were shown to be mediated by the same enzyme-protein (Prieto et al. 1997). In work contemporaneous with Peck (1960), the mechanism of action of thiosulfate reductase had already been postulated to be similar to that of rhodanese (Kaji and McElroy 1959).

The original demonstration of APS reductase in *T. thioparus* used the methylviologen-dependent assay of APS cleavage to AMP and sulfite (Peck 1960), but in the oxidation of thiosulfate the reaction proceeds in the oxidative direction (Eq. 15.4), forming APS. This is the thermodynamically favorable direction of the reaction. Later work showed that APS formation by the reductase could be coupled to the reduction of ferricyanide or to cytochrome c (Peck et al. 1965; Lyric and Suzuki 1970), thereby showing the thermodynamic feasibility of APS as an intermediate in the oxidation pathways for sulfite and thiosulfate.

APS reductase from different sources shows considerable differences in stability of activity. The enzyme from *Desulfovibrio desulfuricans* was stable, but that from *T. thioparus* was labile and could not be purified (Peck et al. 1965). In contrast, APS reductase from *T. denitrificans* was purified to homogeneity and was remarkable heat-stable (Bowen et al. 1966; Taylor 1989).

Further Evidence for the APS Pathway of Thiosulfate Oxidation in *Thiobacillus thioparus*

The requirement for phosphate for the complete oxidation of thiosulfate by *T. thioparus* had already been shown, and arsenate had been shown to replace phosphate in catalyzing thiosulfate oxidation (Vishniac 1952; Vishniac and Santer 1957; Santer et al. 1960). The first evidence for a phosphosulfur compound as an oxidation intermediate was obtained by Vishniac and Santer (1957). They showed that *T. thioparus* incubated with ^{35}S-labeled sulfide and ^{32}P-labeled orthophosphate produced at least one doubly labeled compound of a mixed anhydride type, $-S-O-PO_3^{2-}$. Subsequently, Santer (1959) demonstrated the transfer of ^{18}O from orthophosphate to the sulfate produced from thiosulfate oxidation by whole cells of *T. thioparus* and showed that this transfer was insensitive to the uncoupling agent 2,4-dinitrophenol. These results showed that oxidative phosphorylation was not involved directly in the ^{18}O transfer process but that a phosphate-sulfate exchange occurred during the oxidation process. This led to Peck's (1960) search for the involvement of sulfur-containing nucleotides during thiosulfate oxidation, arguing that the ^{18}O exchange was occurring because of the formation of a $-S-O-PO_3^{2-}$ intermediate (Santer 1959; Kelly 1982).

Armed with the background study of Santer (1959) and the enzymatic evidence for the APS pathway, Peck and Stulberg (1962) undertook to demonstrate the ^{18}O exchange reactions in cell-free extracts of *T. thioparus* during thiosulfate oxidation. This helped provide direct evidence that the APS-reductase pathway was the mechanism by which sulfate was formed. They demonstrated in two stages the reactions that together result in the transfer of ^{18}O from orthophosphate into sulfate during sulfite oxidation by APS reductase (Fig. 15.2). Using APS produced from sulfite and AMP by cell-free extracts (with ferricyanide as electron acceptor), they showed that ADP-sulfurylase catalyzed the exchange of the sulfite group for an ^{18}O-labeled orthophosphate group to produce ADP in which the bridge oxygen between the phosphate groups was ^{18}O. Adenylate kinase catalyzed the conversion of this ADP into AMP, in which the $-PO_3^{2-}$ group contained the single ^{18}O atom. Next Peck and Stulberg (1962) showed that incubating cell-free extracts with reduced glutathione, thiosulfate, orthophosphate, and the ^{18}O-labeled AMP resulted in the production of ^{18}O-labeled sulfate. The amount of ^{18}O recovered in sulfate by these experiments was close to the theoretical amount predicted by the scheme shown in Figure 15.2, thereby supporting the APS pathway for thiosulfate cleavage and sulfite oxidation in *T. thioparus* (Peck 1962; Peck and Stulberg 1962; Kelly 1982).

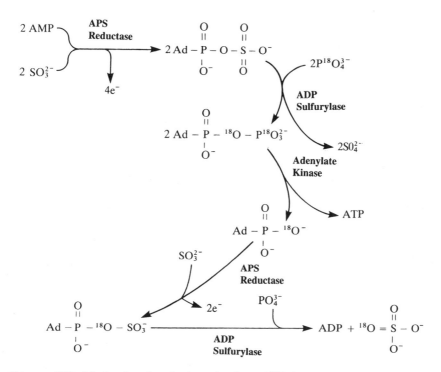

FIGURE. 15.2. Mechanism for the introduction of ^{18}O into sulfate formed during sulfite oxidation by the activities of APS reductase and ADP sulfurylase. Modified from Peck and Stulberg (1962).

Energy Conservation by Substrate-Level Phosphorylation and Its Coupling to Carbon Dioxide Fixation

A further significant contribution from Peck's work was the demonstration of the importance of APS not only as an intermediate in the thiosulfate oxidation pathway but as a high-energy compound from which ATP could be formed (Eqs. 15.4–15.7). This showed for the first time that substrate-level phosphorylation, analogous to that occurring during glycolysis, might be coupled to inorganic sulfur compound oxidation. Direct evidence for the coupling of this thiosulfate-dependent phosphorylation to autotrophic CO_2-fixation was provided by Peck and co-workers. When either thiosulfate or sulfite was oxidized by cell-free extracts of *T. thioparus*, sulfate production and phosphate esterification occurred at a ratio of about two, consistent with Equation 15.7 (Peck and Fisher 1962). The cell-free system producing ADP and ATP from sulfite and AMP could use the esterified phosphate to drive ribose 5-phosphate-dependent fixation of CO_2 by means of the enzymes phosphoribulokinase and ribulose 1,5-bisphosphate carboxylase also present in the crude extracts (Johnson and Peck 1965).

The importance of substrate-level phosphorylation in energy conservation by thiobacilli was confirmed in intact organisms using *Thiobacillus* (now *Halothiobacillus*) *neapolitanus* (Kelly and Syrett 1963, 1964a, 1964b, 1966a; Kelly and Wood 2000a). Suspensions of *Halothiobacillus neapolitanus* oxidizing thiosulfate were shown to produce ATP by both substrate-level and oxidative phosphorylation, with the former being insensitive to uncouplers of oxidative phosphorylation such as dinitrophenol (Kelly and Syrett 1964a, 1966a). The comparative effects of uncoupling agents on $^{14}CO_2$-fixation and ATP during the oxidation of either thiosulfate or sulfide led Kelly and Syrett (1963, 1966a; Kelly 1982) to the conclusion that substrate-level phosphorylation was less important during sulfide oxidation. This suggested that more electron-transport-linked phosphorylation occurred during the conversion of sulfide to sulfate, presumed to be coupled to the initial oxidation of the sulfide to sulfite. The quantitative importance of substrate-level phosphorylation by the APS pathway during the oxidation of different sulfur substrates has still not been resolved, but it was suggested that as much as 40% of the high-energy phosphate generated from thiosulfate by *H. neapolitanus* was by this mechanism (Hempfling and Vishniac 1967; Peck 1968). It was suggested that this estimate may be too high for most species in which the APS pathway occurs, in which oxidative phosphorylation may be the predominant energy-conserving process (Kelly 1990). In some of those species, however, the oxidation of sulfur, either as the element itself, or derived from sulfide or from the sulfane-atoms of thionates, may be catalyzed by an oxygenase, the "sulfur-oxidizing enzyme"

(EC 1.13.11.18) (Suzuki 1965a, 1965b; Charles and Suzuki 1966; Suzuki and Silver 1966; Emmel et al. 1986; Kelly 1999):

$$S_8 + 8O_2 + 8H_2O = 8H_2SO_3 \qquad (15.10)$$

This enzyme is commonly found in aerobic sulfur-oxidizing lithotrophs, but its importance as an alternative to a hydration–dehydrogenation electron-transport-linked energy-generating mechanism is still disputed (Kelly 1999):

$$S_8 + 24H_2O = 8H_2SO_3 + 32H^+ + 32e^- \qquad (15.11)$$

$$32H^+ + 32e^- + 8O_2 = 16H_2O \qquad (15.12)$$

Reaction 15.12 would be catalyzed by the electron-transport chain, with coupled phosphorylation, and all the oxygen in the sulfite product would be derived from water (in contrast to the oxygenase, in which two thirds of the oxygen atoms in sulfite come from dioxygen). Overall, Equations 15.11 and 15.12 produce the same result as Equation 15.10.

If thiosulfate is oxidized by a combination of sulfur-oxygenase and APS reductase pathways, 1 ATP is formed by the APS pathway (from the 2 SO_3^{2-} ions oxidized; Eq. 15.7). The oxidative step leading to 2 APS releases 4 e^- (Eq. 15.4). If these are coupled to electron transport, via flavin (Kappler and Dahl 2001), to oxygen with a P/O ratio of one, a maximum of 2 ATP could be formed (Kelly 1982, 1990). One third of the ATP would then arise from the substrate phosphorylation steps of the APS pathway. This simplistic approximation ignores the metabolic requirement for NAD(P)H generation (Kelly 1982), which would also have to be generated from the e^- released by the APS reductase. Meeting that need for NAD(P)H could result in most of the metabolic ATP arising from substrate-level phosphorylation. In contrast, Equations 15.11 and 15.12 plus the APS pathway (Eq. 15.7) would provide an additional 4[H] to couple to electron transport. This could meet the need for NAD(P) reduction or potentially produce an additional 2 ATP mol^{-1} thiosulfate oxidized, thereby reducing the contribution of substrate-level phosphorylation to a minimum of 20% of the total ATP produced.

Some chemolithotrophs cannot use the sulfur oxygenase during growth because they grow under anoxic conditions. *T. denitrificans* completely oxidizes thiosulfate during anaerobic growth, coupling oxidation to nitrate (or N_2O) reduction. It must, therefore, employ Equation 15.11 for the sulfane oxidation and a modified Equation 15.12 with denitrification rather than oxygen reduction for electron transport and phosphorylation (Kelly 1999). Marked differences in growth yields among thiosulfate-oxidizing chemolithotrophs can be ascribed to both the presence or absence of a functional sulfur-oxygenase and the presence or absence of the APS pathway (Timmer-ten Hoor 1981; Kelly 1982, 1990, 1999; Dopson et al., 2002). It is clear that the end of the path first trodden by Peck in 1960 has not yet been followed to its final destination.

Oxidation of the Sulfane-Sulfur of Thiosulfate and the Resynthesis of the Sulfonate Group of Thiosulfate from Sulfane-Sulfur by Thiobacilli

A further significant observation made by Peck was to confirm the discriminative metabolism of the two unequal sulfur atoms of thiosulfate, resulting in the preferential conversion of the sulfonate ($-SO_3^-$) group to sulfate (Peck and Fisher 1962). The oxidation states of the two atoms of sulfur in thiosulfate ($^-S-SO_3^-$) are -2 and $+5$, respectively (Varaivamurthy et al. 1993), easily allowing the production of the elemental sulfur (oxidation state 0) often seen in *Thiobacillus* cultures. When *T. thioparus* oxidized thiosulfate labeled with ^{35}S in the sulfane atom ($^{-35}S-^{32}SO_3^-$) in the presence of an unlabeled elemental sulfur trap, accumulation of the ^{35}S in the sulfur trap was observed (Skarzynski and Ostrowski 1958; Peck 1962; Peck and Fisher 1962). Sulfate is, however, normally produced from both sulfur atoms of thiosulfate (Santer et al. 1960; Peck 1962; Peck and Fisher 1962; Kelly and Syrett 1966b). This was confirmed for *H. neapolitanus*, which can catalyze the complete oxidation of thiosulfate to sulfate, without sulfur accumulation, although initially sulfate formation from the sulfonate group commenced before that from the sulfane-sulfur (Kelly and Syrett 1966a; Kelly and Wood 1994).

The possible roles in sulfur (and sulfide or sulfane-) oxidation of a sulfur dioxygenase or of electron-transport-linked hydration/dehydrogenation are outlined above, but the fate of the sulfite product may be more complex than previously considered. Vishniac and Santer (1957) showed that ^{35}S-labeled sulfide was rapidly oxidized first to thiosulfate (and polythionates) and then to sulfate by *T. thioparus*. This observation was incorporated into the original Peck scheme (Eqs. 15.13–15.17) by Peck and Fisher (1962), who realized that the complete oxidation of thiosulfate (after reductive scission to sulfite and sulfide; Eq. 15.3) could be explained if there was recycling of sulfide to produce thiosulfate:

$$2SH^- + 2O_2 = S_2O_3^{2-} + H_2O \qquad (15.13)$$

The implication of Equation 15.13, not explored by Peck and colleagues, was that thiosulfate molecules arising in this way would derive both the sulfane- and sulfonate-sulfur atoms from the sulfane-atom of the initial thiosulfate. This resynthesis from the sulfane atom was proved by showing that when *H. neapolitanus* oxidized sulfane-labeled thiosulfate ($^{-35}S-^{32}SO_3^-$), thiosulfate labeled in both sulfur atoms ($^{-35}S-^{35}SO_3^-$) was formed in progressively greater amounts as oxidation proceeded (Kelly and Syrett 1966b;

Kelly and Wood 1994). Only thiosulfate labeled in both atoms was present toward the end of the oxidation process and was derived exclusively from the sulfane-sulfur of the initial thiosulfate supplied (Kelly and Syrett 1966b; Kelly and Wood 1994). No label from sulfonate-labeled thiosulfate ($^{-32}S-^{35}SO_3^-$) appeared in the sulfane-group of thiosulfate in such experiments, and no ^{35}S-label was found in the chemically detectable thiosulfate present in the final stages of $^{-32}S-^{35}SO_3^-$ oxidation (Kelly and Syrett 1966b).

The thiosulfate reductase/rhodanese/APS reductase system is thus supported by evidence from direct enzyme assay, whole-cell metabolism and energetics, and ^{35}S-labeling experiments and provides a robust hypothesis to explain thionate oxidation and energy conservation in at least some chemolithotrophs.

The APS Pathway of Sulfite Oxidation: Into the Twenty-First Century

It is now accepted that sulfite oxidation by lithotrophic bacteria is catalyzed either by direct oxidation to sulfate (using a sulfite: acceptor oxidoreductase) or by the indirect APS pathway, with both systems occurring in many cases in the same organism (Kappler and Dahl 2001). The APS reductase pathway can contribute to inorganic sulfur oxidation by a great diversity of organisms and has been detected in α-, β-, and γ-Proteobacteria; green sulfur bacteria; Gram-positive bacteria; and Archaea (Kappler and Dahl 2001; Borodina et al. 2002). We can conclude that in showing the APS reductase pathway to be a possibility in one obligate chemolithotroph, Peck (1960) started a search that established the APS pathway as one of central metabolic importance in many lithotrophic sulfur bacteria. Among the sulfur-chemolithotrophs, the oxidative APS reductase pathway probably provides enhanced energetic capabilities, as ATP can be conserved at the substrate level as well as by electron-transport phosphorylation (Kelly 1982, 1999; Kappler and Dahl 2001). Given the relatively poor bioenergetics of inorganic sulfur oxidations, this means of draining all available energy from sulfite oxidation could have survival advantage. Moreover, APS reductase itself is revealed as an ancestral enzyme, functioning in the central energy metabolism not only of sulfur- and sulfate-reducing bacteria and Archaea but also in sulfur dissimilation by photolithotrophs and in a huge diversity of chemolithotrophic bacteria and Archaea (Kletzin 1994; Hipp et al. 1997).

Conclusion

In this retrospective analysis of our current understanding of chemolithotrophic sulfur compound oxidation, I have sought to show how the five research papers (and two reviews) on *T. thioparus* published by Peck and co-workers in the 1960s provided the basis of a unitary hypothesis applying to bacteria and Archaea of extremely diverse phylogeny. Before 1960, virtually all hypotheses about sulfur oxidation involved polythionate pathways of greater or lesser complexity (Tamiya et al. 1941; Vishniac and Santer 1957; Kelly 1968). Peck (1960) showed five enzymes in thiobacilli that appeared to be directly involved in converting sulfide and thiosulfate to sulfate; and by the end of the 1960s, several more enzymes had been implicated in thionate metabolism: rhodanese, thiosulfate dehydrogenase (tetrathionate synthesizing), sulfite dehydrogenase, and sulfur oxygenase. Today we are beginning to assemble a molecular understanding of some of these enzymes, but even now the total number of enzymes unequivocally established to function in inorganic sulfur oxidation (including the thiosulfate-oxidizing complex of *P. versutus* and *Paracoccus pantotrophus*) is small (Kelly and Wood 1994; Friedrich 1998; Kelly et al. 2001). Peck's contribution in providing a basic biochemical framework for inorganic sulfur oxidation must be recognized as one of the milestones in twentieth-century biochemistry.

References

Borodina E, Kelly DP, Schumann P, et al. 2002. Enzymes of dimethylsulfone metabolism and the phylogenetic characterization of the facultative methylotrophs *Arthrobacter sulfonivorans* sp. nov., *Arthrobacter methylotrophus* sp. nov., and *Hyphomicrobium sulfonivorans* sp. nov. Arch Microbiol 177:173–83.

Bowen TJ, Happold FC, Taylor BF. 1966. Studies on adenosine 5'-phosphosulfate reductase from *Thiobacillus denitrificans*. Biochim Biophys Acta 111:566–76.

Charles AM, Suzuki I. 1966. Mechanism of thiosulfate oxidation by *Thiobacillus novellus*. Biochim Biophys Acta 128:510–21.

Dopson M, Lindstrom EB, Hallberg KB. 2002. ATP generation during reduced inorganic sulfur compound oxidation by Acidithiobacillus caldus is exclusively due jto electron transport phosphorylation. Extremophiles 6:123–9.

Emmel T, Sand W, König WA, Bock E. 1986. Evidence for the existence of a sulphur oxygenase in *Sulfolobus brierleyi*. J Gen Microbiol 132:3415–20.

Friedrich CG. 1998. Physiology and genetics of sulfur-oxidizing bacteria. Adv Microbial Physiol 39:235–89.

Hempfling WP, Vishniac W. 1967. Yield coefficients of *Thiobacillus neapolitanus* in continuous culture. J Bacteriol 93:874–8.

Hiraishi A, Nagashima KVP, Katayama Y. 1998. Phylogeny and photosynthetic features of *Thiobacillus acidophilus* and related acidophilic bacteria: its transfer to the genus *Acidiphilium* as *Acidiphilium acidophilum* comb. nov. Int J Syst Bacteriol 48:1389–98.

Hipp WM, Pott AS, Thum-Schmitz N, et al. 1997. Towards a phylogeny of APS reductase and sirohaem sulfite reductases in sulfate-reducing and sulfur-oxidizing bacteria. Microbiology 143:2891–902.

Kaji A, McElroy WD. 1959. Mechanism of hydrogen sulfide formation from thiosulfate. J Bacteriol 77:630–7.

Kappler U, Dahl C. 2001. Enzymology and molecular biology of prokaryotic sulfite oxidation. FEMS Microbiol Lett 203:1–9.

Kelly DP. 1968. Biochemistry of oxidation of inorganic sulphur compounds by microorganisms. Aust J Sci 31:165–73.

Kelly DP. 1982. Biochemistry of the chemolithotrophic oxidation of inorganic sulphur. Phil Trans Roy Soc Lond Ser B 298:499–528.

Kelly DP. 1988. Oxidation of sulphur compounds. Soc Gen Microbiol Symposium 42:65–98.

Kelly DP. 1989. Physiology and biochemistry of unicellular sulphur bacteria. In: Schlegel HG, Bowien B, editors. Autotrophic bacteria. New York: Springer-Verlag. p 193–216.

Kelly DP. 1990. Energetics of chemolithotrophs. In: Krulwich TA, editor. Volume 12, The bacteria. New York: Academic Press. p 479–503.

Kelly DP. 1999. Thermodynamic aspects of energy conservation by chemolithotrophic sulfur bacteria in relation to the sulfur oxidation pathways. Arch Microbiol 171:219–29.

Kelly DP, Syrett PJ. 1963. Effect of 2:4-dinitrophenol on carbon dioxide fixation by a *Thiobacillus*. Nature 197:1087–9.

Kelly DP, Syrett PJ. 1964a. Inhibition of formation of adenosine triphosphate in *Thiobacillus thioparus*. Nature 202:597.

Kelly DP, Syrett PJ. 1964b. The effect of uncoupling agents on carbon dioxide fixation by a *Thiobacillus*. J Gen Microbiol 34:307–17.

Kelly DP, Syrett PJ. 1966a. Energy coupling during sulphur oxidation by *Thiobacillus* sp. strain C. J Gen Microbiol 43:109–18.

Kelly DP, Syrett PJ. 1966b. [35S]Thiosulphate oxidation by *Thiobacillus* strain C. Biochem J 98:537–45.

Kelly DP, Wood AP. 1994. Whole organism methods for inorganic sulfur oxidation by chemolithotrophs and photolithotrophs. In: Peck HD, LeGall J, editors. Inorganic microbial sulfur metabolism. Volume 243, Methods in enzymology. p 510–20.

Kelly DP, Wood AP. 2000a. Reclassification of some species of *Thiobacillus* to the newly designated genera *Acidithiobacillus* gen. nov., *Halothiobacillus* en. nov. and *Thermithiobacillus* gen. nov. Int J Syst Evol Microbiol 50:511–16.

Kelly DP, Wood AP. 2000b. Confirmation of *Thiobacillus denitrificans* as a species of the genus *Thiobacillus*, in the β-subclass of the Proteobacteria, with strain NCIMB 9548 as the type strain. Int J Syst Evol Microbiol 50:547–50.

Kelly DP, McDonald IR, Wood AP. 2000. Proposal for the reclassification of *Thiobacillus novellus* as *Starkeya novella* gen. nov., comb. nov., in the a-subclass of the Proteobacteria. Int J Syst Evol Microbiol 50:1797–802.

Kelly DP, Rainey FA, Wood AP. 2001. The genus *Paracoccus*. In: Dworkin M, Falkow N, Rosenberg H, et al. editors. The prokaryotes. 3rd (electronic) ed. New York: Springer-Verlag. link.springer-ny.com.

Kelly DP, Shergill, JK, Wood AP. 1997. Oxidative metabolism of inorganic sulfur compounds by bacteria. Antonie Leeuwenhoek 71:95–107.

Kletzin A. 1994. Sulfur oxidation and reduction in Archaea—sulfur oxygenase/reductase and hydrogenase from the extremely thermophilic and facultatively anaerobic archaeon *Desulfurolobus ambivalens*. Syst Appl Microbiol 16:534–43.

London, J, Rittenberg SC. 1964 Path of sulfur in sulfur and thiosulfate oxidation by thiobacilli. Proc Nat Acad Sci USA 52:1183–90.

Lyric RM, Suzuki I. 1970. Enzymes involved in the metabolism of thiosulfate by *Thiobacillus thioparus*. II. Properties of adenosine 5'-phosphosulfate reductase. Can J Biochem 48:344–54.

Moreira D, Amils R. 1997. Phylogeny of *Thiobacillus cuprinus* and other mixotrophic thiobacilli: proposal for *Thiomonas* gen. nov. Int J Syst Bacteriol 47:522–8.

Nandi DL, Westley J. 1998. Reduced thioredoxin as a sulfur-acceptor substrate for rhodanese. Int J Biochem Cell Biol 30:973–7.

Nathansohn A. 1902. Über eine neue Gruppe von Schwefelbakterien und ihren Stoffwechsel. Mitt Zool Stn Neapel 15:655–80.

Peck HD. 1960. Adenosine 5'-phosphosulfate as an intermediate in the oxidation of thiosulfate by *Thiobacillus thioparus*. Proc Nat Acad Sci USA 46:1053–7.

Peck HD. 1962. Comparative metabolism of inorganic compounds. Bacteriol Revs 26:67–94.

Peck HD. 1968. Energy-coupling mechanisms in chemolithotrophic bacteria. Ann Rev Microbiol 22:489–518.

Peck HD, Fisher E. 1962. The oxidation of thiosulfate and phosphorylation in extracts of *Thiobacillus thioparus*. J Biol Chem 237:190–7.

Peck HD, Stulberg MP. 1962. ^{18}O studies on the mechanism of sulfate formation and phosphorylation in extracts of *Thiobacillus thioparus*. J Biol Chem 237:1648–52.

Peck HD, Deacon TE, Davidson JT. 1965. Studies on adenosine 5'-phosphosulfate reductase from *Desulfovibrio desulfuricans* and *Thiobacillus thioparus*. Biochim Biophys Acta 96:429–47.

Prieto JL, Perez Castiniera JR, Vega JM. 1997. Thiosulfate reductase from *Chlamydomonas*. J Plant Physiol 151:385–9.

Rainey FR, Kelly DP, Stackebrandt E, et al. 1999. A re-evaluation of the taxonomy of *Paracoccus denitrificans* and a proposal for the combination *Paracoccus pantotrophus* comb. nov. Int J Syst Bacteriol 49:645–51.

Santer M. 1959. The role of ^{18}O phosphate in thiosulfate oxidation by *Thiobacillus thioparus*. Biochim Biophys Res Commun 1:9–12.

Santer M, Margulies M, Klinman N, Kaback R. 1960. Role of inorganic phosphate in thiosulfate metabolism by *Thiobacillus thioparus*. J Bacteriol 79:313–20.

Schlegel HG. 1975. Mechanisms of chemo-autotrophy. In: Kinne O, editor. Volume 2, Marine ecology. London: Wiley. p 9–60.

Silver M, Kelly DP. 1976a. Rhodanese from *Thiobacillus* A2: catalysis of reactions of thiosulphate with dihydrolipoate and dihydrolipoamide. J Gen Microbiol 97:277–84.

Silver M, Kelly DP. 1976b. Thin layer chromatography of oxidised and reduced lipoate and lipoamide and their persulfides. J Chromatog 123:479–81.

Silver M, Howarth OW, Kelly DP. 1976. Rhodanese from *Thiobacillus* A2: determination of activity by proton nuclear magnetic resonance spectroscopy. J Gen Microbiol 97:285–8.

Skarzynski B, Ostrowski W. 1958. Incorporation of radioactive sulphur by *Thiobacillus thioparus*. Nature 182:933–4.

Starkey RL. 1935. Products of the oxidation of thiosulfate by bacteria in mineral media. J Gen Physiol 18:325–49.

Suzuki I. 1965a. Oxidation of elemental sulfur by an enzyme system from *Thiobacillus thiooxidans*. Biochim Biophys Acta 104:395–71.

Suzuki I. 1965b. Incorporation of atmospheric oxygen-18 into thiosulfate by the sulfur-oxidizing enzyme of *Thiobacillus thiooxidans*. Biochim Biophys Acta 110: 97–101.

Suzuki I, Silver M. 1966. The initial product and properties of the sulfur-oxidizing enzyme of thiobacilli. Biochim Biophys Acta 122:22–33.

Tamiya H, Haga K, Huzisige H. 1941. Zur Physiologie der chemoautotrophen Schwefelbakterien. Acta Phytochem Tokyo 12:173–225.

Taylor BF. 1989. Thermotolerance of adenylylsulfate reductase from *Thiobacillus denitrificans*. FEMS Microbiol Lett 59:351–4.

Thauer, RK, Jungermann K, Decker K. 1977. Energy conservation in chemotrophic bacteria. Bacteriol Rev 41:100–80.

Timmer-ten Hoor, A. 1981. Cell yield and bioenergetics of *Thiomicrospira denitrificans* compared with *Thiobacillus denitrificans*. Antonie Leeuwenhoek 47: 231–43.

Trudinger PA. 1967. The metabolism of inorganic sulphur compounds by thiobacilli. Rev Pure Appl Chem 17:1–24.

Varaivamurthy A, Manowitz B, Luther GW, Jeon Y. 1993. Oxidation state of sulfur in thiosulfate and implications for anaerobic energy metabolism. Geochim Cosmochim Acta 57:1619–23.

Vishniac W. 1952. The metabolism of *Thiobacillus thioparus*. I. The oxidation of thiosulfate. J Bacteriol 64:363–73.

Vishniac W, Santer M. 1957. The thiobacilli. Bacteriol Revs 26:168–75.

Suggested Reading

Johnson EJ, Peck HD. 1965. Coupling of phosphorylation and carbon dioxide fixation in extracts of *Thiobacillus thioparus*. J Bacteriol 89:1041–50.

16
Reduction of Metals and Nonessential Elements by Anaerobes

LARRY L. BARTON, RICHARD M. PLUNKETT, and BRUCE M. THOMSON

Microorganisms have an important role in geochemical cycles by facilitating oxidation and reduction of many inorganic compounds. Of special interest is microbial reduction of oxidized compounds, as these may be required by microorganisms at trace levels or may have no known metabolic role other than to serve as electron acceptors for bacteria. Reduction by anaerobic bacteria has been demonstrated to transform toxic elements by various processes, including volatilization and formation of insoluble precipitates. Although there are various examples of organisms in the Archaea domain that detoxify metals and metalloids, this review is restricted to anaerobic members of the Bacteria domain. The organisms involved in the reduction processes are characterized as anaerobes or facultative anaerobic bacteria that use molecular hydrogen, lactate, pyruvate, or acetate as their electron donor. Our focus is on the biochemical and physiologic processes associated with dissimilatory reduction of metals, metalloids, and nonessential elements by bacteria growing under anaerobic conditions.

Diversity of Bacteria and Reactants

There is increasing recognition of the great diversity in bacterial types and metabolism, and these organisms have considerable impact on geochemical cycles. Because sulfate-reducing bacteria use many different electron donors and acceptors, scientists for some time have been attracted to these anaerobes. Insight into the bioenergetic strategies used by the sulfate-reducing bacteria in metabolism of inorganic compounds was provided by Peck (1993) and the sulfate reducers serve as useful models for many of the inorganic reactions conducted by other anaerobes. At concentrations needed to support bacterial growth, many of the metals, metalloids, and nonessential compounds amenable to microbial reduction are toxic. Therefore, anaerobic bacteria must have detoxifying systems to enable them to exist in the environment as well as the appropriate respiratory-linked enzymes to charge the plasma membrane for the life processes. It is envisioned that in most instances there are separate processes in bacteria for metal resistance and reduction. There are many examples of anaerobic bacteria that reduce metal cations (Table 16.1) and oxyanions of metals, metalloids, and nonessential elements (Table 16.2). For the most part, bacterial cells from stationary phase have been used in these assays to follow the various reduction activities.

The electrochemical activities of these reactions can be established by examining the single electron half cell reactions. The standard Gibbs energy of formation ($\Delta G°$) is given in Table 16.3 for some of the reactions. The half reactions were calculated using data published by the National Bureau of Standards (Wagman et al. 1982). The species used in calculating the redox couples are those predicted to be most stable in aqueous solutions near pH 7.0, based on thermodynamic stability diagrams (Eh-pH or pe-pH diagrams) prepared by Brookins (1997) and Langmuir (1977). The fourth column in Table 16.3 is the theoretical redox potential for each redox couple under equilibrium conditions in aqueous solution at pH 7.0. This approach is described by Stumm and Morgan (1996) and is used to compare different redox couples. To calculate Eh(W) for reactions involving solid or gas phases, a concentration of $1\,\mu M$ was used in these calculations.

In many instances, substrate phosphorylation is not coupled to oxidation of the electron donor source by the bacteria; therefore, growth will result from oxidative phosphorylation with electrons energizing the plasma membrane for ATP production according to the chemiosmotic system. A list of bacteria displaying dissimilatory reduction where growth is coupled to reduction of metal/metalloid electron acceptors is given in Table 16.4.

In the pursuit of mechanisms for electron transfer, various cytochromes and enzymes have been isolated and purified. A list of proteins implicated in the electron transfer processes with reduction of electron acceptor is given in Table 16.5. Although reductases have considerable specificity for reduction, it is apparent that the low-potential multiheme cytochromes interface with numerous different electron acceptors.

Reduction of Elements Required as Nutrients at Trace Levels

Selenium and molybdenum are required as trace elements by sulfate-reducing bacteria for the synthesis of redox proteins and dehydrogenases (Fauque et al. 1991). Molybdate at high concentrations inhibits the growth of sulfate-reducing bacteria by interfering with sulfate metabolism. Sublethal concentrations of molybdate are reduced by *Desulfovibrio desulfuricans* with the formation of black extracellular, amorphous MoS_2 (Tucker et al. 1997). An electron diffraction analysis was used to

TABLE 16.1. Reduction of metal cations by nongrowing cells or cell-free extracts of anaerobic microorganisms[a].

Metal	Atomic number	Organisms	Reference
Mn(IV)	25	*Acinetobacter calcoaceticus*	Nealson and Saffarini (1994)
		Acinetobacter johnsonii	Nealson and Saffarini (1994)
		Bacillus sp. SG-1	Nealson and Saffarini (1994)
		Bacillus strain 29	Nealson and Saffarini (1994)
		Desulfobacterium autotrophicum	Lovley (1995)
		Desulfomicrobium baculatum	Lovley (1995)
		D. desulfuricans	Lovley (1995)
		D. acetoxidans	Nealson and Saffarini (1994)
		Pseudomonas fluorescens	Nealson and Saffarini (1994)
		G. metallireducens	Lovley et al. (1993a)
		Stain SES-3[b]	Laverman et al. (1995)
		Veillonella atypica	Wolfolk and Whiteley (1962)
Fe(III)	26	*Deinococcus radiodurans* R1[c]	Fredrickson et al. (2000)
		Desulfobacter postgatei	Lovley et al. (1993b)
		Desulfobacterium autotrophicum	Lovley et al. (1993b)
		Desulfobulbus propionicus	Lovley et al. (1993b)
		Desulfovibrio baculatus	Lovley et al. (1993b)
		Desulfovibrio baarsii	Lovley et al. (1993b)
		D. desulfuricans	Lovley et al. (1993b)
		Desulfovibrio sulfodismutans	Lovley et al. (1993b)
		D. vulgaris	Lovley et al. (1993b)
		D. acetoxidans	Lovley et al. (1993b)
		Geothrix fermentans	Coates et al. (1999)
		G. metallireducens	Lovley et al. (1993a)
		Veillonella atypica	Wolfolk and Whiteley (1962)
Co(III)	27	*Geobacter ferrireducens*	Caccavo et al. (1996)
		G. sulfurreducens	Caccavo et al. (1994)
		Shewanella alga	Gorby et al. (1998)
Pd(II)	46	*D. desulfuricans*	Lloyd et al. (1998)
		Clostridium sp.	Francis et al. (1994)
U(VI)	92	*Deinococcus radiodurans* R1[c]	Fredrickson et al. (2000)
		Desulfovibrio baculatus	Lovley et al. (1993b)
		Desulfovibrio baarsii	Lovley et al. (1993b)
		D. desulfuricans	Tucker et al. (1998)
		Desulfovibrio sulfodismutans	Lovley et al. (1993b)
		D. vulgaris	Lovley et al. (1993b)
		G. metallireducens	Lovley et al. (1993a)
		S. oneidensis[d]	Lovley et al. (1991)
		Veillonella atypica	Wolfolk and Whiteley (1962)
Np(V)	93	*S. oneidensis*[d]	Lloyd et al. (2000)

[a] Arrangement is according to the atomic number of the reacting element.
[b] Currently *Sulfurospirillum barnesii* strain SES-3 (Stolz et al. 1999).
[c] Requires anthraquinine-2,6-disulfonate.
[d] Formerly *Shewanella putrefaciens* (Venkateswaran et al. 1999).

TABLE 16.2. Reduction of anions by nongrowing cells or cell-free extracts of anaerobic microorganisms[a].

Electron acceptor	Atomic number	Organisms	Reference
Vanadate	23	*Saccharomyces cerevisiae*	Bisconti et al. (1997)
		V. atypica	Wolfolk and Whiteley (1962)
		D. desulfuricans	Wolfolk and Whiteley (1962)
		Chemolithotroph isolate	Yurkova and Lyalikova (1991)
Chromate	24	*D. vulgaris*	Wang and Shen (1997)
			Tucker et al. (1998)
		D. radiodurans R1	Fredrickson et al. (2000)
		G. metallireducens	Lovley et al. (1993a)
Arsenate	33	*Desulfotomaculum auripigmentum*	Newman et al. (1997b)
		Desulfovibrio strain Ben-RA	Macy et al. (2000)
		V. atypica	Wolfolk and Whiteley (1962)
		Micrococcus aerogenes[b]	Wolfolk and Whiteley (1962)
Selenium	34	*V. atypica*	Wolfolk and Whiteley (1962)
Selenite		*Rhodospirillum rubrum*	Kessi et al. (1999)
		V. atypica	Wolfolk and Whiteley (1962)
		D. desulfuricans	Wolfolk and Whiteley (1962)
		Clostridium pasteurianum	Wolfolk and Whiteley (1962)
Selenite and		*Enterobacter cloacae*	Losi and Frankenberger (1997)
Selenate		*D. desulfuricans*	Tomei et al. (1995)
		W. succinogenes	Tomei et al. (1992)
Molybdate	42	*D. desulfuricans*	Chen et al. (1998)
			Tucker et al. (1997, 1998)
		V. atypica	Wolfolk and Whiteley (1962)
Tc(VI)	43	*D. radiodurans* R1[c]	Fredrickson et al. (2000)
		G. sulfurreducens	Lloyd et al. (2000)
		G. metallireducens	Lloyd et al. (2000)
		D. desulfuricans	Lloyd et al. (1999)
		Thiobacillus thiooxidans	Lyalikova and Khizhnyak (1996)
		Thiobacillus ferroxidans	Lyalikova and Khizhnyak (1996)
		Rhodobacter shpaeroides	Lloyd et al. (2000)
		Paracoccus denitrificans	Lloyd et al. (2000)
Tellurate	52	*V. atypica*	Wolfolk and Whiteley (1962)
Tellurite	52	*V. atypica*	Wolfolk and Whiteley (1962)
		D. desulfuricans	Wolfolk and Whiteley (1962)
		C. pasterianum	Wolfolk and Whiteley (1962)
		M. aerogenes[b]	Wolfolk and Whiteley (1962)
Iodate	53	*D. desulfuricans*	Councell et al. (1997)
Osmium	76	*V. atypica*	Wolfolk and Whiteley (1962)

[a] Arrangement is according to the atomic number of the reacting element.
[b] Currently *Peptococcus aerogenes*.
[c] Requires anthraquinine-2,6-disulfonate.

TABLE 16.3. Half reactions for important reduction by anaerobes.

Redox couple[a]	Reaction	ΔG^{ob} (kJ/mol)	Eh(W)[c] (mV)
Mo(V)/Mo(VI)	$1/2\ Mo_3O_{8(s)} + 2H_2O = 3/2\ HMoO_4^- + 5/2\ H^+ + e^-$	133.7	−181
S(-II)/S(VI)	$1/8\ HS^- + 1/2\ H_2O = 1/8\ SO_4^{2-} + 9/8\ H^+ + e^-$	27.70	−178
Fe(II)/Fe(III)	$Fe^{2+} + 3H_2O = Fe(OH)_{3(s)} + 3H^+ + e^-$	93.79	85
Cu(O)/Cu(II)	$1/2\ Cu_{(s)} + 1/2\ H_2O = 1/2\ CuO_{(s)} + H^+ + e^-$	53.71	142
Se(O)/Se(IV)	$1/4\ Se_{(s)} + 3/4\ H_2O = 1/4\ SeO_3^{2-} + 3/2\ H^+ + e^-$	85.40	175
Te(O)/Te(IV)	$1/4\ Te_{(s)} + 3/4\ H_2O = 1/4\ HTeO_3^- + 5/4\ H^+ + e^-$	68.70	194
As(III)/As(V)	$1/2\ H_3AsO_3 + 1/2\ H_2O = 1/2\ HAsO_4^{2-} + 3/2\ H^+ + e^-$	81.16	220
U(UI)/U(VI)	$1/2\ UO_{2(s)} = 1/2\ UO_2^{2+} + e^-$	39.10	227
Te(IV)/Te(VI)	$1/2\ HTeO_3^- + 1/2\ H_2O = 1/2\ HTeO_4^- + H^+ + e^-$	78.96	404
Se(IV)/Se(IV)	$1/2\ SeO_3^{2-} + 1/2\ H_2O = 1/2\ SeO_4^{2-} + H^+ + e^-$	82.81	443
Pd(O)/Pd(II)	$1/2\ Pd_{(s)} + 1/2\ H_2O = 1/2\ PdO_{(s)} + H^+ + e^-$	86.51	481
Cr(III)/Cr(VI)	$1/6\ Cr_2O_{3(s)} + 5/6\ H_2O = 1/3\ CrO_4^{2-} + 5/3\ H^+ + e^-$	131.36	552
Mn(II)/Mn(IV)	$1/2\ Mn^{2+} + H_2O = 1/2\ MnO_{2(s)} + 2H^+ + e^-$	118.61	574
N(O)/N(V)	$1/10\ N_{2(g)} + 3/5\ H_2O = 1/5\ NO_3^- + 6/5\ H^+ + e^-$	120.53	680
Cl(-1)/Cl(VII)	$1/8\ Cl^- + 1/2\ H_2O = 1/8\ ClO_4^- + H^+ + e^-$	133.90	972
Cl(-I)/Cl(V)	$1/6\ Cl^- + 1/2\ H_2O = 1/6\ ClO_3^- + H^+ + e^-$	139.11	1025

[a] Thermodynamic data for Mo and Te species are from Brookins (1997).
[b] Thermodynamic data are from Wagman et al. (1982).
[c] Eh(W) is calculated for aqueous solution at pH 7.0. All soluble species are $1\,\mu mol/L$.

conclude that Mo(VI) was produced as a $MoS_{2(s)}$ precipitate. It is likely that in the reduction of Mo(VI) in molybdate to Mo(IV) in molybdenum sulfide that there is a Mo(V) intermediate. A brown extracellular soluble Mo(V)-S complex was reported when Mo powder was added to a culture of *D. desulfuricans* (Chen et al. 1998). Although the Mo(V)-S complex may be the intermediate that is reduced to MoS_2, this remains to be established. At this time, the reduction of molybdate has not been coupled to bacterial cell growth.

Selenium, in the form of selenate or selenite, is toxic to *D. desulfuricans* (Tomei et al. 1995) and *Wolinella succinogenes* (Tomei et al. 1992) at elevated levels. At sublethal levels of 0.1–1.0 mM selenite or 10 mM selenate, minimal levels of growth is observed with both *D. desulfuricans* and *Wolinella succinogenes*. With both selenate and selenite, colloidal elemental selenium (Se^0) is produced inside the cell and released into the culture fluid after cell death. This reduction of Se(VI) and Se(IV) by these anaerobes is not coupled to growth and proceeds by mechanisms that have not yet been identified. Selenite and selenate reduction with formation of elemental selenium by these nonrespiratory processes serve to detoxify the environment for future bacteria and may be important for the geochemical cycle of selenium.

TABLE 16.4. Growth coupled to reduction of metals and metalloids.

Electron acceptor	Organism	Reference(s)
Chlorate	Isolate	Malmqvist and Welander (1992)
Perchlorate	Strain perc lace	Herman and Frankenberger (1999)
	Strain GR-1	Rikken et al. (1996)
Chromate	*Desulfotomaculum reducens*	Tebo and Obraztsova (1998)
Mn(IV)	*Bacillus* sp.	Nealson and Saffarini (1994)
	D. reducens	Tebo and Obraztsova (1998)
	D. acetoxidans	Nealson and Saffarini (1994)
	G. metallireducens	Nealson and Saffarini (1994)
	S. oneidensis	Nealson and Saffarini (1994)
Fe(III)	*Aeromonas hydrophila, Bacillus infernus, Deferribacter thermophilus, Desulfuromonas acetexigens, D. acetoxidans, Desulfuromonas chloroethenica, Desulfuromonas palmitatis, Desulfuromusa bakii, Desulfuromusa kysingii, Desulfuromusa succinoxidans, Ferribacterium limneticus, Ferrimonas balearica, Geobacter chapelli, Geobacter grbicium, Geobacter hydrogenophilus, G. metallireducens, G. sulfurreducens, Geothrix fermentans, Geovibrio ferrireducens, Peleobacter carbinolicus, Peleobacter venetianus, Pleobacter propionicus, S. alga, S. oneidensis, Shewanella saccharophila, S. barnesii, Thermoterrabacterium ferrireducens, Thermotoga maritima, Thermus* sp. SA-01, and *T. ferrooxidans*	Lovley (2000)
	D. reducens	Tebo and Obraztsova (1998)
Co(III)	*G. sulfurreducens*	Caccavo et al. (1994)
Arsenate	*Bacillus selenitireducens* strain MLS10	Blum et al. (1998)
	Bacillus arsenicoselenatis strain E1H	Blum et al. (1998)
	C. arsenatis	Kraft and Macy (1998)
	Desulfomicrobium strain Ben-RB	Macy et al. (2000)
	D. auripigmentum strain OREX-4	Newman et al. (1997)
	Sulfospirillum arsenatis strain MIT-13	Newman et al. (1998)
	S. barnesii strain SES-3	Laverman et al. (1995), Stolz et al. (1999)
Selenate	*B. arsenicoselenatis* strain E1H	Blum et al. (1998)
	B. selenitireducens strain MLS-10	Blum et al. (1998)
	Sulfurospirillum arsenophilum strain MIT-13	Stolz et al. (1999), Stolz and Oremland (1999)
	S. barnesii strain SES-3	Oremland et al. (1994) Stolz et al. (1997)
	Thauera selenatis	Rech and Macy (1992)
Uranium	*Desulfovibrio* strain UFZ B 490	Pietzsch et al. (1999)
	D. reducens	Tebo and Obraztsova (1998)

TABLE 16.5. Reduction of metals and metalloids by proteins from anaerobic bacteria.

Chemical reduced	Protein	Organism	Reference(s)
Vanadate	Cytochrome c_7	*D. acetoxidans*	Lojou et al. (1998b)
Chromate	Cytochrome c_3	*D. vulgaris* Hildenborough	Lovley and Phillips (1994)
	Cytochrome c_7	*D. acetoxidans*	Lojou et al. (1998b)
Mn (IV)	Outer membrane protein	*S. oneidensis*[a]	Beliaev and Saffarini (1998)
	Cytochrome c_7	*D. acetoxidans*	Lojou et al. (1998b)
Fe(III)	Outer membrane protein	*S. oneidensis*[a]	Beliaev and Saffarini (1998)
	Cytochrome c_3	*D. desulfuricans* Norway	Lojou et al. (1998a)
	Cytochrome c_3 (M_r 26,000)	*D. desulfuricans* Norway	Lojou et al. (1998a)
	Cytochrome c_3	*D. gigas*	Lojou et al. (1998a)
	Cytochrome c_3	*D. vulgaris* Hildenborough	Lojou et al. (1998a, 1998b)
	Cytochrome c_{553}	*D. vulgaris* Hildenborough	Lojou et al. (1998a)
	Cytochrome c_7	*D. acetoxidans*	Lojou et al. (1998a, 1998b)
Arsenate	Arsenate reductase	*C. arsenatis*	Kraft and Macy (1998)
Selenite	Hydrogenase I	*C. pasteurianum*	Yanke et al. (1995)
	Nitrite reductase	*T. selenatis*	DeMoll-Decker and Macy (1993)
	Selenite reductase	*T. selenatis*	Rech and Macy (1992) Schröder et al. (1997)
Selenate	Cytochrome c_3	*D. vulgaris* Hildenborough	Abdelous et al. (2000)
	Selenate reductase	*T. selenatis*	Schröder et al. (1997)
	Selenate reductase	*S. barnesii*	Stolz et al. (1997) Stolz et al. (1999), Stolz and Oremland (1999)
Tellurite	Hydrogenase I	*C. pasteurianum*	Yanke et al. (1995)
Uranyl	Cytochrome c_3	*D. vulgaris* Hildenborough	Lovley et al. (1993c)

[a] Formerly *S. putrefaciens* (Venkateswaran et al. 1999).

Cytochromes as Oxidoreductases

Several proteins were reported to function as enzymes for the dissimilatory reduction of metals and nonessential elements. As listed in Table 16.4, the most frequently reported proteins involved in metal reduction are the cytochromes from sulfate-reducing bacteria. The focus on these cytochromes supports the initial papers by Lovley and colleagues in which they reported that reduced cytochrome c_3 from *Desulfovibrio vulgaris* Hildenborough reduces uranyl salts (Lovley et al. 1993a) and chromate (Lovley and Phillips 1994).

As reviewed by Lojou et al. (1998a, 1998b), cytochrome (cyt) c_3 and cyt c_7 from the sulfate-reducing bacteria are effective in reducing metals. The cyt c_3 from *D. vulgaris* and from *Desulfovibrio gigas* have a molecular mass of about 13 kDa, and each has four heme groups per molecule. The *D. desulfuricans* strain Norway has two different proteins known as cyt c_3: One is a tetraheme of approximately 13 kDa and the other is a dimer with a molecular mass of 26 kDa and eight hemes per molecule. Cyt c_7 is a triheme that has been purified from *Desulfuromonas acetoxidans* and has a molecular mass of 9.8 kDa. With cyt c_3, cyt c_3 (M_r 26 kDa), and cyt c_7, the iron in each heme is coordinated by two histidine residues. In addition the potentials for the various hemes in each of these cytochromes is in the −100 to −400 mV/NHE range. The second-order rate constants for electron transfer to Fe(III) ammonium citrate decreases in the following range: cyt c_7 (from *D. acetoxidans*) > cyt c_3 (M_r 26 kDa, from *D. desulfuricans* Norway) > cyt c_3 (from *D. vulgaris*) > cyt c_3 (from *D. desulfuricans* Norway) > cyt c_3 (from *D. gigas*) (Lojou et al. 1998a). This is not simply a feature of the redox potential of each cytochrome, because the most positive heme reduction potential, as listed in decreasing order, is as follows: cyt c_3 (M_r 26 kDa, from *D. desulfuricans* Norway) > cyt c_7 (from *D. acetoxidans*) > cyt c_3 (from *D. desulfuricans* Norway) > cyt c_3 (from *D. gigas*) > cyt c_3 (from *D. vulgaris*). As a consequence of three-dimensional structure analysis of several cytochromes, the location of each of the hemes in the cytochrome is known, and for identification purposes, they have been designated as heme I, II, III, or IV. For the reduction of Fe(III) by cyt c_3, the first to be reduced is heme IV, which is in an exposed edge of the cytochrome molecule and is surrounded by various positively charged amino acid residues. This feature of metal reduction may also be the case for cyt c_7, because it has a heme IV; and although cyt c_7 is lacking a heme II moiety, this is not essential for metal reduction. Another feature for metal reduction is the global charge on the cytochrome that can repel the metal ion. The isoelectric point of Cyt c_7 is pH 7.8, which facilitates reduction of chromate, in part, because the cytochrome molecule will not repel chromate.

Not all cytochromes from sulfate-reducing bacteria reduce Fe(III) or other metals. *D. vulgaris* produces a cyt c_{553}, which has a molecular mass of 9 kDa, midpoint redox potential of 0 mV, and a single heme and the iron atom is coordinated by histidine methionine. It is unclear at this time if the inability of this cyt c_{553} to reduce metals is due to lack of a bishistidinyl iron coordination or to some other factor, such as steric hindrance owing to orientation of heme in the protein.

Since both *Geobacter metallireducens* and *Geobacter sulfurreducens* have triheme cyt c_7 proteins (Champine et al. 2000, Afkar and Fukumori 1999; Seeliger et al. 1998), it is appropriate to consider that these electron carriers function in metal reduction in a manner similar to that reported for the cyt c_7 of *D. acetooxidans*. The role of other cytochromes in *Geobacter* has not been explored with respect to metal reduction.

Models for Cellular Reduction

Fe(III) Reduction

For bacterial cells to use metals in dissimilatory reduction, there must be an appropriate set of enzymes and electron carriers arranged in the surface layers of bacteria. A model for Fe(III) reduction was proposed for *G. sulfurreducens* (Lovley 2000), and it involves a 41 kDa *c*-type cytochrome, a 9 kDa cytochrome, and an 89 kDa *c*-type cytochrome, which are positioned in the outer membrane, periplasm, and plasma membrane, respectively. The oxidation of compounds in the cytoplasm reduce a membrane-bound NADH dehydrogenase and electrons are passed along to the outer membrane, where Fe(III) is reduced. In iron-reducing bacteria, Fe(II) accumulates outside the cells. This is fortuitous, because if large quantities of Fe(III) were reduced, as in the case with dissimilatory iron reduction by an intracellular process, the cell would need to establish a procedure to handle accumulation of Fe(II), which could otherwise reach toxic levels. In addition, Beliaev and Saffarini (1998) showed that proteins in the outer membrane of *Shewanella oneidensis* (formerly *Shewanella putrefaciens*, Venkateswaran et al. 1999) are used for reduction of Fe(III) and Mn(IV). This model for electron transfer through a cytochrome cascade is useful in discussing iron and manganese reduction.

U(VI) Reduction

The reduction of U(VI) by sulfate-reducing bacteria appears to occur at the cell surface of the anaerobes, since the reduced products of these elements accumulates in the environment outside of the cell. The proteins of cyt c_3 and cyt c_7 have been demonstrated to function as nonspecific metal dehydrogenases; however, these cytochromes are found in the periplasm and not in the outer membrane. Thus, if it were analogous to Fe(III) reduction, uranyl ions would most appropriately be reduced by a cytochrome in the outer membrane of the sulfate reducers. As demonstrated by Laishley and Bryant (see Chapter 18) cytochromes are located in the outer membrane of certain sulfate reducers; however, their role in reduction of U(VI) remains to be demonstrated.

As(V) Reduction

Chrysiogenes arsenatis is the only known organism capable of using acetate as the electron donor and arsenate as the terminal electron acceptor for growth. This reduction of arsenate to arsenite is catalyzed by an inducible respiratory arsenate reductase, which has been isolated and characterized by Kraft and Macy (1998). Arsenate reductase (Arr) from *C. arsenatis* is a molybdenum-containing protein that contains acid-labile sulfur and zinc cofactor constituents. It consists of an $\alpha_1\beta_1$-heterodimer of 123 kDa, and the two subunits have molecular weights of 87 and 29 kDa, respectively. The enzyme is specific for arsenate and is unable to reduce nitrite, sulfate, or selenate. Arr is able to couple the reduction of arsenate to oxidation of benzyl viologen with an apparent K_m of 0.3 mM and a V_{max} of 7 nmoles per minute per milligram protein. Experiments have determined that 86% of the arsenate reductase activity is in the periplasm and 10% is bound to the plasma membrane. A respiratory role for arsenate reductase has been proposed for *C. arsenatis*; however, the composition of such a respiratory-coupled chain has not been established.

A second example of a membrane-bound arsenate reductase was isolated from *Sulfurospirillum barnesii* and was determined to be a $\alpha_1\beta_1\gamma_1$-heterotrimic enzyme complex (Newman et al. 1998). The enzyme has a composite molecular mass of 100 kDa, and α-, β-, and γ-subunits have masses of 65, 31, and 22, respectively. This enzyme couples the reduction of As(V) to As(III) by oxidation of methyl viologen, with an apparent K_m of 0.2 mM. Preliminary compositional analysis suggests that iron-sulfur and molybdenum prosthetic groups are present. Associated with the membrane of *S. barnesii* is a *b*-type cytochrome, and the arsenate reductase is proposed to be linked to the electron-transport system of the plasma membrane.

The recently isolated *Desulfotomaculum* strain Ben-RB is able to grow using lactate as a substrate and arsenate as the sole electron acceptor (Macy et al. 2000). It has been proposed that arsenate reductase is associated with the respiratory chain of this organism, because >98% of the arsenate reductase bound to the plasma membrane.

Se(VI) Reduction

Thauera selenatis is able to use selenate as a terminal electron acceptor for anaerobic respiration, and a selenate reductase enzyme was isolated (Schröder et al. 1997). The enzyme complex exhibiting selenate reductase activity is composed of an $\alpha_1\beta_1\gamma_1$-trimer, and subunit masses were calculated at α, 96 kDa; β, 40 kDa; and γ, 23 kDa. This reductase is a metal-containing protein that contains molybdenum, iron, and acid-labile sulfide. The sulfur-iron molar values suggest the presence of at least two [Fe-S] centers. The selenate reductase in *T. selenatis* has a high specificity for selenate and does not reduce sulfate, chlorate, nitrate, or nitrite. Enzyme kinetics were calculated when reduced benzyl viologen is the electron donor, and the apparent K_m is 16 μM selenate, with a V_{max} of 40 μmole of selenate reduced per minute per milligram protein. The enzyme can also couple the oxidation of methyl viologen to selenate reduction, although at a relatively low specific activity. The enzyme is localized in the periplasm of the cell, but apparently is a segment of the respiratory chain, because it has an association with a *b*-type cytochrome. When selenate is the only available electron acceptor, it is reduced to selenite without subsequent reduction to elemental selenium. A role for components of the denitrification pathway for metabolism of selenite is suggested, because when nitrate and selenate are both present, selenite does not accumulate but is reduced to Se^0. In addition, it can be considered that the positioning of the selenate reductase in the periplasm may assist in reducing the toxic of selenite to the cells.

Respiratory selenate reduction was studied in *S. barnesii* strain SES-3, and preliminary information regarding a membrane-bound dissimilatory selenate-reductase complex was reported (Stolz et al. 1997; Stolz and Oremland 1999). The selenate-reductase complex is a heterotetramer composed of subunits with masses of 82, 53, 34, and 21 kDa, respectively. The presence of molybdenum as a prosthetic group was suggested from experiments using inhibition by tungsten; however, chemical analysis of the enzyme has not been conducted. The enzyme does not appear to be specific for selenate but also reduces nitrate, fumarate, and thiosulfate. In addition, *S. barnesii* SES-3 has distinct reductases for nitrate, fumarate, and arsenate. The apparent K_m for selanate is $12\,\mu M$. Although the mechanism for selenate reduction has not been resolved for *S. barnesii* strain SES-3, it appears that the mechanism of electron transfer is distinct from that found in *T. selenatis*.

Selenium respiration was reported for two moderately halophilic and alkaliphilic bacteria; however, the mechanism of respiration is unresolved. The physiologic properties of *Bacillus arsenicoselenatis* strain E1H and *Bacillus selenitireducens* growing on selenate and selenite were described by Blum et al. (1998). *Bacillus arsenicoselenatis* is able to grow by dissimilatory reduction of selenate Se(VI), to selenite, Se(IV). Although *B. arsenicoselenatis* is unable to grow using selenate as the electron acceptor, it will grow with selenite reduced to elemental selenium, Se^0. These two strains of *Bacillus* grow in co-culture, with selenate being reduced to Se^0. Enzymes for these reactions have not been purified.

Conclusion

We are at the discovery stage for determining the ability of various bacteria to reduce metals and nonessential compounds. Mechanisms for these reductions generally have not yet been established, and it is apparent that much is unknown. A number of questions pertaining to reduction are raised: Which elements and compounds are reduced at the cell surface? Why are some of the compounds not reduced at the cell surface but become reduced at the plasma membrane or in the cytoplasm? What is the nature of the nonenergetic reactions in the cytoplasm of the bacterial cell? What are the physiologic substrates for the cytochromes and which reactions occur because of substitution of chemicals due to similar structural features?

Certainly, considerable flexibility and adaptability of electron flow is expected in bacteria, and many new strains are expected to be found that obtain energy from these chemical reductions. The natural gene flow over the years in the anaerobic ecosystems has produced microorganisms of considerable physiologic diversity. These anaerobic organisms continue to provide numerous biochemical challenges in the areas of anaerobic reduction of metals, metalloids, and nonessential elements by microorganisms.

Acknowledgments. Research discussed here in the laboratories of LLB and BMT was supported, in part, by grants from the U.S. Department of Energy (WERC Consortium and NABIR Program).

References

Abdelouas A, Gong WL, Lutze W, et al. 2000. Using cytochrome c(3) to make selenium wires. Chem Mat 12:1510–2.

Afkar E, Fukumori Y. 1999. Purification and characterization of triheme cytochrome c_7 from the metal-reducing bacterium, *Geobacter metallireducens*. FEMS Microbiol Lett 175:205–10.

Beliaev AS, Saffarini DA. 1998. *Shewanella putrefaciens* mtrB encodes an outer membrane protein required for Fe(III) and Mn(IV) reduction. J Bacteriol 180: 6292–7.

Bisconti L, Pepi M, Mangani S, Baldi F. 1997. Reduction of vanadate to vanadyl by a strain of *Saccharomyces cerevisiae*. Biometals 10:239–46.

Blum JS, Bindi AB, Buzzelli J, et al. 1998. *Bacillus arsenicoselenatis*, sp. nov., and *Bacillus selenitireducens*, sp. nov.: two halophiles from Mono Lake, California that respire oxyanions of selenium and arsenic. Arch Microbiol 171:19–30.

Brookins DG. 1997. Eh-pH diagrams for geochemistry. Berlin: Springer-Verlag.

Caccavo F, Coates JD, Rossello-Mora RA, et al. 1996. *Geobacter ferrireducens*, a physiologically distinct dissimilatory Fe(III)-reducing bacterium. Arch Microbiol 165:3752–9.

Caccavo F, Lonergan DJ, Lovley DR, et al. 1994. *Geobacter sulfurreducens* sp. Nov., a hydrogen- and acetate-oxidizing dissimilatory metal-reducing microorganism. Appl Environ Microbiol 60:3752–9.

Champine JE, Underhill B, Johnson JM, et al. 2000. Electron transfer in the dissimilatory iron-reducing bacterium *Geobacter metallireducens*. Anaerobe 6: 187–96.

Chen G, Ford TE, Clayton CR. 1998. Interaction of sulfate-reducing bacteria with molybdenum dissolved from sputter-deposited molybdenum thin films and pure molybdenum powder. J Colloidal Interface Sci 204:237–46.

Coates JD, Ellis DJ, Gaw CV, Lovley DR. 1999. *Geothrix fermentans* gen.nov., sp nov., a novel Fe(III)-reducing bacterium from a hydrocarbon-contaminated aquifer. Int J Syst Bacteriol 49:1615–22.

Councell TB, Lada ER, Lovley DR. 1997. Microbial reduction of iodate. Water Air Soil Pollution 100:99–106.

DeMoll-Decker H, Macy JM. 1993. The periplasmic nitrite reductase of *Thauera selenatis* may catalyze the reduction of selenite to elemental selenium. Arch Microbiol 160:241–7.

Fauque G, LeGall J, Barton LL. 1991. Sulfate-reducing and sulfur-reducing bacteria. In: Shivley JM, Barton LL, editors. Variations in autotrophic life. New York: Academic Press. p 271–338.

Francis AJ, Dodge CJ, Lu F, et al. 1994. XPS and XANES studies of uranium reduction by *Clostridium* sp. Environ Sci Technol 28:636–9.

Fredrickson JK, Kostandarithes HM, Li SW, et al. 2000. Reduction of Fe(III), Cr(VI), U(VI), and Tc(VII) by *Deinococcus radiodurans* R1. Appl Environ Microbiol 66:2006–11.

Gorby YA, Caccovo F Jr, Bolton H. 1998. Microbial reduction of cobalt^{III}EDTA$^-$ in the presence and absence of manganese(IV) oxide. Environ Sci Technol 32: 244–50.

Herman DC, Frankenberger WT. 1999. Bacterial reduction of perchlorate and nitrate in water. J Environ Qual 28:1018–24.

Kessi J, Ramuz M, Wehrli E, et al. 1999. Reduction of selenite and detoxification of elemental selenium by the phototrophic bacterium *Rhodospirillum rubrum*. Appl Environ Microbiol 65:4734–40.

Kraft T, Macy JM. 1998. Purification and characterization of the respiratory arsenate reductase of *Chrysiogenes arsenatis*. Eur J Biochem. 255:647–53.

Laverman AM, Blum JS, Schaefer JK, et al. 1995. Growth of strain SES-3 with arsenate and other diverse electron acceptors. Appl Environ Microbiol 61:3556–61.

Lloyd JR, Nolting H-F, Solé VA, et al. 1999. Reduction of technecium by *Desulfovibrio desulfuricans*: biocatalyst characterization and use in a flow through bioreactor. Appl Environ Microbiol 65:2691–6.

Lloyd JR, Yong P, Macaskie LE. 1998. Enzymatic recovery of elemental palladium by using sulfate-reducing bacteria. Appl Environ Microbiol 64:4607–9.

Lloyd JR, Yong P, Macaskie LE. 2000. Biological reduction and removal of Nb(V) by two microorganisms. Environ Sci Technol 34:1297–301.

Lojou E, Bianco P, Bruschi M. 1998a. Kinetic studies on the electron transfer between various c-type cytochromes and iron (III) using a voltametric approach. Electrochim Acta 43:2005–13.

Lojou E, Bianco P, Bruschi M. 1998b. Kinetic studies on the electron transfer between bacterial c-type cytochromes and metal oxides. J Electroanal Chem 452:167–77.

Losi ME, Frankenberger WT. 1997. Reduction of selenium oxyanions by *Enterobacter cloacae* strain SLD1a-1: reduction of selenate to selenite. Environ Toxicol Chem 16:1851–8.

Lovley DR. 1995. Microbial reduction of iron, manganese, and other metals. Adv Agr 54:175–231.

Lovley DR. 2000. Fe(III) and Mn(IV) reduction. In: Lovley DR, editor. Environmental microbe-metal interactions. Washington, DC: ASM Press. p 3–30.

Lovley DR, Phillips EJP. 1994. Reduction of chromate by *Desulfovibrio vulgaris* and its c_3 cytochrome. Appl Environ Microbiol 60:726–8.

Lovley DR, Giovannoni SJ, White DC, et al. 1993a. *Geobacter metallireducens* gen nov sp nov, a microorganism capable of coupling the complete oxidation of organic compounds to the reduction of iron and other metals. Arch Microbiol 159:336–44.

Lovley DR, Phillips EJP, Gorby YA, Landa E. 1991. Microbial reduction of uranium. Nature 350:413–6.

Lovley DR, Roden EE, Phillips EJP, Woodward JC. 1993b. Enzymatic iron and uranium reduction by sulfate-reducing bacteria. Marine Geol 113:41–53.

Lovley DR, Widman PK, Woodward JC, Phillips EJP. 1993c. Reduction of uranium by cytrochrome c_3 of *Desulfovibrio vulgaris*. Appl Environ Microbiol 59:3572–6.

Macy JM, Santini JM, Pauling BV, et al. 2000. Two new arsenate/sulfate-reducing bacteria: mechanism of arsenate reduction. Arch Microbiol 173:49–57.

Malmqvist A, Welander T. 1992. Anaerobic removal of chlorate form bleach effluents. Water Sci Technol 25:237–42.

Nealson KH, Saffarini D. 1994. Iron and manganese in anaerobic respiration: Environmental significance, physiology and regulation. Annu Rev Microbiol 48: 311–43.

Newman DK, Ahmann D, Morel FMM. 1998. A brief review of microbial arsenate reduction. Geomicrobiology 15:255–68.

Newman DK, Beveridge TJ, Morel FMM. 1997a. Precipitation of arsenic trisulfide by *Desulfotomaculum aurigmentum*. Appl Environ Microbiol 63:2022–8.

Newman DK, Kennedy EK, Coats JD, et al. 1997b. Dissimilatory arsenate and sulfate reduction in *Desulfotomaculum auripigmentum* sp. nov. Arch Microbiol 168:380–8.

Oremland RS, Blum JS, Culbertson CW, et al. 1994. Isolation, growth, and metabolism of an obligately anaerobically, selenate-respiring bacterium, strain SES-3. Appl Environ Microbiol 60:3011–9.

Peck HD Jr. 1993. Bioenergetic strategies of sulfate-reducing bacteria. In: Odom JM, Singleton R Jr, editors. The sulfate-reducing bacteria: contemporary perspectives. New York: Springer-Verlag. p 41–76.

Pietzsch K, Hard BC, Babel W. 1999. A *Desulfovibrio* sp. capable of growing by reducing U(VI). J Basic Microbiol 39:365–72.

Rech SA, Macy JM. 1992. The terminal reductases for selenate and nitrate respiration in *Thauera selenatis* are two distinct enzymes. J Bacteriol 174:7316–20.

Rikken GB, Kroon AGM, van Ginkel CG. 1996. Transformation of (per)chlorate into chloride by a newly isolated bacterium: reduction and dismutation. Appl Microbiol Biotechnol 45:420–6.

Schröder I, Rech S, Kraft T, Macy JM. 1997. Purification and characterization of the selenate reductase form *Thauera selenatis*. J Biol Chem 272:23765–8.

Seeliger S, Cord-Ruwisch R, Schink B. 1998. A periplasmic and extracellular *c*-type cytochrome of *Geobacter sulfurreducens* acts as a ferric iron reductase and as electron carrier to other acceptors or to partner bacteria. J Bacteriol 180:3686–91.

Stolz JF, Oremland RS. 1999. Bacterial respiration of arsenic and selenium. FEMS Microbiol Rev 23:615–27.

Stolz JF, Ellis DJ, Blum JS, et al. 1999. *Sulfurospirillum barnesii* sp. nov. and *Sulfurospirillum arsenophilum* sp. nov., new members of the *Sulfurospirillum* clade of the ε-Proteobacteria. Int J Syst Bacteriol 49:1177–80.

Stolz JF, Gugliuzza T, Blum JS, et al. 1997. Differential cytochrome content and reductase activity in *Geospirillum barnesii* strain SeS3. Arch Microbiol 167:1–5.

Stumm W, Morgan JJ. 1996. Aquatic chemistry. 3rd ed. New York: Wiley.

Tebo BM, Obraztsova AY. 1998. Sulfate-reducing bacterium with Cr(VI), U(VI), Mn(IV), Fe(III) as electron acceptors. FEMS Microbiol Lett 162:193–8.

Tomei FA, Barton LL, Lemanski CL, Zocco TG. 1992. Reduction of selenate and selenite to elemental selenium by *Wolinella succinogenes*. Can J Microbiol 38:1328–33.

Tomei FA, Barton LL, Lemanski CL, et al. 1995. Transformation of selenate and selenite to elemental selenium by *Desulfovibrio desulfuricans*. J Indust Microbiol 14:329–36.

Tucker MD, Barton LL, Thomson BM. 1997. Reduction and immobilization of molybdenum by *Desulfovibrio desulfuricans*. J Environ Quality 26:1146–52.

Tucker MD, Barton LL, Thomson BM. 1998. Reduction of Cr, Mo, Se, and U by *Desulfovibrio desulfuricans* immobilized in polyacrylamide gels. J Ind Microbiol Biotechnol 20:13–19.

Venkateswaran K, Moser DP, Dollhopf ME, et al. 1999. Polyphasic taxonomy of the genus *Shewanella* and description of *Shewanella oneidensis* sp. nov. Int J Syst Bacteriol 49:704–24.

Wagman DD, Evans WH, Parker VB, et al. 1982. The NBS tables of chemical thermodynamic properties. J Phys Chem Ref Data 11:1–293.

Wang YT, Shen H. 1997. Modeling Cr(VI) reduction by pure bacterial cultures. Water Res 31:727–32.

Wolfolk CA, Whiteley HR. 1962. Reduction of inorganic compounds with molecular hydrogen by *Micrococcus lactilyicus*. I. Stoichiometry with compounds of arsenic, selenium, tellurium, transition and other elements. J Bacteriol 84:647–58.

Yanke LJ, Bryant RD, Laishley EJ. 1995. Hydrogenase of *Clostridium pasteurianum* functions as a novel selenite reductase. Anaerobe 1:61–7.

Yurkova NA, Lyalikova NN. 1991. New vanadate-reducing facultative chemolithotrophic bacteria. Microbiology 59:672–7.

Suggested Reading

Langmuir D. 1997. Aqueous environmental geochemistry. Upper Saddle River, NJ: Prentice-Hall.

Lyalikova NN, Khizhnyak TV. 1996. Reduction of heptavalent technetium by acidophilic bacteria of the genus *Thiobacillus*. Microbiology 65:468–73.

Macy JM, Lawson S. 1993. Cell yield [Y(M)] of *Thauera selenatis* grown anaerobically with acetate plus selenate or nitrate. Arch Microbiol 160:295–8.

17
Chemolithoautotrophic Thermophilic Iron(III)-Reducer

Juergen Wiegel, Justin Hanel, and Kaya Aygen

Importance of Fe(III) Reduction and Diversity of Thermophilic Fe(III) Reducers

The anaerobic reduction of metal ions, including ferric iron (Fe(III)) to ferrous iron (Fe(II)) by microorganisms, has been observed for >80 years (Harders 1919) and was reviewed several times from various viewpoints (Jones 1986; Lovley 1991, 1995; Nealson and Saffarini 1995). In general terms the process can be described by:

$$Fe^{3+} + electron \Rightarrow Fe^{2+} \tag{17.1}$$

and, if hydrogen is the electron donor, the reaction is exergonic:

$$2Fe^{2+} + H_2 \Rightarrow 2Fe^{2+} + 2H^+ \text{ with a } \Delta G^{\circ\prime} \text{ of } 228.3\,kJ/mol. \tag{17.2}$$

The subsequent formation of magnetite from Fe^{2+} and $Fe(OH)_3$ is also exergonic with a $\Delta G^{\circ\prime}$ of $-60\,kJ/mol$ (Stum and Morgan 1981).

However, only recently has dissimilatory Fe(III) reduction by microorganisms been recognized as a widespread and ecologically and biogeochemically important reaction (for details on the history of recognized nonmicrobial and microbial Fe(III) reduction see e.g., Lovley 1991). Presently, there is no doubt about the ecological importance of microbial Fe(III) reduction in the cycling of metals and mineralization of organic material. Iron, the third most abundant element in the Earth's crust (McGeary and Plummer 1997), plays important roles in the cycling of biomass (Ponnamperuma 1972). Furthermore, biotic dissimilatory iron reduction has become of increasing interest in the development of bioremediation strategies for decontamination of subsurface sites from hazardous waste material, including aromatic compounds (Lovley et al. 1989; Lovley and Lonergan 1990) as well as in novel concepts on the evolution of microbial life (Lovley 1991; Nealson and Saffarini 1995; Vargas et al. 1998; Slobodkin et al. 1999a). Important iron ores are the various ancient Banded Iron Formation deposits, which include the two major types: Algoma, thought to have been formed by submarine volcanic activities, and the Superior type from the Proterozoic Age. Although it is generally believed that microorganisms were involved in the formation of some of the later formations, it has not been unequivocally established whether microbial or geochemical processes have been the major source of the iron precipitation (Guilbert and Park 1986; Lovley 1991).

The extreme negative values for the carbon isotope shifts (up to around δ-60 in some of the banded iron(II)-formations), strongly indicate that microbial processes were involved in the formation of these deposits (Canfield et al. 2000). One explanation of the low values is that the bacteria involved could have used carbon sources that were already metabolized, e.g., by methanogens, and thus already had undergone a significant isotope shift (i.e., $\delta^{12/13}C$ values for methane are already negative; however, the values for the associated CO_2/carbonates are positive, between +5 to +15‰) (Coleman et al. 1993; Horita and Berndt 1999).

Bacteria capable of Fe(III) reduction belong to a wide range of groups in the phylogenetic tree (Lovley 1991, 1995; Lonergan et al. 1996; Slobodkin 1999b). These include obligate anaerobes, aerobic bacteria under oxygen limitation (Short and Blakemore 1986), facultative anaerobes, sulfate and sulfur reducers (Coleman et al. 1993; Roden and Lovley 1993), obligately autotrophic aerobic acidophiles such as *Thiobacillus thiooxidans* and the heterotrophic *Acidophilium cryptum* (Küsel et al. 1999), halotolerant marine bacteria such as *Shewanella alga* (Rosello-Mora et al. 1994), and nonhalotolerant freshwater bacteria such as *Geobacter metalireducens* (Lovley 1995). The diversity also includes a wide range from cold tolerant (formerly called "psychrotrophs") iron reducers to iron-reducing hyperthermophiles such as *Pyrobaculum islandicum* (Vargas et al. 1998; Kashefi and Lovley 2000). Some of the Fe(III) reducers are able to use Fe(III) as the terminal electron acceptor for the complete mineralization of organic compounds, including acetate (Küsel et al. 1999; Roden and Lovley 1993), whereas others such as *Desulfitobacterium dehalogenans* only partially oxidize the carbon sources forming, e.g., acetate, (Wiegel, unpublished results). Fe(III)-reducing bacteria have been isolated or indicated in many different ecological niches with various pH, temperatures, salt concentrations, and contaminations of hazardous waste compounds. However, a larger number of Fe(III) reducers has been isolated from oil and hydrocarbon-contaminated or oil-associated sites (Greene et al. 1997; Tuccillo et al. 1999; Slobodkin et al. 1999a; Coates et al. 1999).

Many novel (facultative) anaerobic Fe(III)-reducing bacteria have been described recently. Most of the mesophilic bacteria belong to the Gram-type negative (Wiegel 1981) proteobacteria with *Shewanella*, *Geobacter*, and *Peleobacter* or major representative. (Lovley 1995). *Shewanella putrefaciens* and *G. metalireducens* are among the most studied Fe(III) reducers. Much less, however, is known about the distribution and diversity of thermophilic and hyperthermophilic Fe(III) reducers. In contrast to the mesophilic iron reducers, the majority of the isolated thermophilic Fe(III)-

reducing species belong to the Gram-type positive branch (Firmicutes). These include the heterotrophic *Bacillus infernus* (Boone et al. 1995), which was the first obligately anaerobic thermophilic Fe(III) reducer isolated; *Thermoterrabacterium ferrireducens* (Slobodkin et al. 1997b); and *Thermoanaerobacter siderophilus* (Slobodkin et al. 1999a) which can grow autotrophically. In contrast, *Deferribacter thermophilus* belongs to the Gram-type negative, or Flexistipes, group (Greene et al. 1997).

Bacillus infernus (Boone et al. 1995), the first described thermophilic Fe(III)-reducer, is despite its phylogenetic position within the generally aerobic *Bacillus* species, an obligate anaerobe. This was the first instance in which the simple division between aerobic spore formers, (*Bacillus* and related bacteria) and the anaerobic spore formers, (clostridia and related bacteria) were disrupted. Since that time an aerobic bacterium, *Thermaerobacter marianensis* was described that was isolated from the Japanese Mariana Trench Chellenger Deep and clearly with the obligately anaerobic clostridia, *Moorella*, *Thermoanaerobacter/Thermoanaerobacterium* species, and the like from the Gram-type positive clostridial branch (Takai et al. 1999). Several strains of *B. infernus* isolated from deep (2700 m) subsurface samples grow in an estimated temperature range of 39–65°C (optimum 60°C) and a pH range of around 6.5 to <9.0 (optimum 7.3) and tolerate up to 0.6 M NaCl.

Thermoterrabacterium ferrireducens (Slobodkin et al. 1997a), the second validly described thermophilic Fe(III) reducer, was isolated in our laboratory from a sample of a run off channel from a hot spring in the Calcite Spring area of Yellowstone National Park. Interestingly, these hot springs are below the Calcite Spring area, which contains oil seeps. The samples from which this strain was obtained contained white streamer material that was covered with black Fe(III)-sulfide. Furthermore, one of the novel Fe(III)-reducing '*Ferribacter thermautotrophicus*' strains and the previously described glycerol utilizing acetogen *Moorella glycerinum* were isolated from the same spring (Slobodkin et al. 1997a). From this site a CO-utilizing Fe(III)-reducing thermophile was also isolated from a sample taken at the same time as our samples (Robb, personal communication 2001). *Thermoterrabacterium ferrireducens* grows between 50 and 74°C (optimum around 70°C) and in the pH^{65C} range of 4.8–8.2 (optimum around 6.1).

Thermoanaerobacter siderophilus (Slobodkin et al. 1999b), a spore-forming bacterium, was isolated from sediments of a newly formed hydrothermal vent of the Karymsky volcano (Kamchatka peninsula). It grows in a temperature range of 39–78°C (optimum around 70°C) and pH range of 4.8–8.2 (optimum 6.3–6.5). Dissimilatory Fe(III) reduction of precipitated amorphous Fe(III) hydroxide/oxide occurs under chemolithoautotrophic and heterotrophic growth conditions. The fermentation products from glucose are ethanol, lactate, hydrogen, and CO_2; less hydrogen is produced when grown in the presence of Fe(III). *Thermoanaerobacter siderophilus* can also reduce Mn(IV). In addition to this species, Slobodkin et al. (1999b) demonstrated that *Thermoanaerobacter wiegelii* (Cook et al. 1996) and *Thermoanaerobacter sulfurophilus* (Bonch-Osmolovskaya et al. 1997) were also dissimilatory Fe(III) reducers, while using peptone as carbon source (other species were not tested).

Deferribacter thermophilus (Greene et al. 1997) is, in contrast to the above obligate anaerobes, a facultative microaerophile that tolerates up to 3% oxygen and is the first thermophilic Fe(III)-reducing bacterium isolated from a petroleum/oil related environment, specifically from production water from an oil field in the North Sea (United Kingdom). Consistent with this and with the observation of several thermophilic Fe(III) reducers in oil deposits by Slobodkin et al. (1999a), magnetite was found to accumulate near petroleum deposits (Lovley 1991). *Deferribacter thermophilus* grows in a relative narrow temperature range of 50–60°C (optimum 60°C) and a pH range of 4.8–8.2 (optimum 6.4). Beside Fe(III), this bacterium can use other metal ions, including Mn(IV), as electron acceptors.

Various thermophilic bacteria and archaea able to reduce (Fe(III). The capability for Fe(III)-reduction appears to be more wide spread among anaerobic thermophilles then originally thought from the isolation of specific iron-reducers discussed above. Slobodkin and Wiegel (1997) showed that in various samples from geothermally heated sites, different hydrogen-oxidizing Fe(III)-reducing microorganisms exist that are able to grow at temperatures up to 90°C. Subsequently, Slobodkin et al. (1999a) isolated several thermophilic/hyperthermophilic bacteria and archaea from deep subsurface petroleum reservoirs of western Siberia. Three of those strains were assigned to the thermophilic *Thermoanaerobacter acetoethylicus* and *Thermoauerobacter brockii*, one strain to the hyperthermophilic bacterium *Thermotoga maritima*, and three strains to the archaeal genus *Thermococcus*. Furthermore, the authors showed that like *Thermotoga subterranae*, which was previously isolated from a French oil well (Jeanthon et al. 1995), the strains were able to reduce Fe(III) heterotrophically, either with peptone or with hydrogen. Between 8 and 20 mmol Fe(III)/l were reduced. This is significantly more than what was demonstrated for the hyperthermophilic bacterium *Thermotoga* and the archaeon *Pyrobacculum* (Vargas et al. 1998). The Fe(III) reduction occurred only in the presence of added peptone or hydrogen (i.e., the utilized medium contained 0.01% yeast extract and 0.2% bicarbonate), indicating that the organisms are able to grow chemolithoheterotrophically as well chemolithoautotrophically with Fe(III) as the terminal electron acceptor. However, it has not conclusively been proven that they truly grow chemolithoautotrophically, i.e., that >50% of the cell carbon came from the inorganic carbon dioxide/carbonate pool. The same is true for the enrichments from geothermal areas described by Slobodkin and Wiegel (1997), which stoichiometrically reduced Fe(III) using hydrogen in ratios of around 2:1.

More recently, there is evidence that many hyperthermophilic anaerobic archaea are able to use soluble, complexed Fe(III) ions and amorphous

Fe(III) hydroxide/oxide as electron acceptors. This indicates that anaerobic Fe(III) reduction was prevalent before the separation into Bacteria and Archaea. This finding may support the hypothesis that Fe(III) respiration is an ancient form of anaerobic respiration and might have been involved in the early evolution of prokaryotic life. The thermophilic iron reducers even include some aerobic bacteria. Bridge and Johnson (1998) demonstrated that the aerobic acidophilic, moderately thermophilic *Acidimicrobium ferrooxidans*, *Sulfobacillus acidophilus*, and *Sulfobacillus thermosulfidooxidans* are also able to reduce Fe(III) with glycerol as carbon and electron donor under oxygen-limiting growth conditions.

In this chapter, we report on additional, recently isolated thermophilic Fe(III) reducers that grow chemolithoautotrophically and reduce Fe(III) stoichiometrically using hydrogen and forming magnetite at unusual high rates. These strains constitute a novel Fe(III)-reducing group, '*F. thermautotrophicus*', gen. nov., sp. nov. In addition, we describe several chemolithoautotrophic Fe(III)-reducing strains that are closely related to well-known heterotrophic species, which neither grow chemolithoautotrophically nor reduce Fe(III) and other metal ions, as far as is known. These bacteria are compared with two previously isolated and well-known thermophiles that can reduce iron heterotrophically but can neither grow autotrophically nor use Fe(III) reduction as a primary energy source.

General Isolation and Growth Conditions for Thermophilic Fe(III) Reducers

The chemolithoautotrophic strains were isolated and routinely grown in anaerobic media under an 80:20 mix of hydrogen and carbon dioxide gases or under nitrogen gas with formate as carbon source and electron donor. The medium typically contained (per liter of deionized water) 0.33 g KH_2PO_4, 0.33 g NH_4Cl, 0.33 g KCl, 0.33 g $MgCl_2$ $2H_2O$, 0.33 g $CaCl_2$ $2H_2O$, 2.0 g $NaHCO_3$, 1 mL vitamin solution (Wolin et al. 1963) and 1.2 mL trace element solution (Slobodkin et al. 1997b). The pH values varied between 7 and 9.0. To this medium, Fe(III) (in the form of amorphous Fe(III) oxide/hydroxide) and/or 9,10-anthraquinone 2,6-disulfonic acid (AQDS) were added as electron acceptors. The Fe(III) concentration in the medium was equal to 90 mM, and the AQDS was at a final concentration of 20 mM. Medium was routinely sterilized by autoclaving at 121°C for 1 h. All cultures were grown in Hungate or Balch tubes at 60°C or at the indicated temperatures, and all transfers were carried out with syringes and needles under anaerobic conditions. For the isolation, usually 100 mL of medium was inoculated with 5–10 g of the samples and incubated statically with mixing only when checked for Fe(III) reduction and magnetite formation. Magnetite formation was checked using magnets and confirmed by X-ray refractive analysis, and Fe(II) formation was determined by employing the 2,2-dipyridyl iron assay of Balashova and Zavarzin (1980).

'*Ferribacter thermautotrophicus*' gen. nov., sp. nov. (Hanel 2000). Based on the report from Slobodkin and Wiegel (1997), three strains were isolated: strain JW/JH-Fiji-2 from the hot spring runoff channel at the soccer field in Savu Savu on Vanu Levu (Fiji) and strains JW/KA-1 and JW/KA-2^T from a runoff channel (Y6) close to the Yellowstone River in the Calcite Springs area of Yellowstone National Park. From the latter, the non-iron-reducing acetogen *Moorella glycerini* had previously been isolated (Slobodkin et al. 1997a). The iron-reducer *Thermoterrabacterium ferrireducens* (Slobodkin et al. 1997b) was isolated from an adjacent hot spring runoff channel (Y7) in the Calcite Spring area. Since the geothermal areas on Fiji and in the Yellowstone National Park belong to different geothermal systems and continental plates, the isolation of the same species indicates that this bacterium is apparently widely distributed. Isolating another thermophilic Fe(III) reducer from the hot springs of Calcite Spring indicates that several thermophilic Fe(III) reducers exist in these systems (see below).

The cells of '*F. thermautotrophicus*' were Gram-type positive rods that usually occurred singly but also in pairs and chains. Their growth temperature ranged from 50 to 75°C, with the optimum around 72°C. The pH^{60C} range for chemolithoautotrophic growth was between 6.5 and 8.5, with the optimum at pH^{60C} 7.3. However, resting cells converted the amorphous Fe(III) hydroxide/oxide stoichiometrically to magnetite in a pH^{60C} range from 5.5 to nearly 11. In the presence of hydrogen and Fe(III), yeast extract, casamino acids, fumarate, and crotonate were utilized in addition to CO_2 and lead to the formation of magnetite. Fumarate, thiosulfate, and the artificial humic acid analog 9,1-anthraquinone 2,4-disulfonic acid were used as electron acceptors beside Fe(III) hydroxide/oxide. Formate, frequently regarded as a pseudoautotrophic substrate could serve as electron donor as well as a carbon source for growth with Fe(III) as the terminal electron acceptor. In fact, formate turned out to be a good substrate for enrichment cultures for these chemolithoautotrophic iron-reducers. It was surprising that, Fe(III) citrate did not serve as an electron acceptor for these strains. This could be because the iron-sequestering system has less affinity for the iron ions than for citrate. Electron micrographs revealed that the strains of '*F. thermautotrophicus*' were always in close proximity to the Fe(III)-hydroxide/oxide precipitation, although less strongly adhered to the Fe(III) particles than *T. ferrireducens* (Slobodkin et al. 1997b). The iron-uptake system is apparently different from that of *T. ferrireducens*, which requires a direct physical contact with the precipitated iron. Further studies are necessary to elucidate the mechanisms of Fe(III) reduction in these bacteria.

The maximal cell density observed under chemolithoautotrophic growth conditions after refeeding twice with hydrogen was 10^8 cells mL^{-1} after

1 week. Under extended incubation times without refeeding, the maximal cell density reached only $2–4 \times 10^7$ cells mL^{-1}; however, this occurs within 24 h owing to a doubling time of about 45 min at 72°C and pH^{60C} 7.3. This density is much less than what can be reached under heterotrophical conditions with *T. ferrireducens* or *B. infernus*. However, the densities reached by Slobodkin et al. (1999a) with the isolates from the deep oil well were also around 4×10^7 cells mL^{-1} when grown autotrophically. Furthermore, our other magnetite-producing isolates (described below) also reached only low densities when grown chemolithoautotrophically with Fe(III) as electron acceptor. The reason for this observed limitation in reaching higher cell densities is not known, but appears to be characteristic for the chemolithoautotrophic Fe(III)-reducing thermophiles.

The most exciting observation was the high reduction rate at neutral pH of around 1.3 µmol Fe(II) formed per hour and milliliter at a cell concentration of $3 \times 10^7 mL^{-1}$. This rate is about 10 times higher than what has been reported for the other species at comparable cell densities. Even at the alkaline pH^{60C} 10, rates of 0.1 µmol $h^{-1} mL^{-1}$ were observed. Based on these properties, we speculate that this bacterium or species with similar properties could have been responsible for or could have been involved in processes leading to the Banded Iron Formation. In comparison, the highest value observed in coal mining-affected freshwater lake sediments supplemented with carbon sources and e-donors (organic or hydrogen) were maximally <1 µmol × mL^{-1} day^{-1} (Küsel and Dorsch 2000).

Chemolithoautotrophic Hydrogen-Oxidizing and Fe(III)-Reducing Strains Related to the Glucolytic Species

Beside the strains constituting *F. thermautotrophicus'* two additional strains were isolated as chemolithoautotrophic Fe(III) reducers, namely JW/JH-1 (from a runoff channel [Y7] close to the Yellowstone River in the Calcite Springs area of Yellowstone National Park) and strain JW/JH-Fiji-1 from a hot spring runoff channel at the soccer field in Savu Savu on Vanu Levu (Fiji). Based on their 16s rRNA sequence analysis, they are closely related to the glycolytic species *Thermobrachium celere* (Engle et al. 1996) and *Clostridium thermopalmarium* as well as *Clostridium thermobutyricum* (Wiegel et al. 1989), respectively (Fig. 17.1). However, neither of the type strains, *T. celere* JW/SL-NZ35^T (DSM 8682, ATCC 700318) as well as the five tested strains of *T. celere* P-6, P-7, P-21, P-32, and P-48 (from our collection), nor *C. thermobutyricum* JW-171K^T (DSM 4928, ATCC 49875) and *C. thermopalmarium* (ATCC 51427) could reduce either soluble Fe(III) citrate or amorphous Fe(III) hydroxide or

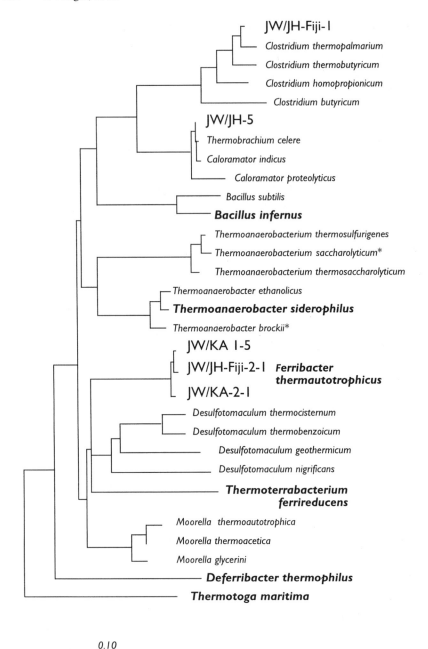

FIGURE 17.1. Phylogenetic tree based on 16S rDNA sequence analysis showing the placement of the novel chemolithoautotrophic Fe(III) reducers. Capital and bold face letters, respiratory Fe(III) reducers; *, tested heterotrophic species that probably reduce Fe(III) predominantly in an assimilatory fashion.

Mn(IV). On the other hand, the Fe(III)-reducing strains JW/JH-1 and JW/JH-Fiji-1, did not grow heterotrophically in the absence of the electron acceptor Fe(III).

In the presence of atmospheric hydrogen, strain JW/JH-1 utilized CO_2, formate (20 mM), glycerol (0.3%, vol/vol), yeast extract (1% wt/vol), tryptone (1% wt/vol), and acetate (20 mM) as carbon sources but not casamino acids (1% wt/vol), glucose (20 mM), cellobiose (20 mM), or pyruvate (20 mM). Strain JW/JH-Fiji-1 exhibited a similar substrate spectrum but did not utilize glycerol or acetate (20 mM). Beside amorphous Fe(III) hydroxide (90 mM), strain JW/JH-1 and strain JW/JH-Fiji-1 utilized AQDS (20 mM) and thiosulfate as electron acceptors in mineral medium, but not Mn(IV) (15 mM), Se(VI) (100 μM), precipitated sulfur (0.1%), sublimated sulfur (0.1%), nitrate, sulfate, crystalline Fe(III) hydroxide (Sigma), or soluble Fe(III) citrate (25 mM). Growing chemolithoautotrophically, i.e., using only H_2/CO_2 as carbon and electron sources, the cell densities were low (3.0×10^7 cells mL^{-1}), similar to those observed for '*F. thermautotrophicus*'. But both chemolithoautotrophic strains grew to higher cell densities on Fe(III) media (2.0×10^8 cells mL^{-1}) in the presence of 1% yeast extract or 1% tryptone. The temperature and pH ranges and optima for growth of the chemolithoautotrophic strains (Table 17.1) differed slightly but are still within possible variations seen for different strains of a species. Strain JW/JH-1 from the Calcite Spring is the first thermophilic Fe(III) reducer for which the oxidation of acetate to CO_2 is reported. Whether the oxidation occurs via the citric acid cycle as shown for *Geobacter sulfurreducens*, however, is not known (Galushko and Schink 2000).

TABLE 17.1. Properties of the novel chemolithoautotrophic strains and the corresponding heterotrophic type strains.

	Thermobrachium celere– like strain JW/JH-1	Thermobrachium celere– type strain	Clostridium thermobutyricum– like strain JW-JH-Fiji-1	Clostridium thermobutyricum– type strain
Motility	–	–	–	+
Spore formation	–	–	–	+
Fe(III) reduction	+	–	+	–
Fermentative Growth	(+)	+	+	+
Temperature ranges (°C)	45–67, optimum 65°C	43–75, optimum 66°C	47–70, optimum 65	26–61.5, optimum 55
pH ranges	6.0–9.5, optimum 8.0	5.4–9.5, optimum 8.2	5.5–9.0, optimum 7.5	5.8–9.0, optimum 7.0
Use of:				
Glucose	–	+	–	+
Tryptone	+	–	+	–
Cellobiose	–	–	–	+
Acetate	+	–	–	–

The stoichiometric ratio for the dissimilatory Fe(III) reduction is 2 mol of Fe(II) formed per 1 mol of hydrogen oxidized. The obtained values for the novel strains were 0.9–1.1 ± 0.2 and 0.8–0.9 ± 0.1 moles of Fe(II) produced for each mole of hydrogen oxidized, depending on the strains tested. These values suggest that for every mole of hydrogen that was used for the reduction of Fe(III), 1 mole of hydrogen was used for anabolic reactions. Similar values have been obtained for *T. siderophilus* (Slobodkin et al. 1999b).

The Overall pH Range for Thermophilic Chemolithoautotrophic Fe(III) Reduction

The observed pH range for thermophilic Fe(III) reduction with the concomitant formation of magnetite is between 5.5 and 11. Although mesobiotic Fe(III) reduction occurs at acidic pH values (e.g., pH 2.5) (Küsel et al. 1999; Küsel and Dorsch 2000), the characteristic formation of magnetite has been observed only in the neutral and alkaline pH range. Magnetite desolves at acidic pH of 5.5 and below (Bell et al. 1987). Nor is the other frequently observed product, the iron carbonate, siderite, stable at these acidic pH values.

Fe(III) Reduction by Known Glycolytic Thermophiles

Slobodkin et al. (1999b) reported that *T. wiegelii* and *T. sulfurophilus* were able to carry out dissimilatory Fe(III) reduction. We tested *Thermoanaerobacterium* strains for dissimilatory Fe(III) reduction during heterotrophic growth. Some limited Fe(III) reduction was observed with *T. brockii* subspecies *finii* and *Thermoanaerobacterium saccharolyticum* strain JW/SL-YS485 but results with *Thermoanaeobacter ethanolicus* JW200[T] were equivocal. To understand why only a limited reduction of Fe(III) was observed, strain JW/SL485 was studied in more detail. When this strain was grown in the presence of the soluble Fe(III) citrate, Fe(III)-was reduced to Fe(II), as evidenced by the color change from black to clear. The amount of Fe(III) reduction was quantified by measuring the Fe(II) accumulation (Balashova and Zavarzin 1980). No magnetite, however, was produced under these conditions. Fe(II) accumulated as cell numbers increased. No significant Fe(III) reduction was observed in a control inoculated with killed cells (Fig. 17.2), nor were Fe(III) oxide/hydroxide and or 9,10-anthraquinone-2,6-disulfonic acid (AQDS) reduced (data not shown). A fermentation analysis (Fig. 17.3) indicated that cultures grown in media containing Fe(III) used all of the provided 0.5% glucose in a shorter period of time while producing less ethanol and slightly more lactate than cells grown in the absence of Fe(III). However, no significant increases in the more oxidized product acetate were observed, as one would expect if Fe(III) served as a respiratory electron acceptor. Furthermore, in the

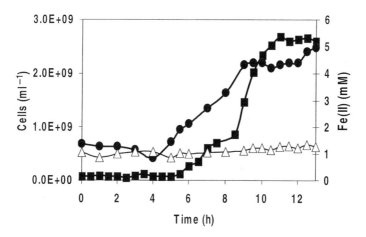

FIGURE 17.2. Fe(III) reduction by *T. saccharolyticum* strain JW/SL-YS48. ■, growth (increase of cells per milliliter); ●, reduction of Fe(III), i.e., increase in Fe(II); △, control, i.e., Fe(II) concentration in the presence of killed cells.

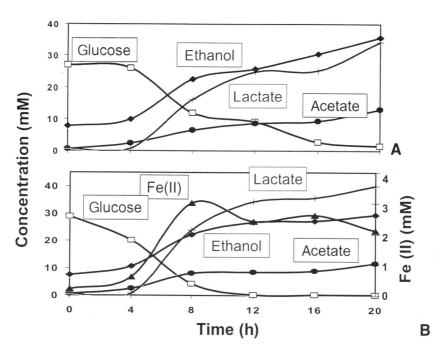

FIGURE 17.3. Fermentation products of *T. saccharolyticum* growing in the presence **A** and the absence **B** of Fe(III) hydroxide/oxide. Medium contained 0.4% yeast extract.

TABLE 17.2. Cell yields and calculated y_{ATP} when grown in the presence and absence of Fe(III) or Fe(II)-supplements[a].

Supplementation	No Fe	Plus Fe(III)	Plus Fe(II)
Biomass (mg/mL) glucose used	0.171	0.254	0.333
(μmol/mL)	14.2	8	18
$y_{glucose}$ (mg/mmoL) y_{ATP}	12.0	31.8	18.5
(g biomass/mol ATP)[a]	5.0	12.0	7.5

[a] The y_{ATP} value was calculated using the theoretical ATP production via the Embden Mayerhoff Parnass pathway, yielding 2 ATP per mole of glucose converted to pyruvate, and 1 ATP per acetate formed from pyruvate: y_{ATP} = micrograms dry weight biomass μmole^{-1} ATP formed; media contained 0.05% (wt/vol) yeast extract.

presence of 0.4% or 0.05% yeast extract, an increased $Y_{glucose}$ was observed compared to a control grown in an identical media lacking the Fe(III) (Table 17.2). Citrate alone, added in the concentrations equal to that of the ferric citrate in previous experiments (25 mM) inhibited growth, suggesting that the citrate was not metabolized as an additional carbon source. Addition of citrate between 1 and 10 mM caused an increase in doubling time; 15, 25, and 35 mM completely inhibited cell growth. When Fe(III) citrate was replaced with Fe(II) citrate, $Y_{glucose}$ yields were also higher than in cultures grown without a supplemental iron species, although still not as high as when grown with Fe(III) citrate (Fig. 17.4).

Based on the following observations, we currently assume that in the heterotrophic strain JW/SL-YS4 85 (and probably in the other tested

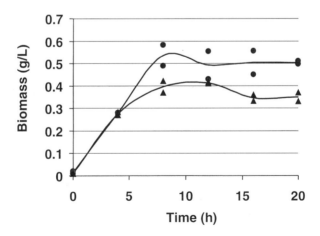

FIGURE 17.4. Growth stimulation of *T. saccharolyticum* by the addition of iron(II) to the growth medium. ●, with Fe(II) supplementation; ▲, without additon of Fe(II).

heterotrophic strains with limited Fe(III) reduction) Fe(III) reduction occurred primarily assimilatory: (1) The addition (and use) of the potential electron acceptor Fe(III) to the growth medium did not lead to an increase in oxidized fermentation products (such as acetate or CO_2), as one would expect if Fe(III) were used as dissimilatory electron acceptor. (2) A moderately increased $Y_{glucose}$ was observed when the reduced iron species, Fe(II) citrate, alone was added. (3) The increased growth yields in the presence of Fe(III) citrate were also observed when nutrient concentrations such as yeast extract were low (<0.1% wt/vol). These observations suggest both a stimulatory effect of Fe(II/III) ions on the overall metabolism as well as that the *T. saccharolyticum* strain JW/SL-YS485 grows preferentially glycolytically but is capable of maximizing its growth yield, especially at low substrate concentrations. Whether the strain gains energy at all via reduction of Fe(III) or that the cells became just more efficient at elevated iron concentrations cannot be decided based on the present experimental results. A similar metabolism was observed in mesophilic fermentative bacteria, such as certain *Vibrio* species (Jones et al. 1984; Lovley 1991). It was calculate that the *Vibrio* use Fe(III) only as a minor electron acceptor and were not observed to conserve energy to support growth (Jones et al. 1984).

The Use of Amorphous Fe(III) Hydroxide/Oxide Versus Soluble Fe(III) Ions Among the Thermophiles

As one may expect from the diversity of microorganisms that can reduce iron, the spectrum ranges from bacteria that can use only amorphous Fe(III) hydroxide/oxide (e.g., *T. ferrireducens*) and apparently require direct contact with the Fe(III) precipitate, as shown by electron micrographs (Slobodkin et al. 1997b), to bacteria that can utilize various forms of Fe(III) ion as precipitated hydroxide or as complexed soluble ions, such as Fe(III) citrate, to bacteria such as *T. saccharolyticum* that can use only soluble Fe(III) citrate but are stimulated by the addition of increased Fe(III) ions. Further studies must to be done to elucidate the nature and which of the bacteria excrete electron mediators (so no direct contact would be required) and which contain cell-wall-bound reductases (which require a direct contact with the Fe(III) precipitate).

Specificity of Reduced Metal Ions Varies Among the Thermophilic Fe(III) Reducers

Whereas the novel "*F. thermautotrophicus*" strains can reduce only Fe(III) among the tested ions, others such as *B. infernus*, *T. siderophilus*, and *D. thermophilus* can use Mn(IV) as terminal electron acceptor, and, similar to the mesophilic Fe(III) reducers, many other metal ions, including uranium. However, the use of various metal ions as electron acceptors has not been investigated in detail among the thermophilic Fe(III) reducers and remains to be done when other strains from diverse locations have been isolated and studied.

Conclusion

Several strains of different chemolithoautotrophic iron-reducing thermophilic bacteria were isolated and partly characterized. They belong to two known species (in different genera) that previously contained only fermentative strains and one novel genus of the Fermicutes. Other thermophilic, glycolytic anaerobic thermophiles tested reduced Fe(III) to an limited degree, apparently primarily in an assimilatory fashion, although $Y_{glucose}$ was increased. The description of novel isolates and the heterotrophic reduction by known glycolytic thermophiles extend the recognized diversity of thermophilic Fe(III) reducers. The isolation of the chemolithoautotrophic strains reported here from distant locations and those by Slobodkin et al. (1999a) from oil reservoirs suggest that among the glycolytic thermophilic anaerobes, the capability of carrying out anaerobic respiratory growth with Fe(III) as an electron acceptor is more widespread than previously assumed. It can be expected that many more diverse thermophilic Fe(III) (and other metal ion) reducers will be isolated. Whether they are indeed prevalent at sites with oil deposits and oil seeps remains to be determined. Furthermore, it is as yet unclear to what extent they can utilize components of crude oil or metabolites originating from oil degradation by other microorganisms.

Although under chemolithoautotrophic growth conditions, cell densities of only $3-5 \times 10^7$ cells per milliliter were observed, the specific rate of Fe(III) reduced per cell unit was about 10 times faster than what had been published for any other Fe(III) reducer. This strengthens the hypothesis that microbially mediated Fe(III) reduction by obligately anaerobic thermophiles could have been an important process on early Earth, when elevated temperatures were predominant (Baross 1998; Kashefi and Lovley 2000), which includes the involvement in the formation of specific Banded Iron Formations. In light of the properties of the above Fe(III) reducers, the theories on the origin and biogeochemistry of Banded Iron Formations should be revisited.

Acknowledgments. The authors are indebted to Anna Louise Reysenbach for the 16S rDNA sequence analysis. The authors also acknowledge Alex Slobodkin who introduced Fe(III) reduction to the laboratory. The initial work was supported by Genencor International.

References

Balashova VV, Zavarzin GA. 1980. Anaerobic reduction of ferric iron by hydrogen bacteria. Microbiology 48:635–9.

Baross JA. 1998. Do the geological and geochemical records of the early Earth support the prediction from global phylogenetic models of a thermophilic ancestor? In: Wiegel J, Adams M, editors. Thermophiles: the keys to molecular evolution and the origin of life? London: Taylor & Francis, pp. 13–18.

Bell PE, Mills AL, Herman JS. 1987. Biogeochemical conditions favoring magnetite formation during anaerobic iron reduction. Appl Environ Microbiol 53: 2610–16.

Bonch-Osmolovskaya EA, Miroshnichenko ML, Chernyh NA, et al. 1997. Reduction of elemental sulfur by moderately thermophilic organotrophic bacteria and description of *Thermoanaerobacter sulfurophilus* sp. nov. Microbiology 66:483–9.

Boone DR, Liu Y, Zhao Z, et al. 1995. *Bacillus infernus* sp. nov., an Fe(III)- and Mn(IV)-reducing anaerobe from the deep terrestrial subsurface. Int J Syst Bacteriol 45:441–8.

Bridge TAM, Johnson DB. 1998. Reduction of soluble iron and reductive dissolution of ferric iron containing minerals by moderately thermophilic iron-oxidizing bacteria. Appl Environ Microbiol 64:2181–6.

Canfield DE, Habicht KS, Thamdrup B. 2000. The archean sulfur cycle and the early history of atmospheric oxygen. Science 288:658–61.

Coates JD, Ellis DJ, Gaw CV, Lovley DR. 1999. *Geothrix fermentans* gen. nov., sp. nov., a novel Fe(III)-reducing bacterium from a hydrocarbon-contaminated aquifer. Int J Syst Bacteriol 49:1615–22.

Coleman ML, Hedrick DB, Lovley DR, et al. 1993. Reduction of Fe(III) in sediments by sulphate-reducing bacteria. Nature 361:436–7.

Cook GC, Rainey FA, Patel BKC, Morgan HW. 1996. Characterization of a new obligately anaerobic thermophile, *Thermoanaerobacter wiegelii* sp. nov. Int J Syst Bacteriol 46:123–7.

Engle M, Li Y, Rainey F, et al. 1996. *Thermobrachium celere* gen. nov., sp. nov., a rapidly growing thermophilic, alkalitolerant, and proteolytic obligate anaerobe. Int J Syst Bacteriol 46:1025–33.

Galushko AS, Schink B. 2000. Oxidation of acetate through reaction of the citric acid cycle by *Geobacter sulfurreducens* in pure culture and in syntrophic coculture. Arch Microbiol 174:314–21.

Greene AC, Patel BKC, Sheehy A. 1997. *Deferribacter thermophilus* gen. nov., sp. nov., a novel thermophilic manganese- and iron-reducing bacterium isolated from a petroleum reservoir. Int J Syst Bacteriol 47:505–9.

Guilbert JM, Park CF Jr. 1986. The geology of ore deposition. New York: Freeman.

Hanel JB. 2000. Thermophilic Iron Reduction. MS-thesis Univ. of Georgia, Athens, GA (US).

Harder EC. 1919. Iron-depositing bacteria and their geologic relations [US Geological Survey Professional Paper 113]. Washington, DC: U.S. Government Printing Office.

Horita J, Berndt ME. 1999. Abiotic methane formation and isotopic fractionation under hydrothermal conditions. Science 285:1055–7.

Jeonthon C, Reysenbach AL, Lharidon S, et al. 1995. *Thermotoga subterranea* sp. nov., a new thermophilic bacterium isolated from a continental oil-reservoir. Arch Microbiol 164:91–7.

Jones JG. 1986. Iron transformation by freshwater bacteria. In: Marshall EC, editor. Volume 9, Advances in microbial ecology. New York: Plenum Press, pp 149–85.

Jones JG, Gardener S, Simon BM. 1984. Reduction of ferric iron by heterotrophic bacteria in lake sediments. J Gen Microbiol 130:45–51.

Kashefi K, Lovley D. 2000. Reduction of Fe(III), Mn(IV), and toxic metals at 100°C by *Pyrobaculum islandicum*. Appl Env Microbiol 66:1050–6.

Küsel K, Dorsch T. 2000. Effect of supplemental electron donors on the microbial reduction of Fe(III), sulfate, and CO_2 in coal mining-impacted freshwater lake sediments. Microb Ecol 40:238–49.

Küsel K, Dorsch T, Acker G, Stackebrandt E. 1999. Microbial reduction of Fe(III) in acidic sediments: isolation of *Acidophilium cryptum* JF-5 capable of coupling the reduction of Fe(III) to the oxidation of glucose. Appl Environ Microbiol 65: 3633–40.

Lonergan DJ, Jentre HL, Coates JD, et al. 1996. Phylogenetic analysis of dissimilatory Fe(III)-reducing bacteria. J Bacteriol 178:2402–8.

Lovley DR. 1991. Dissimilatory Fe(III) and Mn(IV) reduction. Microbiol Rev 55: 259–87.

Lovley DR. 1995. Microbial reduction of iron, manganese, and other metals. Adv Agron 54:175–231.

Lovley DR, Lonergan DJ. 1990. Anaerobic oxidation of toluene, phenol, and *p*-cresol by the dissimilatory iron-reducing organisms, GS-15. Appl Environ Microbiol 56:1858–64.

Lovley DR, Beadecker M, Lonergan D, et al. 1989. Oxidation of aromatic contaminants coupled to microbial iron reduction. Nature 339:297–9.

McGeary D, Plummer C. 1997. Physical geology: earth revealed. New York: McGraw-Hill.

Nealson KH, Saffarini DA. 1995. Iron and manganese in anaerobic respiration: environmental significance, physiology, and regulation. Annu Rev Microbial 48: 311–43.

Ponnamperuma FN. 1972. The chemistry of submerged soils. Adv Agron 24:29–96.

Roden EE, Lovley DR. 1993. Dissimilatory Fe(III) reduction by the marine microorganism *Desulfuromonas acetoxidans*. Appl Environ Microbiol 59:734–42.

Rosello-Mora RA, Caccavo F Jr, Osterlehner K, et al. 1994. Isolation and taxonomic characterization of a halotolerant, facultatively iron-reducing bacterium. Syst Appl Microbiol 17:569–73.

Short KA, Blakemore RP. 1986. Iron respiration-driven proton translocation in aerobic bacteria. J Bacteriol 167:729–31.

Slobodkin A, Wiegel J. 1997. Fe(III) as an electron acceptor for H_2 oxidation in thermophilic anaerobic enrichment cultures from geothermal areas. Extremophiles 1:106–9.

Slobodkin A, Reysenbach A-L, Mayer F, Wiegel J. 1997a. Isolation and characterization of the homoacetogenic thermophile *Moorella glycerini* sp. nov. Int J Syst Bacteriol 47:969–74.

Slobodkin A, Reysenbach A-L, Strutz N, et al. 1997b. *Thermoterrabacterium ferrireducens* gen. nov., sp. nov., a thermophilic anaerobic dissimilatory Fe(III)-reducing bacterium from a continental hot spring. Int J Syst Bacteriol 47:541–7.

Slobodkin A, Tourova TP, Kuznetsov BB, et al. 1999b. *Thermoanaerobacterium siderophilus* sp. nov., a novel dissimilatory Fe(III)-reducing, anaerobic, thermophilic bacterium. Int J Syst Bacteriol 49:1471–8.

Slobodkin AI, Jeanthon C, L'Haridon S, et al. 1999a. Dissimilatory reduction of Fe(III) by thermophlic bacteria and archaea in deep subsurface petroleum reservoirs of western Siberia. Curr Microbiol 39:99–102.

Stum W, Morgan JJ. 1981. Aquatic chemistry. New York: Wiley.

Takai K, Inoue A, Horikoshi K. 1999. *Thermaerobacter marianensis* gen. nov., sp. nov., an aerobic extremely thermophilic marine bacterium from the 11 000 m deep Mariana Trench. Int J Syst Bacteriol 49:619–28.

Tuccillo ME, Cozzarelli IM, Herman JS. 1999. Iron reduction in the sediments of a hydrocarbon-contaminated aquifer. Appl Geochem 14:655–67.

Vargas M, Kashefi K, Blunt-Harris EL, Lovley DR. 1998. Microbiological evidence for Fe(III) reduction on early Earth. Nature 395:65–7.

Wiegel J. 1981. Distinction between the Gram reaction and the Gram type of bacteria. Int J Syst Bacteriol 31:88.

Wiegel J, Kuk S, Kohring GW. 1989. *Clostridium thermobutyricum* sp. nov., a moderate thermophile isolated from a cellulolytic culture, that produces butyrate as the major product. Int J Syst Bacteriol 39:199–204.

Wolin EA, Wolin MJ, Wolfe RS. 1963. Formation of methane by bacterial extracts. J Biol Chem 238:2882–6.

18
Electron Flow in Ferrous Biocorrosion

E.J. Laishley and R.D. Bryant

In 1934, von Wolzogen Kuhr and van der Vlugt reported the involvement of the sulfate-reducing bacteria (SRB) in anaerobic corrosion on metal surfaces. They proposed the classical depolarization theory to explain their observations, which assumes that an initial galvanic cell on the metal surface in an aqueous environment forms a protective film of molecular or atomic hydrogen on the metal surface, essentially neutralizing any further electrochemical reaction. It was surmised that the SRB disrupted this delicate balance by using this hydrogen as an energy source, causing further metal oxidation to maintain the initial electrochemical state. Since their initial report, considerable research has been conducted to determine how the SRB promote corrosion. The experimental data show that the products of SRB metabolism may singly, or in combination, accelerate corrosion (i.e., iron sulfides, H_2S, elemental sulfur, phosphorous compounds, and organic acids) (Pankhania 1988).

Evidence has accumulated that suggests that the activity of the SRB hydrogenase may be directly involved in depolarizing cathodic hydrogen from the metal surfaces and, therefore, they may play an important role in anaerobic corrosion (Cord-Ruwisch and Widdel 1986; Pankhania 1988; Belay and Daniels 1990; Bryant et al. 1991; van Ommen Kloeke et al. 1995). There are other nonSRB genera, such as clostridia and *Shewanella* spp. (Semple and Westlake 1987) found in the corrosion pits on oil pipelines, that possess hydrogenases and have the ability to produce sulfide from partially reduced inorganic sulfur compounds. Studies on the hydrogenase involvement in microbial-induced corrosion have centered on the use of wild and (or) mutant SRB cells (Hardy 1983; Cord-Ruwisch and Widdel 1986; Daumas et al. 1988; Pankhania 1988), but no one has examined the action of cell free hydrogenase on metal surfaces.

This review describes our unique assay system to investigate the role of cell-free hydrogenases in anaerobic biocorrosion, which resulted in proposing a new biocorrosion model.

Phosphate/Mild Steel Cathodic Hydrogen-Generating System

To study the effect of cell-free hydrogenase enzymes on mild steel surfaces, an appropriate assay system was needed. In this regard, we used a Warburg respirometer apparatus to develop an assay in which three mild steel rods (6×12 mm) were immersed in 100 mM KH_2PO_4 buffer (pH 7) in the center compartment of a double sidearm Warburg flask. The flasks were incubated at 37°C under an atmosphere of nitrogen, and gas evolution was monitored by standard manometric techniques. The gas evolved from the interaction between the phosphate buffer and mild steel rods was determined to be hydrogen, and a white precipitate in the assay solution was identified as vivianite, $Fe_3(PO_4)_2$ (Bryant and Laishley 1990) (Fig. 18.1). The stoichiometry of this reaction resulted in an Fe:H_2 ratio close to the predicted electrochemical reaction ratio of 1:1. Apparently, the phosphate reacted with the mild steel surface to produce hydrogen gas and vivianite chemically. In this neutral environment, the phosphate buffer also acted as a source of protons in the cathodic depolarization reaction, resulting in hydrogen evolution as outlined in the following equation: $3Fe + 4H_2PO_4^- \rightarrow Fe_3(PO_4)_2 + 2HPO_4^{2-} + 3H_2$ (Bryant and Laishley 1993). Thus the rate of hydrogen gas evolution was an indirect measurement of the corrosion rate of the mild steel rods. In the control flask containing mild steel rods and

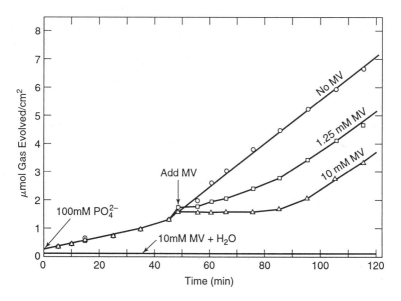

FIGURE 18.1. The effect of phosphate and methyl viologen (*MV*) on hydrogen evolution from a mild steel surface.

water, no soluble iron or gas was detected over the time course of the experiment.

The Effect of Artificial and Natural Electron Carriers on Cathodic Hydrogen Evolution from Metal

The phosphate/mild steel hydrogen–generating assay system was employed to assess the effect of the low-potential electron carrier methyl viologen, a substitute for hydrogenase's physiologic carriers, on hydrogen evolution. Figure 18.1 shows that the addition of the methyl viologen at different concentrations immediately stopped the evolution of hydrogen for varying periods of time. The oxidized methyl viologen (colorless) became reduced (blue color) by preferentially siphoning cathodic-generated electrons from the mild steel instead of allowing the electrons to form hydrogen gas ($2e^- + 2H^+ \rightarrow H_2$). The reduction of methyl viologen continued until a reductive equilibrium was reached; then the flow of cathodic electrons combined with the protons generated by ongoing chemical reaction between the phosphate and the metal to form hydrogen. Thereafter, the rate of hydrogen evolution eventually approached that of the control rate containing no methyl viologen.

Physiologic electron acceptors flavin mononucleotide (FMN) and flavin adenine dinucleotide (FAD) produced similar effects on cathodic hydrogen evolution from mild steel as achieved with methyl viologen (Bryant and Laishley 1990). These experimental results showed that the mild steel rods reacting with phosphate can preferential act as electron donors for the reduction of low-potential electron carriers. All hydrogenases catalyze a reversible reaction for the formation and oxidation of hydrogen, which requires low-potential electron carriers for the enzyme activity (Church et al. 1988; Fauque et al. 1988).

Effect of Hydrogenases on Cathodic Hydrogen Evolution from Mild Steel Metals

Desulfovibrio vulgaris Hildenborough periplasmic [Fe] hydrogenase was isolated by the procedure described by Badziog and Thauer (1980) to study the effect of this enzyme on mild steel, employing our Warburg corrosion assay system. The flask's main compartment contained the phosphate buffer, mild steel rods, and 10 mM methyl viologen (MV), and one of the side arms contained the periplasmic [Fe] hydrogenase (4 μg protein). The reactions were carried out under an atmosphere of nitrogen at 37°C. In the main compartment, the methyl viologen became reduced (blue) owing to the cathodically generated electrons from the phosphate/mild steel chemical reaction; concurrently, no hydrogen evolution was observed.

TABLE 18.1. The effect of *D. vulgaris* Hildenborough periplasmic [Fe] hydrogenase on mild steel corrosion.

Reactions[a]	Fe^{2+} (μmol)	Corrosion fold increase[b]
1: Plus MV (minus phosphate buffer)	0.0	—
2: Plus MV	0.2	—
3: Plus MV Plus [Fe] hydrogenase (4μg protein)	3.1	15.5

[a] All reactions contained mild steel rods; reactions 2 and 3 contained 100 mM phosphate buffer (pH 7), and in reaction 1 distilled water replaced the buffer. MV = 20 μmol.
[b] Compared to reaction 2 [Fe^{2+}].

When the methyl viologen reached its reductive equilibrium, noted by the start of hydrogen evolution, the enzyme in the side arm was tipped into the main compartment, and the reaction stopped after 10 min. The solution in the main compartment of the flasks were analyzed for ferrous ion concentration, which is a measurement for metal corrosion.

Table 18.1 shows that the interaction of phosphate with mild steel promoted cathodic electron reduction of methyl viologen as another 0.2 μmol of Fe^{2+} was detected after 10 min (reaction 2), being corrected for the amount of Fe^{2+} before this time period. In a comparable control flask, (reaction 1), where distilled water replaced the phosphate buffer, no color change occurred in the solution and no Fe^{2+} was detected. When the periplasmic [Fe] hydrogenase was added to the reduced methyl viologen plus mild steel rods/phosphate (reaction 3), significant enzymatic biocorrosion took place over the next 10 min, as 15.5-fold more Fe^{2+} was determined compared to reaction 2. The tremendous increase in Fe^{2+} was attributed to the preferential reduction of methyl viologen by the chemically generated cathodic electrons from the metal and subsequent linking of this reduced electron carrier with the periplasmic [Fe] hydrogenase to activate its enzymatic activity to form copious amounts of hydrogen ($2MV + 2H^+ \rightarrow 2MV^+ + H_2$).

The importance of the hydrogenase enzyme was also demonstrated in experiments in which mixed cultures of SRB recently isolated from the production water of two pipeline systems in Alberta were circulated through two Robbins Devices (McCoy et al. 1981) at a flow rate of $4 L min^{-1}$. Both loops showed detectable SRB attached to corrosion coupons. One population of the SRB had a high level of hydrogenase activity, which correlated well with the subsequent high corrosion of the metal coupons; the other population of SRB had low levels of hydrogenase and low levels of iron loss detected (Bryant et al. 1991).

We were also able to clearly demonstrate the *Clostridium pasteurianum* [Fe] hydrogenase 1 was a corrosive enzyme, accelerating the dessolution of iron and hydrogen production from mild steel surfaces (Bryant and Laishley 1990). We believe hydrogenases are intimately involved in the metal biocorrosion process.

The Role of Iron Regulation in Metal Biocorrosion

We observed in many instances, newly isolated SRB cultures from corroded areas in oil pipelines originally tested positive for hydrogenase activity by a commercial test kit but lost this enzyme activity after being maintained in the laboratory by subculturing into enriched SRB culture medium (Bryant et al. 1991). This observation suggested the hydrogenase(s) of SRB may be subject to induction/repression control mechanism. We found the periplasmic [Fe] hydrogenase from *D. vulgaris* Hildenborough was regulated by ferrous iron availability. The synthesis of this enzyme during growth was regulated by ferrous ion concentration; low iron (<5.0 ppm Fe^{2+}) caused a significant derepression of the periplasmic [Fe] hydrogenase, whereas the presence of Fe^{2+} at >5.0 ppm repressed the enzyme's expression (Bryant et al. 1993). The significance of this finding may explain the lack of correlation in some instances between SRB cell counts and oil pipeline corrosion in field tests (i.e., high cell counts, low corrosion; low cell counts, high corrosion) (Bryant et al. 1991).

Other iron-containing proteins were found to be regulated by [Fe^{2+}] in *D. vulgaris* Hildenborough. We were able to determine that two high molecular weight cytochromes (62.5 and 77.5 kDa) as demonstrated by sodium dodecyl sulfate–polyacrylamide gel electrophoresis (SDS-PAGE) and heme-specific staining were derepressed and located in the outer membrane (OM) under low iron growth conditions (<5 ppm Fe^{2+}) (van Ommen Kloeke et al. 1995). The isolated OM containing these cytochromes and a small amount of contaminating periplasmic [Fe] hydrogenase accelerated hydrogen evolution from the mild steel rods/phosphate assay system compared to the control reaction minus the OM fraction (Fig. 18.2) (van Ommen Kloeke et al. 1995). This increased hydrogen production reflected the enhanced cathodic flow of electrons from the metal surface via the cytochromes/[Fe] hydrogenase system present in the OM fraction compared to the control phosphate/mild steel reaction evolving hydrogen at constant rate over the same time period. This suggested the cytochromes(s) were preferentially syphoning the generated cathodic electrons and interacting with the periplasmic [Fe] hydrogenase, resulting in an enhanced acceleration of hydrogen in a similar manner previously reported for the electron carrier methyl viologen. In another experiment, where the phosphate buffer was replaced with distilled water, the OM fraction addition showed no hydrogen evolution, even though it contained the cytochromes and periplasmic [Fe] hydrogenase. Again, phosphate was shown to initiate this anodic/cathodic reaction on mild steel, which could be significantly amplified by the OM fraction containing the heavy molecular weight cytochromes and [Fe] hydrogenase.

This discussion has centered on the importance of iron in regulating the cellular protein components that play crucial roles in the microorganisms' ability to affect biocorrosion of metals. However, it should also be noted

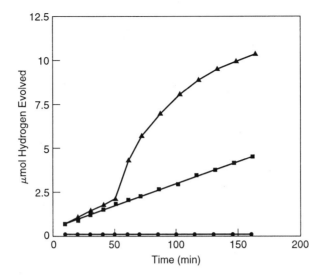

FIGURE 18.2. The effect on the phosphate buffer/mild steel hydrogen evolution system by the isolated OM fraction (100 μg protein), from *D. vulgaris* Hildenborough grown on 5 ppm Fe^{2+}. ■, control: mild steel rods and phosphate buffer (pH 7); ◆, mild steel rods, phosphate buffer (pH 7) and OM added at 50 min; ●, mild steel rods, water and OM added at 50 min (minus phosphate buffer).

that iron has a chemotactic function in facilitating the physical attachment of *D. vulgaris* to the metal. The experiment in Figure 18.3 shows that under low iron concentration growing conditions (<5 ppm Fe^{2+}) there was a linear attachment of cells to the iron coupons >60 min after coupons were suspended (and subsequently removed in staggered time intervals) in freshly inoculated cultures that were 24 h old. There was an exponential rise in cellular attachment between 1 and 18 h. This linear attachment of cells to metal coupons in the first 60 min was not observed when a comparable experiment was run under high iron concentration conditions (>5 ppm Fe^{2+}). However, there was an increase, albeit small relative to the low iron condition, in cellular attachment to the coupons between 1 and 18 h.

This small increase in attachment to the cells after 1 day was likely because the iron level in culture decreased with time as a result of the metabolic production of sulfide, which complexed with the iron to produce the iron sulfide precipitate, effectively lowering the dissolved iron concentration toward that of the low iron condition. It would seem that when organisms are placed in low-iron environments, they will use their chemotactic response/biologic inductions mechanisms to ensure that the iron requirements of the cells' high iron requiring proteins (i.e., cytochromes, ferredoxin, hydrogenases) are satisfied.

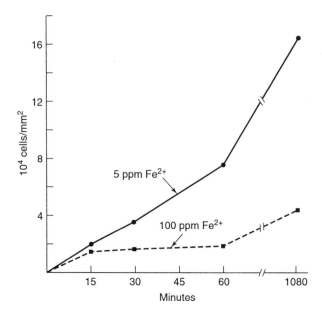

FIGURE 18.3. *Desulfovibrio vulgaris* Hildenborough attachment kinetics to mild steel coupons under various Fe^{2+}. Metal coupons (1.5×6.0 cm) were suspended into 24-h growing cultures under various Fe^{2+} concentrations and removed at the times indicated. Coupons were washed with distilled water, dried, fixed with 5% gluteraldehyde, and analyzed by scanning electron microscopy. The bacterial count at each datum point represents an average of 10 random sites on the coupon, counted from scanning micrographs and equated to a number (10^4 cells) per unit area (mm^2) metal.

Conclusion

The present biocorrosion model (Odom 1993) implies that the SRB hydrogenase would have to be located on the exterior surface of the OM to use the surface hydrogen film on the metal. However, hydrogenases have been found only in the periplasm, cytoplasmic membrane, and cytoplasm of SRB (Badziog and Thauer 1980; van Ommen Kloeke et al. 1995), which raises questions about the current cathodic depolarization theory. Our present findings showed that Fe^{2+} concentration regulated *D. vulgaris* Hildenborough periplasmic [Fe] hydrogenase, OM high molecular weight cytochromes, chemotactic response to metal surfaces, and preferential electron siphoning from mild steel via electron carriers needed for [Fe] hydrogenase reductive activity, resulting in accelerated biocorrosion of the metal surface.

Thus we propose a novel biocorrosion model (Fig. 18.4) (van Ommen Kloeke et al. 1995) whereby the derepressed OM cytochrome(s) remove electrons from the cathodic site on the mild steel surface and couple with

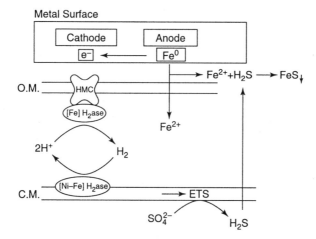

FIGURE 18.4. Proposed biocorrosion model for cathodic electron depolarization of mild steel by *D. vulgaris* Hildenborough. *CM*, cytoplasmic membrane; *HMC*, high molecular weight cytochrome; *[Fe] H₂ase*, iron hydrogenase; *[NiFe] H₂ase*, nickel iron hydrogenase; *ETS*, electron transport system.

the inducible periplasmic [Fe] hydrogenase to produce hydrogen gas with simultaneous release of Fe^{2+} from the anodic site. We visualize that this hydrogen gas is produced in the periplasmic space to diffuse into the cytoplasmic membrane, where it is oxidized by the constitutive [NiFe] hydrogenase; the electrons are passed though an electron-transport system to the terminal acceptor SO_4^{2-}, which is reduced to H_2S and evolved from the cell. To determine if this biocorrosion is universal among SRB, more work of a similar nature will be required on other species of SRB.

Acknowledgments. This work was supported by the Natural Sciences and Engineering Research Council of Canada and British Petroleum Canada.

References

Badziog W, Thauer RK. 1980. Vectorial electron transport in *Desulfovibrio vulgaris* (Marburg) growing on hydrogen plus sulfate as sole energy source. Arch Microbiol 125:167–74.

Belay N, Daniels L. 1990. Elemental metals as electron sources for biological methane formation from CO_2. Antonie Leeuwenhoek 57:1–7.

Bryant RD, Laishley EJ. 1990. The role of hydrogenase in anaerobic biocorrosion. Can J Microbiol 36:259–64.

Bryant RD, Laishley EJ. 1993. The effect of inorganic phosphate and hydrogenase on the corrosion of mild steel. Appl Microbiol Biotechnol 38:824–7.

Bryant RD, Jansen W, Boivin J, et al. 1991. Effect of hydrogenase and mixed sulfate-reducing bacterial populations on the corrosion of steel. Appl Environ Microbiol 57:2804–9.

Bryant RD, van Ommen Kloeke F, Laishley EJ. 1993. Regulation of the periplasmic [Fe] hydrogenase by ferrous iron in *Desulfovibrio vulgaris* (Hildenborough). Appl Environ Microbiol 59:491–5.

Church DL, Rabin HR, Laishley EJ. 1988. Role of hydrogenase 1 of *Clostridium pasteurianium* in the reduction of metronidazole. Biochem Pharmacol 37:1525–34.

Cord-Ruwish R, Widdel F. 1986. Corroding iron as a hydrogen source for sulphate-reduction in growing cultures of sulphate-reducing bacteria. Appl Microbiol Biotechnol 25:169–74.

Daumas S, Massiani Y, Crousier J. 1988. Microbiological battery induced by sulphate-reducing bacteria. Corros Sci 28:1041–50.

Fauque G, Peck HD Jr, Moura JJG, et al. 1988. The three classes of hydrogenases from sulfate-reducing bacteria of the genus *Desulfovibrio*. FEMS Microbiol Rev 54:299–344.

Hardy JA. 1983. Utilization of cathodic hydrogen by sulphate-reducing bacteria. Br Corros J 18:190–3.

McCoy WF, Bryers JD, Robbins J, Costerton JWC. 1981. Observations on biofilm formation. Can J Microbio 27:910–17.

Odom JM. 1993. Industrial and environmental activities of sulfate-reducing bacteria. In: Odom JM, Singleton R Jr, editors. The sulfate-reducing bacteria: contemporary perspectives. New York: Springer-Verlag. p 189–210.

Pankhania I. 1988. Hydrogen metabolism in sulphate-reducing bacteria and role in anaerobic corrosion. Biofouling 1:27–47.

Semple KM, Westlake DWS. 1987. Characterization of iron-reducing *Alteromonas putrefaciens* strain oil field fluids. Can J Microbiol 33:366–71.

van Ommen Kloeke F, Bryant RD, Laishley EJ. 1995. Localization of cytochromes in the outer membrane of *Desulfovibrio vulgaris* (Hildenborough) and their role in anaerobic biocorrosion. Anaerobe 1:351–8.

von Wolzogen Kuhr CAH, van der Vlugt LS. 1934. Graphication of cast iron as an electrochemical process in anaerobic soils. Water 18:147–65.

Index